THE ANATOMY OF MOUNTAIN RANGES

PRINCETON SERIES IN GEOLOGY AND PALEONTOLOGY

EDITED BY ALFRED G. FISCHER

THE ANATOMY OF MOUNTAIN RANGES

JEAN-PAUL SCHAER AND JOHN RODGERS, EDITORS

PRINCETON UNIVERSITY PRESS

PRINCETON, NEW JERSEY

Copyright © 1987 by Princeton University Press
Published by Princeton University Press,
41 William Street, Princeton, New Jersey 08540
In the United Kingdom: Princeton University Press,
Guildford, Surrey

This book has been composed in Linotron Times Roman and Helvetica

Clothbound editions of Princeton University Press books
are printed on acid-free paper, and binding materials
are chosen for strength and durability.
Paperbacks, although satisfactory for personal collections,
are not usually suitable for library rebinding

Printed in the United States of America by
Princeton University Press
Princeton, New Jersey

CONTENTS

CONTRIBUTORS

A. Baud
Institut et Musée de Géologie
Palais de Rumine
CH-1005 Lausanne, Switzerland

Asger Berthelsen
Institut for almen Geologi
Øster Voldgade 10
DK-1350 Copenhagen K, Denmark

R. Caby
Centre Géologique et Géophysique
Université des Sciences et Techniques
    du Languedoc
34060 Montpellier Cedex, France

W. Frank
Geologisches Institut
Universität Wien
Universitätsstrasse 7
A-1010 Vienna, Austria

K. Honegger
V. Trommsdorff
Institut für Mineralogie und Petrographie
ETH-Zentrum
CH-8092 Zürich, Switzerland

M. Julivert
F. J. Martínez
Universitat Autònoma de Barcelona
Bellaterra, Spain

H. P. Laubscher
B. Meier
M. Schwander
Universität Basel
CH-4056 Basel, Switzerland

François Mégard
Centre Géologique et Géophysique
Université des Sciences et Techniques
    du Languedoc
34060 Montpellier Cedex, France

A. G. Milnes
Swiss Federal Institute of Technology
Zürich, Switzerland

John Rodgers
Department of Geology and Geophysics
Yale University
New Haven, CT 06520
United States

Jean-Paul Schaer
Université de Neuchâtel
Institut de Géologie
CH-2000 Neuchâtel 7, Switzerland

Brian F. Windley
Department of Geology
The University
Leicester LE1 7RH, England

John Suppe
Department of Geological and Geophysical
        Sciences
Princeton University
Princeton, NJ 08540
United States

Part I

GENERAL

Chapter 1

INTRODUCTION: COMPARATIVE ANATOMY IN GEOLOGY

JEAN-PAUL SCHAER

Université de Neuchâtel

Students working in small universities are more and more absorbed by multiple teaching tasks to which are added burdensome management duties. This situation generates a relative isolation that keeps them away from the dynamic impulses of modern research. To enable them to remain in contact with the most active research groups, the French-speaking universities of Switzerland and the University of Bern have joined together to provide postgraduate courses in Earth Science. Every year a set of lectures is organized, taking stock of one modern and important topic.

During the first months of 1982, the University of Neuchâtel was responsible for setting up a series of lectures on the theme "Comparative Anatomy of Mountain Ranges." Guest speakers from Switzerland and abroad presented for the benefit of an attentive audience the results of their research, which covered orogenic belts on almost all parts of the earth (excluding Australia) and of all ages. The course brought out the great complexity of structures, reflecting the diversity of materials involved and the specifics of structural evolution.

Mountain ranges such as the Alps, and even more the Andes, Appalachians, and Himalayas, cover such vast areas that it is almost impossible for a single person to be familiar with all their specific details. The problems of enormous space and limited time thus prevent most researchers from acquiring a firsthand knowledge that extends over many mountain ranges. Because of this limitation, few authors in Earth Science venture into the field of comparative anatomy. Thanks to his relatively broad knowledge of our globe, John Rodgers, who acted as coordinator for the lecture series, could embark in that direction. His comments in the next article must be read with this in mind. To him and to all those who took part in our efforts, we extend our deep gratitude.

Part of the material expounded at Neuchâtel was elaborate enough for publication, and it is gathered here. We regret not being able to include other contributions that brought out specifics of other ranges but also analogies. We hope the material published will prove useful to a wide audience and pave the way for future papers in comparative anatomy in Earth Science.

It may be a cause for wonder that the course on anatomy of mountain ranges was given in Neuchâtel. One must not forget, however, that this small academic town benefits from a strong tradition in the field. Louis Agassiz taught biology here during the past century. More recently, Emile Argand first and then his student C. E. Wegmann contributed much to impregnating Earth Science with this line of thought.

We felt it would be interesting and instructive to sketch some aspects of the lives and thoughts of these researchers in order to highlight the richness and strength of this method of investigation. From this standpoint, our aim would remain unfulfilled if we did not first recall the man who founded the discipline, the great *Georges Cuvier* (1769–1832), who was born in Montbéliard, a French city (at that time attached to the Duchy of Württemberg) not far from the Swiss border.

As soon as he had given its *lettres de noblesse* to comparative anatomy using anatomical structure and its evolution as a basic principle, Cuvier set about to make it the foundation for all animal classification. By systematic comparison of the architecture of living organisms and their fossil remains, he demonstrated that the past must be considered when one is studying the organization of the entire animal world. From the beginning of his research he insisted on the importance of the connections that must exist between the various organs in order to make of them a living unit. He understood that certain functions have such a determining influence on the organism that they can easily be used as a general guide to classification; thus, for him, the nervous system gave the phyla, the respiratory and circulatory organs gave the classes.

In geology to this day, the classifications of mountain ranges are based mostly on positional criteria (ocean-continent, continent-continent collision ranges; intracratonic chains; and so forth). One may ask whether there are other parameters with enough influence to impose some structural specificities while excluding others (presence of large granitic batholiths, evolution of sedimentary basins, deformation style, and metamorphic regimes, for example). For Cuvier, the anatomical characteristics that distinguish groups of animals are the proof that different species have not altered since their creation. Can the structural analogies that have been described between young ranges and old ones be taken as proof that our earth and its continents have been evolving for 2,000 or 3,000 Ma (Ma = million years) in an at least similar if not identical fashion? It took Darwin and those who followed him to integrate into comparative anatomy all the riches of ontogeny, embryology, and, above all, phylogeny, which, because it encompasses lines of descent, spreads into the field of paleontology. At this level of research, both the zoologist and the paleontologist know that their documentation is incomplete, because the softer parts of animals very rarely leave any fossil traces and because it is virtually impossible to know all the elementary mutations which, when accumulated along an evolutionary line, can produce perhaps the modifications leading to new species. The geologist studying mountain ranges is put into an even more difficult position since he can only contemplate a very thin slice of the crust (15 km in favorable zones). It is true that, for the past few years, modern geophysics has been bringing forth new information about the specifics of deep zones; this information holds the promise of fruitful comparisons. Although the task is complex, it is important to remain optimistic and to follow Cuvier who stated, in the preface to his work ''Recherches sur les ossements fossiles des quadrupèdes,'' ''As antiquarian of a new species, I had to learn how to decipher and restore these monuments, how to recognize and bring together in their primitive order the scattered and mutilated fragments of which they are composed'' (Cuvier 1812). This is precisely the object of our study when we set out to classify, analyze, and understand mountain ranges by using the mutilated fragments left to us by successive erosion periods.

Jean Louis Rodolphe Agassiz (1807–1873), born at Môtier in the Swiss canton of Fribourg, was the son of a clergyman whose family originated from the villages of Orbe and

Bavois in the canton of Vaud. Agassiz was educated first at Bienne and then for two years at Lausanne Academy. He was a student at the Universities of Zurich, Heidelberg, and Munich, and he finished his studies with a Doctor of Philosophy degree at Erlangen and a degree in medicine at Munich (1830). At the age of 22 he had already published an important paper about fish brought back from Brazil by a German expedition. He dedicated this work to Cuvier, and in his accompanying letter to Cuvier the young author expressed both his admiration for the great teacher and his ambition to study natural sciences even in the face of his precarious financial situation.

Agassiz received a very encouraging reply from Cuvier, informing him that the results described in his publication would be incorporated into the second edition of ''Le Règne animal.'' Agassiz returned to Switzerland for a few months and then left for Paris, passing through Germany where, with the help of his draftsman Dinkel, he studied fossil fish in a number of museums. In Paris he met Cuvier, who was at first somewhat reserved toward him but was soon won over by the young naturalist's enthusiasm and knowledge and by the quality of his work. The friendship between the two men soon became so close that Cuvier gave Agassiz all the documents he had collected concerning fossil fish. With this gift the teacher indicated clearly that he was abandoning his own research on the subject.

Agassiz was scientifically more active than ever, yet he found time to make contact with such interesting people as Alexander von Humboldt, who was in Paris at the time as the personal representative of his king, Frederick William III of Prussia, to Louis Philippe of France. Humboldt loved to encourage young talent and was generous with his time and money. He made Agassiz his protégé and passed on to him his own scientific conception of the world, also impressing upon his young disciple the importance of public relations for someone who is going to make a career in science. Humboldt's example and teaching played a decisive part in Agassiz's education, for it proved to be the basis of his success in America, where his career depended as much on charm and public relations as on scientific knowledge.

In the last phase of his training in Paris, Agassiz was profoundly influenced by Cuvier, so much so that for the rest of his life he considered him his only teacher, the one who enlarged his scientific concepts by explaining to him that precise analysis and patience with facts must always take precedence over theoretical ideas, interpretations, and synthesis. Unfortunately, this period of training was very short, less than six months, because Cuvier died in May 1832. Agassiz thus found himself alone with his enormous projects and as always without financial help. Thanks to the generosity of some citizens of Neuchâtel, enough money was gathered to offer him the position of Professor of Natural Science in their town. Because it gave him the opportunity to live near his family again in a country that he knew and loved, he accepted the offer, arrived in Neuchâtel in September 1832, and wasted no time in getting down to work. In addition to his teaching, he took in the same year a leading role in the founding of the local Society of Natural Science, which immediately impressed the scientific world by the high quality of its publications. Agassiz also became Director of the Museum of Natural History, and an extraordinary scientific activity inflamed the little provincial town, where the population at that time was no more than 5,000. In those days Neuchâtel, as a principality of the King of Prussia, was still dependent on Berlin. For Agassiz this was most fortunate, since it was again possible for him to benefit from Humboldt's generous protection.

During the period Agassiz lived in Neuchâtel, his scientific activity was superlative in both quality and quantity. The spiritual inheritance from Cuvier is present throughout Agassiz's work. For him, meticulous observation remained the basis of sound research, but not carried out blindly, in view of the intensity and diversity of the living world. Major efforts were to be directed toward subjects that offered the best opportunity to improve knowledge. Certain classes of fossils, like the extremely well-preserved echinoids, were ideal material for a better understanding of the past world and its organization (Agassiz et al. 1838–1842). A sum of characteristics was often a more reliable criterion for classifying organisms than was any specific function, however well-defined. But Agassiz also insisted that study of a part, or a function, when carried out in all its detail, would always promote discoveries relating to the entire system. When one has a good knowledge of some important organs, it becomes possible to imagine or even to describe elements that should coordinate them even if they have not been preserved. The constraint of organized life offers a better approach to truth than does an uncontrolled imagination (Agassiz 1833–1843).

Emile Argand (1874–1940) was born in Geneva. His life and work have been admirably described by his teacher Maurice Lugeon (1940). Here only some striking aspects of the career of this exceptional man can be presented. All those who knew him were impressed by his amazing visual memory; he was able to draw a map of any part of the world from memory with amazing accuracy. Certainly, he had strengthened this gift when, on leaving college, he did his apprenticeship as a draftsman. Before devoting his life to geology, he studied medicine for 2 years, until the age of 23, and it was probably then that he received a solid grounding in anatomy. Throughout his life Argand stressed the importance of geometry and precise drawing, and several of his papers are accompanied by beautiful sections, admirably drawn not just for themselves but to enable the reader to have an exact picture of volumes which, as they are constantly modifying with time, imply permanent work within this supplementary dimension. His geometric analysis is governed by the cylindrical principle, though not as rigidly as once thought, since it is softened and guided by axial variations that show the relative plasticity of materials and the heterogeneity of obstacles. Axial variations in the vertical plane are above all important for deciphering deep structures to a depth of 20 to 40 km. After 1912, probably about 1914, Argand discovered Alfred Wegener's work, and its broad vision reinforced his own enthusiasm from then on. From his research on the topography and geometry of the Central Pennine Alps, he developed a powerful technique that enabled him to understand the architecture, the deformations, and the movements not only of the Alps but also of the continents, which fell into an order that is perfectly coherent both for the time and the future. Already in his paper "Sur l'arc des Alpes occidentales" (Argand 1916), nearly all the "argandian" message is clearly given. The analysis focuses on the core of the alpine structure, showing how the superficial character of the deformation of the Austro-Alpine nappes contrasts with the plastic style of the Pennine folds, which in turn impose their style on deeper zones.

Argand then took a great interest in recent ranges, especially those of East Asia, because they show side by side and simultaneously the successive stages through which more evolved chains have passed. Thus comparative anatomy allowed Argand to go back in time, and it led to a concept of tectonic embryology (Figure 1-1). Even if continental drift is not directly referred to, the whole evolution of the geosynclinal domain is already visualized as the consequence of tangential movements affecting rigid blocks, which as they approach deform plastically the large marine zone separating them.

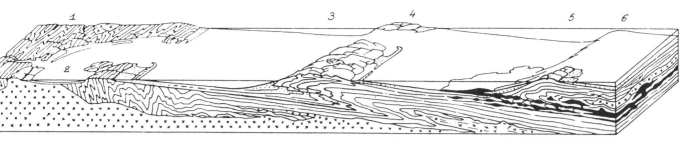

FIGURE 1-1. Embryonic tectonics (after Argand, 1916). 1: Foreland. 2: Epicontinental sea. 3: Geosyncline. 4: Frontal cordillera or geanticline. 5: Geosyncline with ophiolites (in solid black). 6: Second geanticline. (With permission of the Geological Society of Switzerland.)

In 1924 Argand's masterwork, *The Tectonics of Asia*, appeared. Its message is a direct continuation of Edward Suess's work, "Das Antlitz der Erde," which the teacher from Vienna had finished a few years earlier, and which presents an analytical geological description of the entire earth. Argand admired this work so much that he referred to it in the first line of his text. But for Argand the analysis and the description of structures had to be pushed still further to include studies of different scales, from continents to thin sections. These static images must be used to discover at each instant the evolution of volumes in time, to create an animated picture. It should be possible to place all observations in the framework of the unitary theory of Wegener's continental drift, a theory full of force, flexibility, and possibilities of invention. All the structures should fall into place as functions of displacement and of the relative plasticity of the deformed materials and preexisting structures. Anatomy, the geometry of structures, was the basic science that permitted one to control imagination. In 1927 the Academy of Science in Paris awarded the Prix Cuvier to Argand for the whole of his work, without underlining, as might have been expected, the spiritual line of descent that joins the zoologist to the geologist, and the essential contribution of the latter to this new form of comparative anatomy.

C. E. Wegmann (1896–1982), one of Argand's few students, analyzed in his Ph.D. thesis the frontal part of the Pennine zone in the Valais. After periods in Zurich and France, he went while still young to the Nordic countries, where he met J. J. Sederholm, whose strong, warm personality had a profound scientific and human influence on him. There he applied with success the principles of geometrical and kinematical analysis studied at Neuchâtel, first in certain sectors of the Caledonian range, then in the Precambrian basement in Finland. Thus Wegmann succeeded in combining the ideas of the new school of Scandinavian petrography with the basic principles of structural analysis that had been developed in the Alps.

In 1932 he went to Greenland, first to the east coast, then to the south. In a short time, he managed to build syntheses of these regions, which numerous subsequent studies have confirmed. During this same period, he published "Zur Deutung der Migmatite" (Wegmann 1935). In this outstanding paper Wegmann presents the structural evolution of different tectonic levels as a function of the physical "climate" prevailing in the deeper levels where important redistributions of material are observed. In 1940, after the death of Argand, Wegmann was asked to replace him as professor at Neuchâtel. In contrast to the three pioneers profiled above, and with the exception of "Zur Deutung der Migmatite," Wegmann did not publish his ideas in major synthetic works. They are scattered through many short notes, which seem at first to be of only regional interest. Wegmann was afraid of any schematization, because it evolves too rapidly into dogmatism that weakens our ability to grasp realities. Argand, without much field experience outside of the Alps, but with the help of his fantastic memory and his great gift for synthesis, was able to grasp the entire surface of the world and present a structural picture that was far in advance of its time. Unlike him, Wegmann needed the constant support of precise observations made during his numerous travels; individual mountain ranges were examined from their surface to their deepest part and from their granite roots to their cover foldings with the aim of combining different types of information into several coherent pictures. The idea of structural levels appears very early in his studies, and several drawings published in Scandinavian newspapers between 1927 and 1930 show different aspects of his constant preoccupation with a threefold division of the earth's crust. In these first sketches, the deformation is still fully considered in the light of Argand's "embryotectonics," and the Penninic style is largely dominant (Figure 1-2). Step by step, the individuality of each zone and especially the deepest one becomes apparent. In Greenland and Scandinavia, deep erosion and the existence of strong axial depressions offer the opportunity to compare, side by side and in continuous sections, deformations from the upper levels with the deepest parts of the crust. Particular attention has to be given to chronology: one must recognize synchronous events in the horizontal as well as in the vertical direction. Very useful as guides in this last approach are basic dikes, since they give a precise time signal preserved from the lowest crust to the surface (Wegmann 1960) and allow one to distinguish between structures, movements, and transformations produced before, during, and after their emplacement (Figure 1-3). They are also one of the best sources of information for studying the kinematical and rheological behavior of the deeper levels, where mobility is a fact that has to be considered even if it does not integrate well into many accepted syntheses.

During his stay in Scandinavia, Wegmann was struck by the complexity of the problems raised by the comparative anatomy of mountain ranges. Geology, which should be preemi-

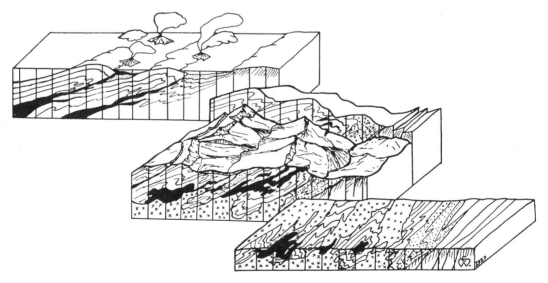

FIGURE 1-2. Wegmann's schematic view of the evolution of an orogenic segment. (Published in Tronderske Blade, 26 April 1930.)

FIGURE 1-3. Greeting card drawn in 1963 by C. E. Wegmann, comparing the surface with the deeper levels in an orogenic domain.

nently a planetary science, suffers more than any other scientific field from political and linguistic problems. To overcome these problems, Wegmann suggested, international meetings that included excursions into typical areas should be encouraged. Efforts should be made to set precise limits for tectonic terms while keeping when possible to the original definitions. In going from the surface to the deeper part of the crust, several structural discontinuities may be crossed. In some cases no major break or change in style can be recognized, but in others, for example salt tectonics, the cover may float over its basement, and movements in and over the salt are no longer related to those of the substratum (Wegmann 1963). Similar decoupling horizons can be produced by rising granitic bodies coming from infrastructural material activated by metamorphic transformation, differentiation, and deformation. The existence of these major discontinuities limits possible extrapolation and leaves room for creative imagination.

Wegmann was deeply impressed by the complex problems that arise in the analysis of the updoming of Fennoscandia following the removal of its ice sheet, especially the interferences between crustal movement and the changing sea level. Recent tectonics is a particularly promising field for comparative anatomy (Wegmann 1955), but it is necessary to increase regional investigations in order to achieve as large an inventory as possible of the different types of recent movements, which should then provide us with a better knowledge of the actual kinematical state of the earth's crust. Comparisons between recent and ancient deformations will be instructive and will help us to visualize the relation between surface and deep-seated evolution. In Wegmann's work everything leads to comparison, taking root in a particular penetrating analysis reinforced by a fertile imagination that surpasses doctrinal representations.

REFERENCES

Agassiz, L., 1833–1843: Recherches sur les poissons fossiles. Neuchâtel.

Agassiz, L., et al., 1838–1842: Monographie d'échinodermes vivantes et fossiles. Neuchâtel.

Argand, E., 1916: Sur l'arc des Alpes occidentales. Eclogae geol. Helvetiae *14*, 145-191.

Argand, E., 1924: La tectonique de l'Asie. 13th Internat. Geol. Congr., Bruxelles 1922, Comptes Rendus, *1*, 171-372.

Cuvier, G., 1812: Recherches sur les ossements fossiles des quadrupèdes. Discours préliminaire. Paris.

Lugeon, M., 1940: Emile Argand. Soc. neuchâteloise. Sci. Nat. Bull. *65*, 25-63.

Wegmann, C. E., 1935: Zur Deutung der Migmatite. Geol. Rundschau *26*, 305-350.

Wegmann, C. E., 1955: Tectonique vivante: vue d'ensemble. Geol. Rundschau *43*, 273-306.

Wegmann, C. E., 1960: Wege in die Tiefen. Geol. Rundschau *50*, 84-94.

Wegmann, C. E., 1963: Tectonic patterns at different levels. Alex L. du Toit Memorial Lectures, *8*, Geol. Soc. South Africa, Trans. *66*, Annexure, 78 p.

Chapter 2

DIFFERENCES BETWEEN MOUNTAIN RANGES

JOHN RODGERS

Yale University

Geologists have been searching for and arguing about the cause of orogeny and mountain ranges for a long time. By the 1930s, there were a large number of competing theories of orogeny, as I learned when Professor C. R. Longwell assigned each member of his advanced class in structural geology one theory to present (I drew Erich Haarmann). The wide variety of theories was most striking, as well as instructive. I learned then, and the lesson has been enforced since in several fields, that when the number of theories to explain one group of phenomena is large, roughly one per theorist, some essential ingredient is lacking; the data are insufficient or some basic idea has not been thought of yet. So it proved in the study of orogeny; the data from the oceans were woefully sparse—if not downright wrong and misleading—until after World War II, and then in the 1960s the integrated theory of plate tectonics provided the set of basic ideas that had been lacking. Orogenic theorists quickly divided into believers and nonbelievers; the believers to be sure were soon arguing about many aspects of the new theory, especially the driving mechanisms, but not about its essentials, whereas the nonbelievers spent most of their time not proposing viable alternatives but trying to disprove plate tectonics by exposing the circular reasoning of some of its proponents (by no means an unhealthy exercise).

One of the characteristics of the older pre–plate-tectonic theories was their persistent attempt to explain all mountain ranges in the same way, different for each theory, of course, and almost always based on the mountain range first studied by the theorist, but the same for all ranges according to that theory. Some plate-tectonic theorists have fallen into the same trap, but the majority recognize, I think, that the new theory permits or even predicts that orogeny will take a number of different forms, and that study of the differences between mountain ranges can be as fruitful as study of their resemblances, which used to be stressed. Thus the idea of a Comparative Anatomy of Mountain Ranges is far more acceptable now than in the old days, when one was tempted rather to examine the comparative anatomy of orogenic theories. Yet differences between mountain ranges can be recognized only if good, detailed descriptions of the mountain ranges are available. Recently there have been several attempts to provide and assemble such descriptions. Such an attempt was the cours romand du 3e cycle en Sciences de la Terre - 1982 - des Universités de la Suisse romande, on which this book is based. If the book has a special virtue, it is that for the most part the articles collected describe less familiar mountain ranges; the Alps were deliberately excluded from the course because they are so familiar to the Swiss students following it. Nevertheless an effort was made to include descriptions of mountain ranges on most of the continents and of every age from early Precambrian to Recent.

What sorts of differences exist and what can they tell us? One obvious difference is in the age and degree of dissection by erosion. Mountain ranges recently uplifted, like the Himalayas and the Andes, or still forming and rising, as in Taiwan, show us the more surficial aspects, although even in these regions erosion has already cut down several kilometers so that the "original" form of the range cannot be seen, except perhaps in southern Taiwan and under water to the south. Mountain ranges like the Hercynian chains of Europe and the Appalachians, the Caledonides of Great Britain and Scandinavia, the Pan-African chains, and especially the earlier Precambrian "ranges" of the Baltic and other shields have lost these surficial aspects to erosion, but the same erosion has exposed their metamorphic and plutonic cores. Indeed it might be argued that most of the differences between ranges are only apparent, the result of dissecting different individuals of a single species to various depths, like learning about the human body by examining it at different levels from the skin to the most internal organs. I believe, however, that mountain ranges in more or less the same degree of dissection—the Himalayas and the Andes, for example, or the Alps and the North American Cordillera—show contrasts that can hardly be the result of age and dissection alone. It is no coincidence that the contrasts are best brought out by comparing ranges from the east-west belt across southern Eurasia with ranges from the circum-Pacific belt (compare the articles by *Frank et al.* and *Mégard* in this book). But there may well be more than just two "species" of ranges to dissect, like learning not only human anatomy but also that of other vertebrates from cats back to fish; only by studying many can we learn if a basic general plan or plans underlie them.

Among the significant differences that distinguish mountain ranges are the presence or absence of particular types of structure or the abundance or paucity of particular types of rock. Thus some ranges are notable for the vast areas of granitoid intrusions that are exposed, but others almost lack them; compare, for example, the Andes and the Alps. Again, chains of large basement-cored uplifts appear on the cratonal platforms in front of several conventional (geosynclinal) mountain ranges, such as the Eastern Rockies of Wyoming and Colorado in front of the main fold-and-thrust belt of the U.S. Cordillera, the High and Middle Atlas in front of the Alpine Rif, and the Timan in front of the northern Urals, but are absent in front of otherwise similar ranges, such as the Canadian Rockies, the Alps, and the Appalachians.

Many mountain ranges include an external zone of strongly deformed but mostly unmetamorphosed sedimentary strata, which have been folded and thrusted over a surface of décollement near the base of the sedimentary cover, above an older basement clearly continuous with that of the nearby craton toward which the folds and thrusts generally verge. Classic examples are the Jura, the Subalpine chains of France, the Helvetic folds and nappes of Switzerland, and also the Valley and Ridge province of the Appalachians. But such belts, despite their common characters, are by no means all alike; several interesting variants are described in this volume. In the Gourma, described by *Caby*, the fold-and-thrust belt at the margin of the Pan-African chain advances in a great convex arc into a preexisting depression—an aulacogen—that extends westward far into the adjacent West African craton. A very similar arrangement is found at each end of the Vilyuysk depression between the main craton of central

Siberia and its southeastern appendage, the Aldan platform and shield; at the southwest end the Baikalian (i.e., approximately Pan-African) chain exhibits a sharply curved arc convex to the north, whereas at the northeast end the much younger (Jurassic) Verkhoyansk Range makes a broader arc convex to the southwest.

On the other hand, in Cantabria, described by *Julivert*, the external fold-and-thrust belt of the Hercynian chain of Iberia forms a sharp bend concave toward the "foreland"; the bend was probably inherited from the original shape of the geosynclinal margin but was considerably tightened and intensified during the later phases of deformation. In Taiwan, as described by *Suppe*, the fold-and-thrust belt is forming today, and its growth can be studied in three dimensions and in time, thanks in part to many oil wells that also provide information on overpressured fluids within the deforming strata. A comparable belt in front of the Himalayas, the Siwalik or Ganges molasse zone, is also active today, as noted by *Frank, Baud, Honegger, and Trommsdorff*. In the Peruvian Andes, described by *Mégard*, there appear to be two such fold-and-thrust belts, the very young (Quechua 3)—if not still active—Subandean belt at the east margin of the chain, verging east toward the Brazilian craton, and a somewhat older (Incaic) belt farther west, verging northeast toward the Cordillera Oriental, which had already been consolidated in still earlier (Peruvian) phases of orogeny.

Such fold-and-thrust belts appear to be lacking in some older mountain ranges, chiefly, perhaps, because, as they are zones of relatively shallow deformation above a décollement, they are readily removed if erosion goes deep enough. Thus such a belt is lacking along most of the eastern front of the Scandinavian Caledonides, but it is still visible in Jämtland in Sweden and especially where the Permian down-faulting of the Oslo graben has preserved a narrow slice across its full original width of 100 km. Similarly, a fragment of a fold-and-thrust belt is preserved northwest of the Grenville front in the Mistassini region of Quebec, and another example may be the Dal group of southern Sweden, caught under the Dalsland Boundary fault mentioned by *Berthelsen*. Further examples are known back throughout the Proterozoic, as mentioned by *Windley*; notable is the Wopmay orogen, nearly 2,000 Ma old, in the northwestern corner of the Canadian shield.

The fold-and-thrust belts of the western Alps and the Appalachians are backed by anticlinoria of basement (the external massifs in the Alps, the Blue-Green-Long axis in the Appalachians), which appear just where metamorphism becomes significant in the cover rocks. Until recently, these anticlinoria were generally regarded as essentially autochthonous or parautochthonous uplifts of the same cratonal basement that underlies the adjacent fold-and-thrust belt, although, at least in parts of the Appalachians, the presence of upside-down unconformities, stacked horizontal thrust sheets involving basement, and large windows suggested that certain parts were allochthonous. Now these suggestions have been strongly confirmed in both mountain ranges by geophysical evidence, such as gravity minima where maxima would be expected and regular subhorizontal reflecting "horizons" continuous with the décollement surfaces under the fold-and-thrust belts but extending partly or entirely beneath the anticlinoria. Thus large segments of these anticlinoria appear to be completely allochthonous, and the rest may be partially so.

A particularly spectacular example is the recently discovered subsurface "COCORP" fault in the southern Appalachians, on which, at a late stage in the orogenic cycle, already metamorphosed, intruded, and consolidated rocks were thrust some 200 km northwestward, with little or no additional internal deformation, but acting as a plunger to produce the intense shortening in the adjacent fold-and-thrust belt of the Valley and Ridge province. Indeed, one might ask whether the already consolidated Cordillera Oriental of Peru has not recently acted or is not now acting as a similar plunger to produce the Subandean fold-and-thrust belt.

Similarly, in the northern Scandinavian Caledonides, as soon as one enters the zone of metamorphism, one encounters a pile of thrust sheets (stacked up during two different orogenic episodes); many of them include both basement and cover. Farther south, a thin basement thrust sheet (the Tännäs augen gneiss nappe of west-central Sweden and its continuations in central Norway) forms one member of a pile of superposed thrust sheets, lying between two others composed only of younger sediments; it must record 100 or 200 km of horizontal displacement, although its rocks still belong to the Baltic craton over which they have been thrust. In places in western Norway, the whole pile of sheets, already stacked up, was then recumbently folded during metamorphism, perhaps more than once. *Milnes* compares the basal thrust of the pile in a nearby part of the range to the southern Appalachian ''COCORP'' fault.

Again, there are significant differences between mountain ranges. The Narcea anticline in Cantabria, described here by *Julivert*, is similar in position to the anticlinoria in the Alps and Appalachians, but it is made of only slightly older sedimentary strata instead of ''crystalline'' basement. Similarly, the comparable Ural-Tau anticlinorium of the Urals is made mostly of the older Riphaean (late Precambrian) sedimentary sequence, the true basement cropping out only locally. Moreover, although all the other anticlinoria and basement thrust sheets mentioned above verge toward the fold-and-thrust belt, the Ural-Tau verges away from it (eastward). Yet there is no reason to doubt that the décollement beneath the fold belt to the west roots under the anticlinorium. Another possible uplift of this sort might be the ''Mitteldeutsche Schwelle'' between the Rhenohercynian zone (the external, thin-skinned, fold-and-thrust belt) and the Saxothuringian zone (the more internal, older, metamorphic belt) on the northern side of the Hercynian orogenic belt in Germany.

In other ranges, such as the North American Cordillera, external massifs or anticlinoria are poorly developed or lacking, unless their place be taken by the ''metamorphic core complexes'' or ''low-angle normal-fault complexes,'' which are, however, much younger than the main orogenic deformations there. In some ranges, on the other hand, crystalline thrust sheets or nappes represent the basement of the craton on the opposite side of the mountain range. Examples are the Austro-Alpine nappes of the Alps; the highest thrust sheets in the western Tien-Shan; perhaps also the allochthonous, katazonal metamorphic complexes in Galicia and northern Portugal, described here by *Julivert*; the Münchberg granulite massif or klippe lying within the Saxothuringian zone in east-central Germany; or the Jotun nappe in Norway.

Transcurrent or wrench faults, with different characteristics according to the materials that were already present, also appear in several of the mountain ranges described in this volume. Much of the material sliced up by both normal and wrench faults in the High Atlas of Morocco, described by *Schaer*, was already consolidated (Hercynian) ''basement,'' but the thick Mesozoic sediments deposited in the resulting pull-apart basins, especially in the eastern part of the range, were intensely deformed during and in large part because of the continued faulting. Such a mechanism probably also applies in the northern Pyrenees and the Greater

Caucasus. *Laubscher* describes a complicated mosaic of plates, jostled by large transcurrent movements, in the northern Andes, and *Meier, Schwander, and Laubscher* show the effects of such movements on strata that in a simpler setting might have produced a more conventional fold-and-thrust belt.

The existence of large transcurrent movements in relatively recent mountain chains (California, Anatolia) raises the question of their presence in older chains, in zones since overprinted by metamorphism, and of the original relations between the various terrains now juxtaposed but originally perhaps far apart. This question of ''exotic terrains'' has been explored particularly in the North American Cordillera but is pertinent in many other ranges, for example, the Avalonian terrain in the northern Appalachians, described here by *Rodgers*. Could the major north-south transcurrent faults that cut across the Touareg shield, described by *Caby*, be separating originally distant terrains?

In the interior metamorphic belts of many ranges, we find remarkable large-scale recumbent folds or nappes, commonly called ''of Pennine type'' from the typical examples in the Pennine zone of the western Alps (similar nappes are exposed in the Tauern window in the eastern Alps). The type examples have cores of older basement, but the basement was so thoroughly metamorphosed during the Alpine orogeny that its rocks were as ductile as the enveloping cover; in other cases, however, such cores are lacking. Examples of such nappes are the Tassendjanet and other nappes in the Pan-African belt, described by *Caby*; the Mondoñedo nappe in Galicia, described by *Julivert*; and the crystalline nappe in the Himalayas, described by *Frank, Baud, Honegger, and Trommsdorff*. *Windley* wonders whether the complicated structure recently described in the Archean greenstone belts is not of the same sort.

In many chains it seems probable that the Pennine-type nappes formed considerably earlier than the more external fold-and-thrust belt of the same mountain chain. Nevertheless, although a general outward progression of deformation can be recognized in many orogenic periods and cycles, it is by no means universal—an instance, once again, of significant differences in mountain ranges and their history.

The kind of regional metamorphism present also differs between mountain ranges. Barrovian or even higher pressure series are widespread in some; high-temperature metamorphism predominates in others. ''Paired belts,'' with a high-pressure belt toward the coast and a high-temperature belt inland, are certainly common around the Pacific, but they are not universal even there and seem to be altogether lacking in many other ranges. In some instances, at least, early high-pressure metamorphism has been followed by low-pressure/high-temperature metamorphism, making the earlier metamorphism difficult to decipher. Again one wonders to what extent age, repeated deformation, and degree of dissection have influenced what we now see. In any case, as *Windley* points out, one can no longer say that blueschist metamorphism and ophiolites are unknown in the Precambrian.

In a number of chains geologists have described volcanic belts, the chemistry of whose rocks suggests that they were once volcanic island arcs like those that now festoon the western margin of the Pacific basin. Examples are the volcanic rocks of the Upper Proterozoic in the western Hoggar, described by *Caby*; the Bronson Hill anticlinorium in the northern Appalachians, described by *Rodgers*; the Dras volcanic belt in the western Himalayas, described by *Frank, Baud, Honegger, and Trommsdorff*; and the volcanic belt in the eastern Coastal Range of Taiwan, described by *Suppe*. The question immediately arises as to whether the basins on either side of these belts, commonly filled with thick bodies of often volcanogenic sediments, plus volcanics, were originally floored by oceanic or by continental crust, and vigorous debate continues over many examples.

As an example of the kinds of complexities we ought to expect to find in ancient mountain chains, I would like to cite the present-day volcanic arc of the Aleutian Islands, whose western part is flanked on both sides by wide areas of oceanic crust, but whose eastern end is superposed on the continental crust of southern Alaska, probably already consolidated by Cretaceous time, though orogeny related to subduction of the Pacific plate continues to be very active there. Thus one and the same volcanic arc is oceanic at one end, continental at the other.

Indeed the whole problem of how much of the core belts of mountain ranges was originally oceanic (whether "true" ocean like the present Atlantic or wide marginal seas like the Bering and Philippine Seas) is in the center of current tectonic debate, as is well illustrated in many of the articles in this book. Thus belts of ultramafic and mafic rocks (continuous or, more often, discontinuous) are variously interpreted as sutures representing ancient oceans or, on the contrary, as zones where mantle material has been injected along intracontinental rifts or transcurrent faults. Again, mountain ranges may well differ greatly in this regard, some being formed essentially from deposits in marginal seas or in old oceans or along their borders and others being largely or entirely intracontinental, the thick sedimentary and volcanic piles of which they are formed having been deposited in rift valleys or other continental basins.

In this connection one thinks also of the famous "greenstone" belts of the Archean, described here by *Windley*. Although commonly thought to be confined to the Archean, both *Windley* and *Berthelsen* point out Proterozoic examples; elsewhere Windley has described a much younger example in the internal zones of the southern Andes (Zone D, discussed by *Milnes*), and they may well be present elsewhere in younger mountain ranges, representing old marginal seas. Thus one comes to the broader question, raised by both *Windley* and *Berthelsen*, of whether orogenic processes have evolved through geological time, so that certain types of mountain ranges or structures (greenstone belts, for example) might be characteristic or more characteristic of the distant past and others (fold-and-thrust belts, for example) of more recent time.

Another broad theoretical question that divides observers and indeed affects the way they view orogenic belts is what motor drives orogeny. The immense forces required to produce the observed deformations in vast masses of solid rock can ultimately be explained only by considering the earth as a heat engine; the heat flowing from the interior to the surface (attested by the well-known if highly nonuniform thermal gradients in the crust) can certainly provide the necessary energy. But some mechanism for converting the heat flow into mechanical energy must be sought. Fortunately, calculations show that the mechanism can have a very low efficiency, only a few percent. The great differences in tectonic theory in our day concern not the ultimate heat source ("original" and radioactive heat in the earth) but this mechanism, and these differences cannot help but color our descriptions of the mountain ranges.

If I may oversimplify, two basic ideas about the mechanism or motor are currently competing. According to one, the heat energy causes differential vertical movements of the crust, and the evident horizontal movements of rock masses recorded in the mountain ranges are simply the result of *gravity* which acts on masses that stand high and makes them slide or flow into lower regions, like immense landslides, or like the ice in the continental ice sheets of

Antarctica and Greenland, which flows or spreads away from the high centers of the sheets by gravity, even if some of the ice is actually moving uphill. According to the other idea, the heat flow, by a mechanism akin to *convection*, causes slow but large horizontal movements of the whole crust or lithosphere, tearing it apart in some areas and smashing it together in others, the smashed-together areas being in general where orogeny occurs and mountain ranges appear. Indeed, some of the apparent differences between mountain ranges may simply reflect the different basic ideas of those who describe them.

By a third mechanism much favored in the past, tangential compression resulted directly from *contraction* during cooling of the earth, as an earlier cooled and hence already contracted crust or lithosphere adjusted itself to the continued cooling of the interior. This theory was first formulated when heat flow from inside the earth was explained *only* as loss of original heat, and hence as true cooling; the discovery that radioactivity creates heat within the earth decreased by an order of magnitude or so the supposed rate of cooling and hence the available tangential compression, which in any case proved insufficient to explain the amount of tangential shortening that geologists were recording all over the world as they studied more and more mountain ranges. Harold Jeffreys therefore, in a famous paragraph in the second edition of his book *The Earth* (paragraph 15.25, repeated with little change in subsequent editions up to 1976), scolded geologists for daring to record and believe in such large amounts of shortening, which were contradicted by the calculations that he had based on the contraction theory. But the obvious conclusion is the reverse; the existence of the shortening proves instead that the calculations rest on the false premise that the compression responsible for the shortening resulted only from contraction during cooling. Thus, by Jeffreys' own argument, the contraction theory is quite inadequate to explain orogeny.

The contrast between the two other ideas—gravity spreading and lateral compression powered by convection—was posed with particular clarity in the debate between David Elliott and William Chapple concerning the deformation of the Canadian Rockies. Both applied sophisticated mathematics to the question, but as they started from somewhat different assumptions, they reached different conclusions. Elliott replied to Chapple's criticisms, but in later lectures and conversations he seemed to be swinging toward Chapple's position. The mechanical analysis of the Taiwan fold-and-thrust belt in the article by *Suppe* supports Chapple's point of view that lateral compression is the prime motor, that the topographical slope called on by Elliott to instigate gravity spreading was produced by that compression. Needless to say, the argument is not over yet.

We hope that the collection of mountain-range descriptions in this volume will add to the data base needed for further debate on these questions and, perhaps, for their ultimate solution.

REFERENCES

Chapple, W. M., 1978: Mechanics of thin-skinned fold-and-thrust belts. Geol. Soc. America Bull. *89*, 1189-1198.

Elliott, David, 1976: The motion of thrust sheets. Jour. Geophys. Res. *81*, 949-963.

Elliott, David, 1980: Mechanics of thin-skinned fold-and-thrust belts: Discussion. Geol. Soc. America Bull. *91*, pt. 1, 185-187.

Chapter 3

TECTONIC FRAMEWORK OF PRECAMBRIAN BELTS

BRIAN F. WINDLEY

University of Leicester

The Precambrian eon can be usefully divided into the Archean and Proterozoic. These eras are separated by the Archean-Proterozoic Boundary (APB), which represents a diachronous, transitional time period of ca. 2,700 to 2,300 Ma B.P.

THE ARCHEAN

The twofold division of Archean rocks into greenstone belts (GB) and granulite-gneiss belts (GGB) is well known (Windley 1984). For the first third of geological history, the same types of rocks were produced, but secular variations were also present. The paragraphs that follow describe the main features of these two types of belts.

Condie (1981) lists 22 factual observations about typical Archean GB. They are volcano-sedimentary basins that formed in environments ranging from ensialic rifts in a gneissic basement, like the Chitradurga belts of southern India (Chadwick et al. 1981), to ensimatic basins between small sialic nuclei, like those in southern Canada (Goodwin and Smith 1980). Two types of volcanic associations are recognized: bimodal (lacking andesites) and calc-alkaline. Early GB may have either type, but late ones are typically calc-alkaline. Ultramafic and mafic volcanics predominate in the lower levels of the successions, with increasing proportions of calc-alkaline and felsic volcanics in the higher levels. Komatiites, especially peridotitic types, are distinctive. Mature sediments are more common in early GB and immature sediments in later belts—an unexpected secular variation (Condie, personal communication). Intrusive plutons range from early gneissic tonalites and trondhjemites to rare post-tectonic high-K granites (Condie 1981).

It is increasingly realized that the structural evolution of GB is far more complicated than is suggested by a layer-cake, time-continuous stratigraphy, and that these belts are not simple downfolds between rising diapirs (Burke et al. 1976). In South Africa the Barberton, Murchison, and Sutherland GB are floored by tectonic contacts and have been thrust onto underlying gneisses (Fripp et al. 1980). Thrust-nappe tectonics has given rise to inverted successions in GB in Zimbabwe (Stowe 1974, Coward et al. 1976); Barberton, South Africa (Williams and Furnell 1979), Ontario, Canada (Poulsen et al. 1980), and the Yilgarn, southern Western Australia (Archibald et al. 1978). Later subhorizontal compression produced refolds with a steep cleavage, and finally the belts were intruded by diapiric plutons.

Granulite-gneiss belts have not been so usefully synthesized. The predominant rock unit (about 80-85% surface area) is a tonalitic to granodioritic orthogneiss in high amphibolite or granulite grade. Usually only about 5 to 10% of the gneisses have a sedimentary parentage. Within the gneisses are layers up to about 1 km thick of metavolcanic tholeiitic amphibolite, which is typically associated with cratonic or continental margin-type metasediments of the quartzite, marble, mica schist association, usually up to only a few meters or tens of meters thick. Rarely, as in the Limpopo belt of southern Africa, they reach 1 km in thickness. Commonly intruded into the above supracrustals are layered cumulate stratiform complexes with predominant anorthosites and leucogabbros (Windley et al. 1981). The precursors of the voluminous orthogneisses were intruded into the supracrustals and the layered complexes. The whole assemblage was then metamorphosed at about 9 to 11 kb and deformed into complex fold interference patterns in the lower part of late Archean (mostly 3,000-2,700 Ma) thickened continental crust.

Some Archean GGB contain relicts of gneiss material 3,600 to 3,800 Ma old (West Greenland, Labrador, the Limpopo belt), but the peak of tonalitic gneiss injection was in the late Archean, approximately coincident with the peak of GB formation. There are some interesting long-term variations: (1) the 3,800-Ma Amitsoq gneisses in West Greenland are more potassic than the 3000-Ma Nuk gneisses; (2) except for one very small occurrence in West Greenland (Griffin et al. 1980), all the very old gneissic material only reached an amphibolite grade of metamorphism, whereas granulites were commonly produced in the late Archean in association with the generation of very thick crust; and (3) the layered anorthositic complexes are only of late Archean age.

It is widely considered that the most important phase of continental crustal growth was in the late Archean. Estimates of the percentage of present crustal mass generated by the APB range from over 50 (Condie 1981) to about 85 (Dewey and Windley 1981).

CRUSTAL CONDITIONS IN THE EARLY PROTEROZOIC

The massive thickening of lithosphere in the late Archean gave rise by the early Proterozoic to extensive, thick, rigid, and stable continental parts of plates underlain by tectospheric roots; this time boundary therefore represents the most important threshold in geological history (Windley 1984, Dewey and Windley 1981). The fact that early Proterozoic lithospheric plates were beginning to respond to deformation, deposition, and intrusion in a mode comparable to that of today is indicated by the following features:

1. Large, stable continental areas showing a high degree of rigidity, as demonstrated by the deposition and preservation of little-deformed cratonic sequences (Witwatersrand, Huronian) and by the intrusion of extensive linear dike swarms (Molson dikes, Canada; Waterberg, Botswana-Transvaal; Umkondo, Zimbabwe; see Windley 1984).

2. Continental shelf-rise sequences on and aulacogen (Milanovsky 1981) re-entrants in continental margins (Aldan shield), often with early alkalic complexes associated with the rifted margin (Wopmay orogen, Hoffman 1980).

3. Linear/arcuate orogenic belts bordering continental plates left as stable interiors (Circum-Ungava geosyncline, Gibb 1975) with fore-arc thrust sheets over basement (Wopmay

orogen, Hoffman 1980), flysch belts (Lapland, Barbey et al. 1982), and accretionary prisms (Manitoba and Montana, Fountain and Desmarais 1980).

4. The construction of island arcs, Andean arcs, and back-arc basins, the igneous rocks of which are chemically similar to modern equivalents. Examples include the Churchill Province of Canada (Moore 1977, Lewry et al. 1981); the Great Bear volcano-plutonic belt in the Wopmay orogen (Hildebrand 1981); Arizona, U.S.A. (Anderson 1977); Svecokarelides (Hietanen 1975, Torske 1977, Löfgren 1979, Loberg 1980); Queensland in Australia (Wilson 1978); Ketilidian in southern Greenland (Berthelsen, this volume); Colorado, U.S.A. (Condie and Nuter 1981); Finnish Lapland (Raith et al. 1982); and Wyoming, U.S.A. (Condie 1982a).

5. Himalayan-type collisional belts (the Svecokarelides of the Baltic shield, Bowes 1980; the Musgrave Range of Australia, Davidson 1973; the Churchill Province of Canada, Gibb et al. 1980), widespread basement reactivation adjacent to linear/arcuate thrust belts involving the generation of high-potash minimum-melting granites (East Labrador Trough; Hepburn batholith in the Wopmay orogen, Hoffman 1980); and slip-line indentation fracture systems in cratonic forelands (Wopmay orogen, Hoffman 1980; Churchill Province-Slave Craton, Gibb 1978; deep-level shear zones in West Greenland, Watterson 1978).

Condie (1982b) grouped early to middle Proterozoic supracrustal successions into the following three lithologic assemblages: (1) quartzite-carbonate-shale, which reflects stable continental margins or intracratonic basins; (2) bimodal volcanics-quartzite-arkose formed in lithosphere-activated continental rifts or aborted mantle-activated rifts; and (3) continuous (tholeiitic and calc-alkaline) volcanics-graywacke, an assemblage similar to that in Archean GB and formed in mantle-activated cratonic rifts or basins associated with converged plate boundaries. The possible inter-relationships between these three types of assemblages and their tectonic environments in a plate-tectonic framework is not yet understood.

It should be emphasized that the APB is in need of major reassessment. Although it does separate features of permobility from those of cratonic stability, as discussed below, it did not prevent the two main types of Archean orogenic belts (GB and GGB) from continuing into the Proterozoic (Tarney and Windley 1981, Windley 1983). Because Proterozoic belts of all types can increasingly be understood in terms of a modern type of plate-tectonic regime, they should enable us to understand better the more problematic mode of origin of comparable Archean belts.

THE WILSON CYCLE IN EARLY PROTEROZOIC TIMES

The APB separates Archean rocks, whose tectonic mode of origin cannot *easily* be related to any modern type of tectonic regime, from early Proterozoic rocks, a great many of which have features directly comparable to those of modern belts formed by plate-tectonic processes. Thus the APB separates rocks that can and cannot be readily ascribed to the Wilson cycle. It is relevant here to enumerate some key features of these early Proterozoic belts.

The Wopmay orogen (2,200-1,800 Ma) in northwestern Canada (Hoffman 1980) is outstanding because it retains evidence of a complete Wilson Cycle: continent-derived alluvial and deltaic clastics with bimodal lavas formed during early rifting; trailing continental margin with terrace (shelf slope) and rise; collapsed carbonate bank followed by continent-directed flysch and molasse; aulacogens projecting into the continental margin; west-dipping followed by oblique east-dipping subduction; paired batholithic belts; the Great Bear volcano-plutonic belt containing high alumina-low TiO_2 andesites, basaltic andesites, and basaltic to dacitic to rhyolitic ash-flow tuffs, which are nearly identical in major and minor element geochemistry to modern continental magmatic arcs (Hildebrand 1981); and Himalayan stage of collision, represented by craton-directed overthrusts in the margin of the belt and slip-line indentation fracture systems in the foreland. This type of geological framework is different from anything found so far in the Archean.

In the western Churchill Province of Canada there is well-documented evidence of 2,000 to 1,800 Ma arc-microcontinent collisional tectonics (Ray and Wanless 1980, Lewry 1981, Lewry et al. 1981). There is a major volcano-plutonic arc, a microcontinent, an intervening telescoped fore-arc basin, a transform fault between the microcontinent and a remnant interarc basin, and a monzogranitic-granodioritic Andean-type batholith.

In Arizona there are four 1,800- to 1,710-Ma volcanic belts flanked by aprons of volcanogenic graywackes and siltstones which show mafic to felsic evolution with time and vertical stratification from low-K tholeiites at the base to calc-alkaline volcanic, fragmental, and clastic types (Anderson 1977). Each volcanic belt shows lateral alkali-silica polarity and has a coeval plutonic batholithic belt, which likewise shows chemical polarity. The features enumerated above have analogues in modern island arcs.

The reason for describing these three belts is to illustrate the point that this kind of sedimentary, igneous, structural, and chemical polarity across the strike of a belt was only possible in the early Proterozoic (in contrast to the Archean) because of the existence of stable cratons or continental margins along which these belts could develop at the interface with what must have been a subducting oceanic plate (given the evidence of the igneous products). The relationships described are essentially the same as those seen today in the Himalayas, which went through a major period of growth of island arcs and Andean-type margins before the final continent-continent collision. Thus in separating a period in which no substantial stable plates formed from one in which they formed on such a scale as to enable continental-margin belts to develop around them, the APB defines the time at which modern-style plate-tectonic processes were able to begin.

GRENVILLE-DALSLANDIAN MOBILE BELT

In the period 1,000 $\pm$ 200 Ma a series of mobile belts formed, some of which are so deeply eroded that their high-grade lower crustal segments are exposed. Prominent in the North Atlantic region is the Grenville-Dalslandian belt. Many controversial interpretations have arisen in an effort to account for the tectonic evolution of the Grenville belt. This is not surprising, not because of anything unique or unusual about the Grenville belt, but simply because very little is known about how the lower parts of modern orogenic belts formed.

Ideas on the origin of the Grenville belt include *in situ* ensialic (no lateral shortening) deformation of cover-basement (Wynne-Edwards 1972), a dextral shear zone (Baer 1977), very slow sideways shunting of continental crust over a hot spot (Wynne-Edwards 1976), and a Wilson Cycle with prominent Tibetan-style late continental collision (Dewey and Burke 1973, Baer 1976, Young 1980). As a result of a conference on the Grenville, Baer and others (1974) stated that ''most experts would agree that the Grenvillian orogeny may be explained in terms of plate tectonics.'' A major constraint on such models is provided by paleomagnetic data. The sequence of loops with amplitudes of 60° and periods of 200 to 300 Ma is consistent with the operation of the Wilson Cycle, four hypotheses being invoked by Irving et al. (1974). In contrast, Dunlop et al. (1980) conclude that the data suggest tentatively that a small ocean between ''Grenvillia'' and Laurentia may have closed after 1,200 Ma by 90° rotation about a pole northeast of Newfoundland, and that the well-dated poles for the Grenville Loop in the North American apparent polar wander (APW) path between 1,050 and 800 Ma rule out the possibility of a collision between ''Grenvillia'' and the rest of the Canadian shield in this time interval. Evidence in the foreland that strongly supports the idea of the operation of a Grenville Wilson Cycle includes the presence of the contemporaneous Keweenawan rift system and its attendant gravity high, the rift triple junction in the Lake Superior region, the belt of alkaline complexes marginal to the northwest side of the Grenville belt, and the Abitibi dike swarm. There is evidence for the existence of oceanic lithosphere in southeast Ontario, where a belt of mafic pillow basalts overlies a mafic-ultramafic metaigneous complex; is overlain by more salic volcanics, sediments, and marbles; and is intruded by granodioritic to granite plutons (Chappell et al. 1975). There is evidence also in the Llano Uplift of Texas, where a serpentinite formed in an ophiolite 1,000 to 2,000 Ma old (Garrison 1981). Here 4 km of low-K tholeiites whose trace-element distributions are similar to those of modern immature emerging arcs are overlain by 3 km of andesite, geochemically comparable to volcanic rocks from mature arc systems.

Farther to the east, according to Young (1980), there are Grenville-age rocks and deformation in southern Norway and East Greenland, northwest Africa, the Channel Islands, the Moines of Scotland, Brittany, and the Nubian and Arabian shields. Young (1980) suggests that the Bou Azzer ophiolite in Morocco may be of Grenville age (1,000 Ma), but this would seem to be incorrect in the light of the new isotopic data of Leblanc and Lancelot (1980), which indicate a history from 788 to 534 Ma. As yet, the geochronological data do not enable us to separate clearly the events of Grenville and Pan-African age; nevertheless some ophiolites may be related to the Grenville orogeny. In particular, serpentinites, gabbros, sheeted dikes, and mafic volcanics 1,200 to 850 Ma old in Egypt are associated with chert and tectonic mélange (Shackleton et al. 1980); these record the presence of a former ocean or marginal basin of Grenville age, but whether it was closed in late Grenville or early Pan-African times is uncertain.

An interesting picture is emerging from the Dalslandian belt (1,000 ± 200 Ma) in southern Sweden and southern Norway where Berthelsen (1980) has recognized several low-angle thrusts on which slabs of crustal thicknesses 200 to 300 km long and 50 km wide have been piled on top of each other, and he interprets this thrust-stacking as a result of a Himalayan-style continental collision. According to Falkum and Petersen (1980), aulacogens about 1,190 Ma old, preserved in the Telemark area of southern Norway, formed during the early stages of this Sveconorwegian orogeny. Evidence elsewhere of crustal tectonics 1,200 to 800 Ma old includes many aulacogens of the Russian plate, the GGB of Antarctica, the Namaqua-Natal mobile belt of South Africa, and the granulite belt on the south coast of Western Australia.

PAN-AFRICAN BELTS

In the period 800 to 500 Ma a series of mobile belts formed in many continents; the Pan-African, Brazilian, and Cadomian belts are particularly well known. As late as 1973 it was widely thought that many of these belts formed ensialically, but there has been considerable success in recent years in establishing geological relationships that point toward a plate-tectonic mode of development. A major problem still remains, however, both in the lack of age constraints with regard to the subdivision of the many belts that formed in this period and in the fact that the older belts overlap with those of Grenville age and the younger ones overlap with those of Caledonian-Appalachian age.

The key results indicating modern-style plate-tectonic activity are listed in Table 3 of Windley (1983) and are summarized in detail in Windley (1984). Such conclusions based on geological and geochemical relationships are further supported by the accumulating paleomagnetic evidence for large and rapid APW from all major continents; because their APW paths cannot all be matched, it follows that large relative continental movements occurred.

One group of rocks deserves special comment. Engel et al. (1980) describe the Pan-African "oceanic-arc complexes" of Egypt, Sudan, and Saudi Arabia as constituting "the largest and one of the few well preserved greenstone belts in the Proterozoic of the world." Serpentinized ultramafic rocks, overlain by gabbros (with gabbro dikes) and pillowed tholeiitic basalts, form an oceanic substrate overlain in turn by a calc-alkaline sequence of andesitic and more felsic metavolcanics of arc type interfingered with coeval immature sediments (tuffaceous wackes and volcanoclastics intruded by sills of basaltic komatiite) and banded iron formations of Algoma type. These rocks are engulfed by synkinematic and younger granites. As Engel et al. (1980) point out, these rocks have geological and geochemical features comparable on the one hand to modern ocean-floor ophiolites and overlying island-arc suites and on the other to many Archean GB (and by implication to early Proterozoic GB, as indicated in this paper). These Pan-African rocks thus provide an important link between modern and early Precambrian sequences.

SUMMARY

Many Proterozoic mobile belts have a sedimentary, igneous, metamorphic, structural, and geochemical record that is entirely compatible with a modern type of plate-tectonic evolution. The massive thickening of crust and lithosphere in the late Archean gave rise by the early Proterozoic to thick and stable continental plates that began to respond to deformation, deposition, and intrusion in a mode comparable to that of today. Orogenic belts developed along the margins of continental plates, and all stages of the Wilson Cycle can be recognized from the early Proterozoic. But like Mesozoic-Cenozoic orogenic belts, each Proterozoic belt has its own signature, dependent on local tectonic conditions at the time and place. Thus overly simple correlations should not be expected.

In contrast, Archean belts formed in a tectonically more mobile regime; thus early cratons did not survive. Orogenic belts tend to lack the linearity, narrowness, and asymmetry of the Proterozoic belts. Special tectonic mechanisms have to be invoked in order to explain the origin of these belts, but such mechanisms may only be variants of their modern counterparts.

REFERENCES

Anderson, P., 1977: The island arc nature of Precambrian volcanic belts in Arizona. Geol. Soc. America Abstr. Programs *10*, 156.

Archibald, N. J., Bettenay, L. F., Binns, R. A., Groves, D. I., and Gunthorpe, R. J., 1978: The evolution of Archean greenstone terrains, Eastern Goldfields Province, Western Australia. Precambrian Res. *6*, 103-131.

Baer, A. J., 1976: The Grenville Province in Helikian times: A possible model of evolution. R. Soc. London, Phil. Trans. *A280*, 499-515.

Baer, A. J., 1977: The Grenville Province as a shear zone. Nature *267*, 337-338.

Baer, A. J., Emslie, R. F., Irving, E., and Tanner, J. C., 1974: Grenville geology and plate tectonics. Geoscience Canada *1*, 54-61.

Barbey, P., Touret, L., Capdevila, R., and Hameurt, J., 1982: Mise en évidence de deux séries paléovolcaniques dans la ceinture protérozoïque de la Tana; conséquences sur l'environnement géotectonique de la ceinture des Granulites de Laponie (Fennoscandie). Acad. Sci. Paris C. R. (2) *294*, 207-210.

Berthelsen, A., 1980: Towards a palinspastic tectonic analysis of the Baltic shield. *In*: Cogné, J., and Slausky, M. (eds), Geology of Europe. Bur. Rech. Géol. Min. Mém. *108*, 5-21; 26th Internat. Geol. Congr., Paris 1980, Colloque *C6*: Geology of Europe, 5-21.

Bowes, D. R., 1980: Correlation in the Svecokarelides and a crustal model. *In*: Mitrofanov, F. P. (ed), Principles and Criteria of Subdivision of Precambrian in Mobile Zones. Nauka, Leningrad, 294-303.

Burke, K., Dewey, J. F., and Kidd, W.S.F., 1976: Dominance of horizontal movements, arc and microcontinental collisions during the later permobile regime. *In*: Windley, B. F. (ed), The Early History of the Earth. Wiley, New York and London, 113-129.

Chadwick, B., Ramakrishnan, M., and Viswanatha, M. N., 1981: The stratigraphy and structure of the Chitradurga region: An illustration of cover-basement interaction in the late Archaean evolution of the Karnataka craton, southern India. Precambrian Res. *16*, 31-54.

Chappell, J. F., Brown, R. L., and Moore, J. M., Jr., 1975: Subduction and continental collision in the Grenville province of southeastern Ontario. Geol. Soc. America Abstr. Programs *7*, 733-734.

Condie, K. C., 1981: Archaean Greenstone Belts. Elsevier, Amsterdam, 434 p.

Condie, K. C., 1982a: Plate-tectonics model for Proterozoic continental accretion in the southwestern United States. Geology *10*, 37-42.

Condie, K. C., 1982b: Early and middle Proterozoic supracrustal successions and their tectonic settings. Am. Jour. Sci. *282*, 341-357.

Condie, K. C., and Nuter, J. A., 1981: Geochemistry of the Dubois greenstone succession: An early Proterozoic bimodal volcanic association in west-central Colorado. Precambrian Res. *15*, 131-156.

Coward, M. P., Lintern, B. C., and Wright, L. I., 1976: The pre-cleavage deformation of the sediments and gneisses of the northern part of the Limpopo belt. *In*: Windley, B. F. (ed), The Early History of the Earth. Wiley, New York and London, 323-330.

Davidson, D., 1973: Plate tectonics model for the Musgrave Block-Amadeus Basin complex of central Australia. Nature Phys. Sci. *245*, 21-23.

Dewey, J. F., and Burke, K.C.A., 1973: Tibetan, Variscan and Precambrian basement reactivation: Products of continental collision. Jour. Geology *81*, 683-692.

Dewey, J. F., and Windley, B. F., 1981: Growth and differentiation of the continental crust. *In*: Moorbath, S., and Windley, B. F. (eds), The Origin and Evolution of the Earth's Continental Crust. R. Soc. London, Phil. Trans. *A301*, 189-206.

Dunlop, D. J., York, D., Berger, G. W., Buchan, K. L., and Stirling, J. M., 1980: The Grenville Province: A paleomagnetic case-study of Precambrian continental drift. *In*: Strangway, D. W. (ed), The Continental Crust and its Mineral Deposits. Geol. Assoc. Canada Spec. Paper *20*, 487-502.

Engel, A.E.J., Dixon, T. H., and Stern, R. J., 1980: Late Precambrian evolution of Afro-Arabian crust from ocean arc to craton. Geol. Soc. America Bull. *91*, pt. 1, 699-706.

Falkum, T., and Petersen, J. S., 1980: The Sveconorwegian orogenic belt, a case of late-Proterozoic plate-collision. Geol. Rundschau *69*, 622-647.

Fountain, D. M., and Desmarais, N. R., 1980: The Wabowden terrane of Manitoba and the pre-Belt basement of southwestern Montana: A comparison. Montana Bur. Mines Geology Spec. Publ. *82*, 35-46.

Fripp, R.E.P., van Nierop, D. A., Callow, M. J., Lilly, P. A., and du Plessis, L. U., 1980: Deformation in part of the Archaean Kaapvaal Craton, South Africa. Precambrian Res. *13*, 241-251.

Garrison, J. R., Jr., 1981: Coal Creek Serpentinite, Llano Uplift, Texas; a fragment of an incomplete Precambrian ophiolite. Geology *9*, 225-230.

Gibb, R. A., 1975: Collision tectonics in the Canadian Shield? Earth Planet. Sci. Lett. *27*, 378-382.

Gibb, R. A., 1978: Slave-Churchill collision tectonics. Nature *271*, 50-52.

Gibb, R. A., Thomas, M. D., and Mukhopadhyay, M., 1980: Proterozoic sutures in Canada. Geoscience Canada *7*, 149-154.

Goodwin, A. M., and Smith, I.F.M., 1980: Chemical discontinuities in Archaean metavolcanic terrains and the development of Archaean crust. Precambrian Res. *10*, 301-311.

Griffin, W. L., McGregor, V. R., Nutman, A., Taylor, P. N., and Bridgwater, D., 1980: Early Archaean granulite facies metamorphism south of Ameralik, West Greenland. Earth Planet. Sci. Lett. *50*, 59-79.

Hietanen, A., 1975: Generation of potassium-poor magmas in the northern Sierra Nevada and the Svecofennian of Finland. U.S. Geol. Surv. Jour. Res. *3*, 631-645.

Hildebrand, R. S., 1981: Early Proterozoic LaBine group of Wopmay Orogen: Remnant of a continental volcanic arc developed during oblique convergence. *In*: Campbell, F.H.A. (ed), Proterozoic Basins in Canada. Geol. Surv. Can. Paper *81-10*, 133-156.

Hoffman, P. F., 1980: Wopmay Orogen: A Wilson cycle of early Proterozoic age in the northwest of the Canadian Shield. *In*: Strangway, D. W. (ed), The Continental Crust and its Mineral Deposits. Geol. Assoc. Canada Spec. Paper *20*, 523-549.

Irving, E., Emslie, R. F., and Ueno, H., 1974: Upper Proterozoic paleomagnetic poles from Laurentia and the history of the Grenville structural province. Jour. Geophys. Res. *79*, 5491-5502.

Leblanc, M., and Lancelot, J. R., 1980: Interprétation géodynamique du domaine pan-Africain (Précambrien terminal) de l'Anti-Atlas (Maroc) à partir de données géologiques et géochronologiques. Canadian Jour. Earth Sci. *17*, 142-155.

Lewry, J. F., 1981: Lower Proterozoic arc-microcontinent collisional tectonics in the western Churchill Province. Nature *294*, 69-72.

Lewry, J. R., Stauffer, M. R., and Fumerton, S., 1981: A Cordilleran-type batholithic belt in the Churchill Province in northern Saskatchewan. Precambrian Res. *14*, 277-313.

Loberg, B.E.H., 1980: A Proterozoic subduction zone in southern Sweden. Earth Planet. Sci. Lett. *46*, 287-294.

Löfgren, C., 1979: Do leptites represent Precambrian island arc rocks? Lithos *12*, 159-165.

Milanovsky, E. E., 1981: Aulacogens of ancient platforms: Problems of their origin and tectonic development. Tectonophysics *73*, 213-248.

Moore, J. M., Jr., 1977: Orogenic volcanism in the Proterozoic of Canada. *In*: Baragar, W.R.A., et al. (eds), Volcanic Regimes in Canada. Geol. Assoc. Canada Spec. Paper *16*, 127-148.

Poulsen, K. H., Borradaile, G. J., and Kehlenbeck, M. M., 1980: An inverted Archean succession at Rainy Lake, Ontario. Canadian Jour. Earth Sci. *17*, 1358-1369.

Raith, M., Raase, P., and Hormann, P. K., 1982: The Precambrian of Finnish Lapland: Evolution and regime of metamorphism. Geol. Rundschau *71*, 230-244.

Ray, G. E., and Wanless, R. K., 1980: The age and geological history of the Wollaston, Peter Lake, and Rottenstone domains in northern Saskatchewan. Canadian Jour. Earth Sci. *17*, 333-347.

Shackleton, R. M., Ries, A. C., Graham, R. H., and Fitches, W. R., 1980: Late Precambrian ophiolitic mélange in the Eastern Desert of Egypt. Nature *285*, 472-474.

Stowe, C. W., 1974: Alpine-type structures in the Rhodesian basement complex at Selukwe. Geol. Soc. London Jour. *130*, 411-425.

Tarney, J., and Windley, B. F., 1981: Marginal basins through geological time. *In*: Moorbath, S., and Windley, B. F. (eds), The Origin and Evolution of the Earth's Continental Crust. R. Soc. London, Phil. Trans. *A301*, 217-232.

Torske, T., 1977: The South Norway Precambrian region; a Proterozoic cordilleran-type orogenic segment. Norsk. Geol. Tidsskr. *57*, 97-120.

Watterson, J., 1978: Proterozoic intraplate deformation in the light of south-east Asian neotectonics. Nature *273*, 636-640.

Williams, D.A.C., and Furnell, R. G., 1979: Reassessment of part of the Barberton type area, South Africa. Precambrian Res. *9*, 325-347.

Wilson, I. H., 1978: Volcanism on a Proterozoic continental margin in northwestern Queensland. Precambrian Res. *7*, 205-235.

Windley, B. F., 1983: A tectonic review of the Proterozoic. Geol. Soc. America Mem. *160*, 1-10.

Windley, B. F., 1984: The Evolving Continents. Wiley, New York and London, 399 p.

Windley, B. F., Bishop, F. C., and Smith, J. V., 1981: Metamorphosed layered igneous complexes in Archean granulite-gneiss belts. Ann. Rev. Earth Planet. Sci. *9*, 175-198.

Wynne-Edwards, H. R., 1972: The Grenville Province. *In*: Price, R. A., and Douglas, R.J.W. (eds), Variations in Tectonic Styles in Canada. Geol. Assoc. Canada Spec. Paper *11*, 263-334.

Wynne-Edwards, H. R., 1976: Proterozoic ensialic orogenesis: The millipede model of ductile plate tectonics. Am. Jour. Sci. *276*, 927-953.

Young, G. M., 1980: The Grenville orogenic belt in the North Atlantic continents. Earth Sci. Rev. *16*, 277-288.

Part II

EUROPE

Chapter 4

A TECTONIC MODEL FOR THE CRUSTAL EVOLUTION OF THE BALTIC SHIELD

ASGER BERTHELSEN

Institut for almen Geologi, Copenhagen

The Precambrian crystalline rocks exposed in the Baltic shield form part of the oldest continental crust in Europe. On its eastern and southern side, the Baltic shield is fringed by Paleozoic and Mesozoic strata that form the cover to the Precambrian basement of the neighboring Russian platform and Fennoscandian border zone southwest of the shield. The northwestern margin of the shield is outlined by the front of the Scandinavian Caledonides.

The Baltic shield is a classic region for studies of Precambrian geology (Sederholm 1907, 1927, 1932; Wegmann 1928; Eskola 1929, 1948; Backlund 1936, 1943). At the dawn of this century, Sederholm demonstrated that there is no essential difference between the ancient orogenic belts of the Precambrian shield and younger mountain chains. Deciphering the palimpsest in the shield rocks, Sederholm (1932) also saw a pattern of successive orogenic belts stamped on the shield.

However, the true length of the complex geological evolution of the Baltic shield was not fully realized until lately, when modern isotope age determinations revealed that it spanned from more than 3,100 Ma (? 3,500 Ma) to about 600 Ma ago, making it at least four times the length of Phanerozoic time.

To outline the main trends in this prolonged and complex crustal evolution and to arrive at a more coherent picture of the tectonic development of the entire Baltic shield, I present below a tectonic division of the shield rocks.

PRINCIPLES OF DIVISION

CRUSTAL AGE AND AGE PROVINCES

The principle underlying this tectonic division is dualistic. Available geochronological and geological information is used to evaluate (1) when, in certain parts of the shield, the first/oldest continental crust was formed and (2) when the youngest large-scale orogenic deformational patterns were developed in particular regions. In this way, the shield is divided into (1) crustal age provinces and (2) structural age provinces. This dualistic type of division is thought to bring out the large-scale tectonic traits of a shield area more clearly than does the principle of structural levels (Bogdanoff 1963).

Ideally, a crustal age province is defined by a limited (ca. 100 Ma) grouping of the ages of formation of its oldest, mantle-derived rocks, which from geological, geochemical, and geochronological reasoning can be considered to signal the first formation of a continental crust in the region. The crustal age of a given province, therefore, generally is notably older than the average age of the crustal rocks occurring within it, even if undeformed post-orogenic and cratonic rocks are disregarded.

Following this principle of division, three Proterozoic crustal age provinces (P_1, P_2, and P_3) have been distinguished in the Baltic shield (see Fig. 4-1). The age spread from the Archean terrain, as well as geological observations, suggests that at least three older crustal age provinces might be found in the so-called Archean nucleus in the northeastern part of the shield, but data from this region are still too scarce to allow a systematic subdivision. In this paper, therefore, the Archean nucleus is provisionally treated as a single "bulk age" crustal province.[1]

STRUCTURAL AGE PROVINCES

The next step in the analysis is to subdivide the shield according to structural criteria. This is generally done by delineating structural provinces. As defined by Stockwell (1982), a structural province is characterized by its overall structural trends and style of folding. Since, however, a structural province may be influenced by more than one major episode of orogenic activity, some ambiguity may arise. Therefore, I have used the age of the youngest regionally developed deformational pattern in a province as the decisive criterion of classification. The provinces distinguished in this way are called structural age provinces.

The deformational patterns used in this work comprise fold and thrust structures revealed by regional structural analysis and dated indirectly by means of isotopic age determinations of deformed and post-deformation rocks. Finally, the borders between the established crustal age provinces and structural age provinces were drawn on the same map. Because nowhere in the shield do the borders between the two types of provinces coincide, a larger number of tectonic units, delimited by either crustal age or structural age boundaries, result (Figure 4-1).

Where the first-formed crust is of Archean age, it is divided, as in Figure 4-1, into tectonic units (A_1-A_4) according to the degree of Proterozoic reworking, while, as already pointed out, where the crustal accretion started at different times during the Proterozoic, it is divided into crustal age provinces designated P and P_1-P_3.

Important tectonic boundaries, all developed during the Proterozoic, are numbered 1–9. They can be interpreted as representing either continent-to-continent collision sutures (1), healed subduction sutures (3, ?4), or infracrustal sutures formed by tectonic underplating or related to crustal peeling accompanied by ductile thickening of the lower lithosphere (2, 5, 6, 7, 8, 9).

[1] By the time the manuscript of this article was revised, new data (e.g., Bernard-Griffiths et al. 1984, Kozlovsky 1984) had become available, showing that this "bulk age" crustal province includes two belt-form early Proterozoic crustal age provinces: the granulite complex proper (marked P on Figure 4-1) and the Kola suture belt (the black belt marked 1 on Figure 4-1).

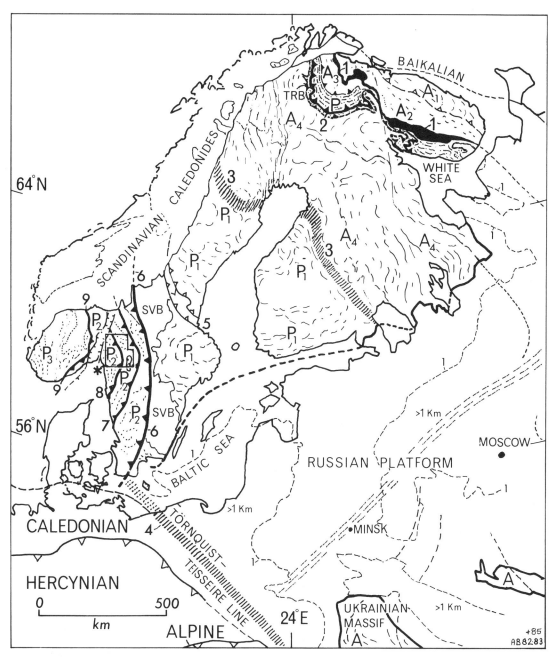

FIGURE 4-1. Tectonic sketch map of the Baltic shield. The broken line marked 1 is 1-km isopach of Phanerozoic cover. The small frame marked by an asterisk indicates the area depicted in the block diagram shown in Figure 4-10.

Aboriginal and Reworked Crust

The tectonic division thus arrived at permits us, when information is optimal, to trace the crustal evolution of a tectonic unit right from its beginning, when the first sialic crust was formed, to its end, when it acquired its last deformational pattern and was in a state of being cratonized.

The division assists us in analyzing the often complex time relations between crust-forming and crust-deforming processes, as well as the relative importance of these processes at various times. Thus it becomes possible to distinguish between tectonic units with aboriginal crust and others with structurally reworked crust.

The term *aboriginal crust* is used here for crust that has not been influenced by any significant tectonic movements since its initial formation. A subsequent supply of sialic material may have occurred, but the deformational pattern related to the initial crustal accretion is still intact. Aboriginal crust may therefore be defined by a correspondence between, or a significant overlap of, the spreads of its crustal and structural ages.

A tectonic unit with *reworked crust* has been exposed to superimposed tectonics, and it is thus characterized by a structural age that is markedly younger than its crustal age. The distinction between aboriginal and reworked crust is important for understanding why tectonic styles with profound geometrical contrasts may develop in neighboring units during the same orogenic episode.

THE AGE ASYMMETRY IN THE BALTIC SHIELD

A closer study of Figure 4-1 reveals that there is a clear asymmetry in the distribution of the crustal age provinces in the Baltic shield. The oldest crustal age province, the Archean age province, occupies the northeastern half of the shield (see footnote 1, p. 32), while in the southwestern half, younger and younger Proterozoic crustal age provinces have been welded onto the southwest margin of that nucleus, specifically (1) the Svecofennian province, P_1, ca. 1,900 Ma old; (2) the crustal age province of southwestern Sweden (Gorbatschev 1980) and southeastern Norway (Berthelsen 1980), P_2, ca. 1,700 Ma old; and (3) the crustal age province of southwestern Norway (Priem and Verschure 1982), P_3, ca. 1,500 Ma old. A similar succession is not to be found in the distribution of the structural age provinces. This is because the structural age provinces that develop in relation to the 1,700 and 1,500 Ma old crustal-forming episodes have been modified by the overprint of the younger, Sveconorwegian structures (ca. 1,200-850 Ma old). The structures of the neighboring Svecofennian tectonic unit are thus cut off by the eastern Sveconorwegian front (marked 6 in Figure 4-1). The age asymmetry in the Baltic shield was recognized by Berthelsen (1980) in connection with a tectonic palinspastic analysis of the southwestern part of the shield. In this analysis the pre-Sveconorwegian setting of the P_2 and P_3 crustal age provinces was reconstructed by making allowance for the crustal shortening caused by the Sveconorwegian folding and thrusting. By this procedure, the age asymmetry became even more evident than as shown in Figure 4-1 (cf. Figure 8 in Berthelsen 1980 and Figure 4-9 in this paper).

In earlier reviews of the tectonic evolution of the Baltic shield (see also Simonen 1980) it has been generally assumed that remnants of pre-Svecofennian and Svecofennian crustal rocks occur within the Sveconorwegian tectonic unit, but except within the easternmost sub-provinces, this assumption has not been confirmed by recent mapping and isotopic age studies (Pedersen et al. 1978, Gorbatschev 1980, Priem and Verschure 1982).

Berthelsen (1980) has presented a plate-tectonic interpretation of the southwestern part of the Baltic shield, according to which the southwestwards to westwards younging is explained by stepwise oceanward shifts of the site of an active subduction zone and corresponding successive periods of crustal accretion. Immediately after a shift, subduction can be imagined to have taken place in an oceanic environment, but with time the crustal accretion above the descending lithospheric slab and associated submarine and island-arc volcanism and sedimentation caused a new sialic crust/crustal age province to be formed and to be welded onto the margin of the preexisting shield. Figure 4-2 shows an interpretative section through the ca. 1,800 Ma old shield.

It is interesting to note that this model of marginal accretionary growth concurs with recent paleomagnetic studies. These studies indicate a consistent APW path for the period 2,700-1,200 Ma for rocks of the same age but from different tectonic units in the shield (Pesonen and Neuvonen 1981). The fact that the Sveconorwegian poles from Sveconorwegian areas (1,200-1,100 Ma) form an exception to this consistency and differ considerably from the poles of rocks of the same age from other parts of the Baltic shield is also in agreement with Berthelsen's (1980) interpretation.

Pesonen and Neuvonen (1981) conclude that the consistency in the APW curves indicates that ensialic tectonics has been dominant during the evolution of the shield. I agree with this conclusion insofar as I consider (1) the Sveconorwegian tectonic evolution (ca. 1,200-850 Ma) in south Norway and southwest Sweden to be entirely of ensialic nature and (2) the previous Proterozoic crustal evolution to have started in oceanic (simatic) environments, but to have affected neighboring sialic environments as well.

THE DISTAL EFFECT OF MARGINAL SUBDUCTION

The effect exerted by a low-dipping Proterozoic interoceanic subduction zone on the neighboring shelf and continental areas of the proto-Baltic shield is here called the distal effect of the subduction. As shown in Figure 4-2, it is assumed that the low-dipping slab not only traveled below the marginal belt, where new continental crust was being formed above it, but also traveled so far that it reached below the inner parts of the shield. With increasing distance from the active site of subduction, the ascending magmas could evolve over a longer and longer period, and contamination with preexisting continental crust would change their composition and delay their final emplacement. This conceptual model explains the overall trend in the geochronological distribution, geochemistry, and time of emplacement of the plutonic rocks of the marginal belts and of the corresponding inner/distal regions in the shield.

The distal effect of a subduction zone involves not only magmatic activity but also tectonic reworking and metamorphic overprints in the preexisting neighboring crustal segments in the shield. Since the width of the region affected by each distal event appears to have been greater than the width of the corresponding newly formed crustal belt or crustal age province

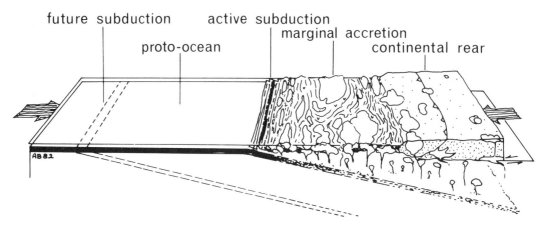

FIGURE 4-2. Schematic diagram showing how an active subduction zone in an oceanic environment causes proximal crustal accretion and deformation at the margin of the continent and accompanying distal deformation and magma emplacement giving rise to late-orogenic to cratonic intrusions in the rear of the existing continent. When the site of active subduction is shifted oceanwards, a new crustal belt will be welded onto the shield, and a corresponding distal effect will influence a broader distal region.

close to the subduction zone, a considerable overlapping of regions affected by the distal effects of successive marginal subduction zones is to be expected.

In a situation where the effects of successive subduction events overlap and thus are superimposed, we may easily get the impression of a very prolonged orogenic evolution if we study a distal region without regard for its wider surroundings. This may be the reason why the Svecokarelian orogenic cycle has been estimated by some authors to span from before 2,000 Ma to around 1,500 Ma—or almost as long as the whole of Phanerozoic time.

The distal-effect concept outlined above was conceived for tectonic reasons by Berthelsen (1980), and related petrological and geochemical problems were discussed by Gorbatschev (1980) and Wilson (1980, 1982). It forms a most important aspect of the model of marginal accretionary growth, because it offers an opportunity of finding indirect evidence for the existence of former active subduction zones. Direct evidence may be difficult to provide, because marginal crustal accretion and welding of new sialic belts onto the shield, in combination with the distal effects of migrating subduction zones, may have effectively healed former plate boundaries and obliterated most traces of the rock types and high-pressure mineral associations considered specific for former converging margins.

THE NORTHERN TECTONIC UNITS OF THE ARCHEAN NUCLEUS (A_1 AND A_2)

Most tectonic units of the Archean nucleus have suffered from Proterozoic tectonic movements, metamorphism, and magmatic activity, but the effect of these different events varies considerably from unit to unit.

In unit A_1, which makes up the northeastern part of the Kola peninsula, the original Archean crust underwent only slight tectonic reworking but was thrust en bloc onto the A_2 unit south of it (Khain and Leonov 1979). This thrusting may even be of late Archean age. In the A_2 unit, Archean fold-and-thrust structures also have been preserved more or less intact, as appears to be the case at least in the Sydvaranger district of northeast Norway (Figure 4-3). Here a complex Archean crustal evolution can be traced. After the initial formation of a granitoid basement of unknown age, there followed emplacement of tonalitic magmas and remobilization of the old basement with accompanying downfolding and metamorphism of a cover sequence of quartzites, shales and metaturbidites, coarse conglomerates, quartzitic schists and quartzites including magnetite-banded quartzites and quartz-banded iron ore, and metavolcanics of andesitic to dacitic compositions. Subsequent large-scale overthrusting resulted in crustal slicing and duplications (Figure 4-3). Emplacement of granitic-monzonitic and ul-

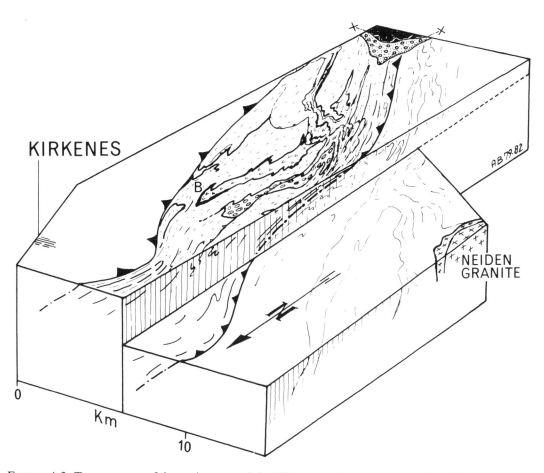

FIGURE 4-3. Tectonogram of the region around the Björnevann iron ore deposit (B) in Sydvaranger (northwest end of unit A_2 on Figure 4-1). Tonalitic to trondhjemitic gneisses (around Kirkenes) have been overthrust onto the Archean supracrustals of the Björnevann Group, which rests with tectonic contact on an old granite-gneiss basement (with downfolded supracrustals), now intruded by the Neiden granite. In the south, the late Archean folded and thrusted rocks are discordantly overlain by the Neverskrukk conglomerate, over which metabasic rocks of the Kola suture belt (tectonic boundary 1 on Figure 4-1) have been thrust. (Based on Bugge and Iversen 1981, with observations by the author 1979–1982.)

tramafic magmas marked the end of this episode. The Neiden granite, dated to about 2,550 Ma, most probably was intruded at the same time. Then followed, after a period of denudation, deposition of conglomerates and grits (the Neverskrukk conglomerate) and intrusion of several generations of pre-Riphean basic dikes (Meriläinen 1976, Bugge 1978a and b, Bugge and Iversen 1981, author's personal observations).

THE KOLA SUTURE

As shown in Figure 4-1, a major collision suture (tectonic line 1) separates the tectonic units designated A$_2$ and A$_3$. This suture, here called the Kola suture, can be followed for about 500 km from the Caledonian front in the west, through Norway, Finland, and the USSR, right to the east coast of the Kola peninsula. No doubt this suture forms one of the major tectonic boundaries in the Baltic shield (Bugge 1978a, Khain and Leonov 1979, Berthelsen 1982b, Berthelsen and Marker 1986). West of the Norwegian-Russian frontier, the Kola suture is outlined by a belt up to 10 km broad of northwards upthrust and overthrust metavolcanics and metasediments, which in places can be seen to overlie the practically undisturbed beds of the Neverskrukk conglomerate with a marked tectonic contact (Figure 4-3). The metavolcanic rocks of the suture belt comprise former tholeiitic basalts, andesites, and dacites to rhyodacites, and talc-chlorite schists point to the former presence of true ultramafic lavas (Bugge 1978a). The metasediments are calcareous schists, dolomite, and phyllite and graphite schists. Gabbroic and ultramafic rocks intrude the sequence.

East of the Norwegian-Russian frontier, the suture belt forms part of a large arcuate structure in the Pechenga region. Due to a combined axial virgation and depression, the total thickness of the overthrust suture sequence here reaches almost 10,000 m. Important copper-nickel mineralizations are associated with layered peridotite phacoliths in phyllites (Väyrynen 1931, Gorbunov 1971).

Until lately, information on the age of the suture belt rocks has been incomplete and in part conflicting. Meriläinen (1976) assumed an Archean age, while Bugge and Råheim (see Bugge and Iversen 1981) obtained Rb/Sr ages suggesting a Proterozoic age for some metavolcanics. As summarized by Berthelsen and Marker (1986), recent data from the Soviet part of the Kola suture suggest an early Proterozoic age (ca. 2,400-1,900 Ma) for the suture belt rocks.

When viewed in relation to the Proterozoic evolution of the neighboring crustal units (A$_2$ and A$_3$ in Figure 4-1), it is obvious that the suture belt rocks must have been formed in a wide rift zone or spreading ocean of Red Sea type (here called the Kola ocean), and they presumably continued to form during the closure of this ocean (Bugge 1978a, Berthelsen 1982b). The ultimate collision between the Archean units north and south of the suture took place around 1,900 Ma—prior to the intrusion of the post-kinematical Vainospää granite (northeast of Inari Lake in Finnish Lapland), which cuts and interrupts the belt. This granite has been dated to about 1,790 Ma (Haapala and Front 1985).

No paleomagnetic check of the tectonic significance of the Kola suture has yet been made. In their recent review of the paleomagnetism of the Baltic shield, Pesonen and Neuvonen (1981) do not present any data for rocks from the tectonic units north of the Kola suture.

THE GRANULITE BELT (A$_3$ AND P)

The tectonic units A$_3$ and P (Figure 4-1) constitute the Granulite belt of Lapland and its continuation north of the White Sea, south of the Kola suture. The southern border of unit P is defined by a marginal low-angle thrust (marked 2 in Figure 4-1). According to Gorbunov (1980), this marginal thrust merges into the White Sea (see also Berthelsen and Marker 1986). The units A$_3$ and P thus cover more or less the same terrain as the so-called Belomorides or Marealbides (derived respectively from the Russian and Latin names for the White Sea—see Figures 4-1 and 4-4).

The Granulite belt is divided into two units (A$_3$ and P) by a zone of blastomylonites: A$_3$ forms a northern "rampart" with Archean crust adjacent to the Kola suture, and P is the granulite complex proper. The latter corresponds to what Kröner (1980) termed the Inari mobile belt, stressing that its tectonic development was ensialic and alien to Wilson Cycle processes.

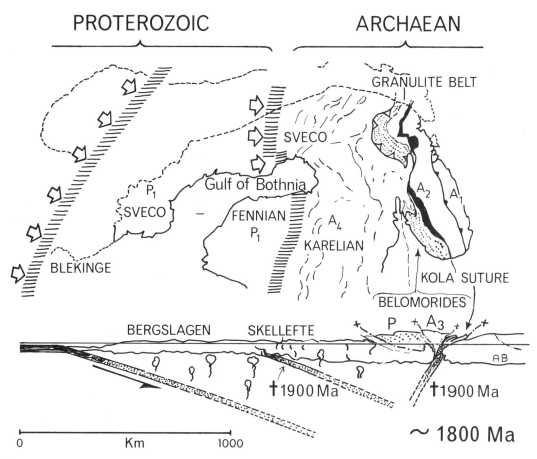

FIGURE 4-4. Sketch showing the possible plate-tectonic situation in the Baltic shield around 1,800 Ma ago, when the Main Svecofennian subduction zone was still active and the Svocofennian crustal age province was being formed. The subduction zone just south of the Skellefte field (1,900 Ma or more) had become inactive and was being healed. The subduction zone that caused the Kola collision suture had also been arrested around 2,100 Ma ago.

Recent reviews of the Granulite belt by Barbey et al. (1984) and Berthelsen and Marker (1986) contradict this view.

The Archean crustal evolution in the northern unit, A_3, seems to have been much like that of neighboring regions in the Archean nucleus, in which a pre-greenstone basement of tonalitic gneisses ca. 3,100 Ma old has been reported (Kröner et al. 1981) and evidence of tectono-thermal episodes around 2,800 and 2,500 Ma has been obtained. Meriläinen (1976) has found evidence of the ca. 2,500 Ma event. However, it was during the Proterozoic, and along with the closing of the Kola ocean, that the Granulite belt acquired its specific tectonic and metamorphic characters (Barbey et al. 1984, Bernard-Griffiths et al. 1984, Berthelsen and Marker 1986). Between 2,000 and 1,900 Ma ago, an early Proterozoic turbidite and mass flow sequence, deposited in a wide back-arc basin south of the northern rampart (A_3), was metamorphosed and altered into the garnet-sillimanite-cordierite-biotite granulites of the granulite complex proper, P, where quantitatively less important orthogneisses stem from tholeiitic to calc-alkaline intrusions.

The tectonic stacking and overthrusting of the rocks of the granulite complex proper may thus be interpreted as a direct consequence of the continental collision in the nearby Kola suture. Along with the large-scale southwestward overthrusting of the complex onto the Karelian tectonic unit (marked A_4 in Figure 4-1) and accompanying upthrusting of the northern rampart, the garnet-sillimanite-cordierite-biotite granulites were partly altered into blasto-mylonitic granulites with platy quartz foliation.

Emplacement of calc-alkaline rocks in the ca. 1,950-1,900 Ma old Archean crust of the northern rampart of the Granulite belt (Berthelsen and Marker 1986) suggests, moreover, that the collision 1,900 Ma ago was preceded by the functioning of a southwest-dipping subduction zone and the development of an andesitic island arc near the southern margin of the Kola ocean (Figure 4-4). The remains of this island arc now form part of the Kola suture belt.

THE KARELIAN TECTONIC UNIT

TECTONICS AND STRATIGRAPHY

The Karelian tectonic unit (A_4 of Figure 4-1), which forms the southwestern part of the Archean nucleus, is limited in the northeast by the marginal overthrust of the Granulite belt (marked 2 in Figure 4-1) and in the southwest by the former continental margin of the Archean proto-shield (marked 3 in Figure 4-1). On the geological map the Karelian tectonic unit therefore comprises (1) Archean gneiss terrains in culminations or as median massifs and (2) the northwest-southeast trending Karelides of Finmarksvidda (Norway), northern Sweden, northern and eastern Finland, and Russian Karelia.

Unless otherwise stated, the following interpretative review of the Karelian tectonic unit is based mainly on Meriläinen (1976), Gaal et al. (1978), Rickard (1979), Simonen (1980), Kröner et al. (1981), and Gaal (1982), and on earlier contributions (see the references given either by the aforementioned or by Eskola 1963).

With the possible exception of the median massifs of Karelia (East Finland and USSR), most of the Archean basement terrain of the Karelian tectonic unit shows evidence of tectonic reworking, metamorphism, and magmatic activity related to the ca. 2,000 to 1,800 Ma old Proterozoic events. Older trends may have become modified by rising granite diapirs or by shearing between crustal blocks, and if intact, they are often hidden by overlying Proterozoic supracrustals. It is therefore difficult to draw a regionally coherent picture of the Archean tectonic pattern. In this respect, the trend of geological boundaries and aeromagnetic anomalies forms an incomplete basis of information. One major feature, the Tana River belt (Barbey et al. 1980), deserves mention however (TRB in Figure 4-1). The Tana River belt surrounds the convex southwest thrust margin of the Granulite belt as a belt 5 to 15 km wide and more than 350 km long (Barbey et al. 1980). It consists of a banded sequence of garnet amphibolites and pyroxene-, amphibole-, and garnet-bearing gneisses of volcanosedimentary origin. The metavolcanic rocks show basaltic to basalto-andesitic compositions of tholeiitic character. In addition, small bodies of hornblendites, pyroxenites, and spinel cortlandtites occur, and banded gabbros and granites intrude the sequence.

Berthelsen and Marker (1986) suggest that the Tana River belt probably represents overthrust parts of the oceanized basement of the former back-arc basin, in which the protoliths of the granulite complex proper (marked P in Figure 4-1) were laid down.

The Archean basement south of the Tana River belt is known to have developed through at least three major "orogenic" episodes. Next to internally complexly built gneiss terrains where tonalitic, trondhjemitic, and granodioritic rocks predominate, a number of schist belts and greenstone belts occur; the latter typically contain komatiites and quartz-banded iron ores. Two greenstone sequences of different age have been distinguished. The oldest greenstones were affected by a tectono-thermal event ca. 2,800 Ma, while the younger (Laponian) sequence postdates the ca. 2,800 Ma event but was affected by a ca. 2,500 Ma event.

In Finland, near the southwestern margin of the proto-shield, the Archean basement is locally overlain by Sariolian metaconglomerates, meta-arkoses, and metavolcanics ca. 2,500 to 2,300 Ma old, while farther to the north it is the Jatulian transgressive sequence, ca. 2,400 to 2,000 Ma old, that forms the unconformable cover to the basement, which here was exposed to sialitic weathering. Within the Archean basement occur a number of layered mafic to ultramafic intrusions, which have been dated 2,450 to 2,430 Ma. The chromite ores at Kemi (Figure 4-5) are associated with pyroxenitic and peridotitic differentiates of one of these intrusions (Isokangas 1978).

The Jatulian sequence is evidence of a widespread transgression onto the proto-shield. The basal (locally uraniferous) Jatulian quartzites are overlain by a marine succession of dolomite and carbonaceous slates and black schists. Hematite-banded iron ores are found locally, and the Jatulian sediments are invariably associated with basic volcanics (albite-rich spilites), sills, and dikes. Because of this latter feature, several occurrences of Jatulian supracrustals have been described as greenstone terrains. There are thus three stratigraphically different greenstone associations in the Karelian tectonic unit.

In a belt approximately 100 km wide along the southwestern margin of the Archean proto-shield, the Jatulian formations are in several places unconformably overlain by a flysch-like sequence several km thick, the Kalevian group, which comprises quartzites, slates, phyllites and mica schists (largely derived from turbidites), tholeiitic metabasalts with metachert interbeds, and lenticular bodies of serpentinized and metasomatically altered dunites. These

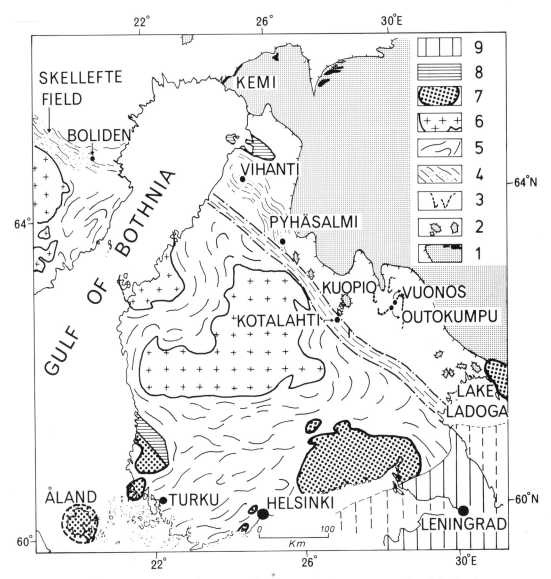

FIGURE 4-5. Sketch map of the region around the former southwestern margin of the Archean proto-shield. 1: Archean basement (locally intruded by gabbros ca. 2,440 Ma old). 2: Mantled domes with cores of reworked Archean. 3: Karelian schists (with ultramafics indicated around Outokumpu). 4: Metavolcanics of the Skellefte field and Vihanti-Pyhäsalmi belt. 5: Svecofennian infra- and supracrustals. 6: Late Svecofennian granite massifs. 7: Post-Svecofennian intrusive complexes (rapakivi, gabbro, and so on). 8: Jotnian conglomerates and sandstone. 9: Phanerozoic cover formations. The trend of the Kotalahti shear zone is also shown.

magmatic rocks constitute the so-called "ophiolites" noted by earlier authors (see, e.g., Eskola 1963). The stratiform but folded and recrystallized copper ores of Outokumpu and Vuonos (Figure 4-5) occur in the Kalevian sequence (Isokangas 1978). Following Eskola (1963), the Jatulian and Kalevian rocks are often referred to collectively as the Karelian schists.

THE EARLY SVECOKARELIAN OROGENIC EVOLUTION

Simonen (1980) states that the deposition of the Kalevian sediments postdated the Jatulian volcanic activity and took place between ca. 2,050 and 1,900 Ma, while Gaal (1982), referring to older lead model ages for Finnish galenas, suggests that the Kalevian sedimentation had already started 2,200 to 2,100 Ma ago. However, the Kalevian sequence comprises both autochthonous and allochthonous units. The latter were thrust eastward or northeastward on top of the Jatulian and autochthonous Kalevian rocks in connection with Svecokarelian tectonic movements. The apparently contradictory age relationships between the Jatulian and Kalevian formations may therefore be resolved by assuming that those Kalevian schists that appear to be "too old" were originally deposited in fault-controlled basins on an attenuated continental margin, while, at the same time, Jatulian deposition and volcanism took place on the continent proper. Later, the Kalevian facies transgressed onto the Jatulian platform, giving rise to the Jatulian-Kalevian unconformity.

The early Kalevian sedimentation off a passive attenuated margin continued apparently from ca. 2,200 to 2,050 Ma ago, and the attenuation of the southwestern continental margin as well as the Jatulian transgression over the Archean continent may therefore have been contemporaneous with the opening of the Kola ocean. It is also tempting to suggest that the change marked by the Jatulian-Kalevian unconformity (ca. 2,000 Ma old) corresponds to the initiation of an active margin off the southwestern border of the continent and the related development of a marginal flysch trough over the neighboring parts of the continent. The Kalevian sedimentation may thus have spanned from an early passive-margin environment to the filling-up of the subsequently formed "synorogenic" marginal trough.

The existence of a former subduction zone, ca. 2,000 Ma old, off the southwestern margin of the proto-shield (approximately line 3 in Figure 4-1) is best shown in northern Sweden (Figures 4-5 and 4-6), where the acid to intermediate volcanics of island-arc type graywackes,

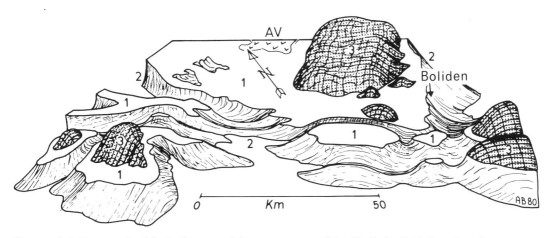

FIGURE 4-6. Interpretive block diagram of the eastern part of the Skellefte field, based on data supplied by Grip (1978) and Lundberg (1980). Note that the late (Revsund) granite intrusions have been omitted and that the upper part (2)—the phyllitic metasediments and basic volcanics—of the Skellefte Group has been removed, so that the lower metavolcanics (1) and the diapirs of Jörn granite (3) outline the structures. AV stands for the terrestrial Arvidsjaur volcanics, which dominate farther to the north.

pelites, and basic pillow lavas of the Skellefte field give way northward to the terrestrial Arvidsjaur volcanics and the Jörn granitoids (Rickard 1979, Lundberg 1980); here porphyry copper-molybdenum deposits have been found in association with early (Jörn-type) Svecokarelian plutonic rocks (Walser and Einarsson 1982). On the Finnish coast of the Gulf of Bothnia (Figure 4-5), one can look for the continuation of the Skellefte volcanic field in the acid to basic effusives, tuffites, and pyroclastics of the Vihanti-Pyhäsalmi belt (Isokangas 1978, Gaal 1982). Both the Skellefte field and the Vitangi-Pyhäsalmi belt contain important sulfide deposits (Rickard and Zweifel 1975, Rickard 1979). The lead model ages of the galenas of the Vihanti belt are 1,975-1,925 Ma (Simonen 1980).

Farther southeast, most direct traces of a former active margin appear to become lost in the subvertical Kotalahti shear belt, which corresponds to Hietanen's (1975) Svecokarelian fault and the Raahe-Ladoga zone of Gaal (1982). It is up to 20 km wide and can be traced for 430 km northwest to southeast across southern Finland (Figure 4-5). The Kotalahti belt houses a number of Ni-Cu occurrences associated with sheet- to pipe-like dioritic, basic, and ultrabasic bodies ca. 1,925 Ma old (Mikkola and Niini 1968, Gaal 1972, 1982, Isokangas 1978).

The Kotalahti shear belt is generally believed to reflect an important break in the structure of the underlying crust and mantle, which probably was inherited from the early (i.e., ca. 2,000 Ma old) Svecokarelian tectonic development. The shear belt, however, was not formed as such until the crust of the former active margin had become at least partially consolidated. Its development therefore should probably be seen in relation to the ca. 1,950 to 1,800 Ma old Svecofennian evolution, which caused a new (P_1) crustal age province to be welded onto the shield (Figure 4-4). A forthcoming paper by Berthelsen and Marker (in prep.) describes the evolution of the Kotalahti (or Raahe-Ladoga) shear zone in relation to other Late Svecokarelian strike-slip megashears in the Baltic shield.

RESULTING TECTONIC STYLES

Within the Karelian unit are found a variety of tectonic styles, depending on the degree to which Archean granite-greenstone–type structures have been preserved or reworked as well as on the intensity of the deformation that post-Archean cover rocks have suffered. Reviewing the crustal development of part of the westernmost Karelian tectonic unit in northern Sweden, Witschard (1980) emphasized the importance and repeated influence of deep-reaching fault zones that caused vertical mechanical segmentation of the crust and resulted in the ultimate formation of a mosaic pattern with more or less rounded nuclei circumscribed by "local" mobile zones (clearly displayed on aeromagnetic maps). Berthelsen and Marker (in prep.) suggest that the north-south trends in the westernmost Karelian unit were developed in connection with important strike-slip movements. Thrusts that displace earlier-formed fold structures (Figure 4-7) supply evidence of local transpressive regimes during the late Svecokarelian strike-slip faulting.

Discussing the Archean crust of Finland, Gaal et al. (1978) related the overall trend of the granite-greenstone terrains to deep-seated fractures that controlled Archean magmatism and even influenced the deformation of the Proterozoic cover. Silvennoinen (1972), in his description of the Jatulian greenstones and metasediments of the Rukantunturi area in northeast

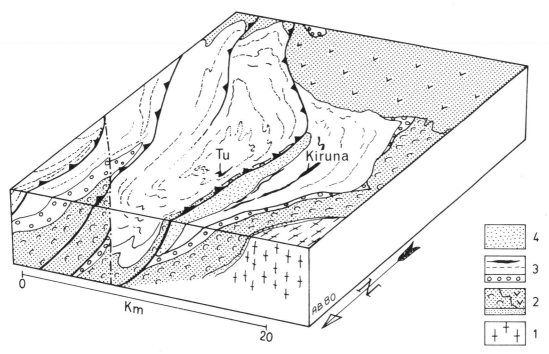

FIGURE 4-7. Tectonogram of the Kiruna region, northern Sweden, based on the Kiruna NV, NO, SV, SO map sheets prepared by the Geological Survey of Sweden (Ser. Af, Nr. 1-4, 1967). Archean basement gneiss (1) is overlain by the Kiruna greenstone with a basal breccia (2), that by the Kurravaara conglomerate and Kiruna porphyries (3), and all took part in the Svecokarelian folding prior to the deposition of the conglomerate, graywackes, arkoses, and sandstones of the Upper Kauki series (4). A younger deformational event (? 1,840 to 1,800 Ma old) has given rise to prominent listric overthrusts that cut earlier folds. This event may be related to transpression caused by large-scale strike-slip movements (Berthelsen and Marker in prep.). The Tuolluvaara (Tu) ore deposit lies in the northern part of a marked axial depression.

Finland, suggested that the deformation of the cover sequence took place by décollement in combination with fault-controlled block movements.

Probably the best known tectonic style of the Karelian tectonic unit is displayed by the mantled domes of the Pitkäranta and Kuopio areas (Eskola 1948, Preston 1954), where partly recrystallized Archean basement rocks form a core surrounded by Jatulian schists. Occurring not far from the former southwestern margin of the proto-shield (cf. Figure 4-5), the Kuopio domes are the most southwesterly outcrops of the Archean basement in this part of Finland. Farther to the north and east, near the large median massif of Finnish and Soviet Karelia, the Archean crust reacted in a more rigid manner and now forms upthrust massifs and detached slices delimited by blastomylonitic shear zones. The major thrust fault depicted by Gaal (1982, Figure 1) northeast of the Kalevian flysch belt affects both Jatulian-Kalevian schists and the Archean basement.

Despite the diversity of its effect within the Karelian tectonic unit, the Svecokarelian tectonic development was everywhere of ensialic nature, and north of the active margin in the southwest, the regional metamorphic imprint was of low grade, reaching only the greenschist and epidote-amphibolite facies. Even though large amounts of basic to acid magmas were added to the crust of the Karelian unit, it largely retained its tectonic integrity and behaved as a semi-stable segment between the strongly tectonized Granulite belt south of the Kola suture in the northeast and the active margin in the southwest.

THE SVECOFENNIAN TECTONIC UNIT (P_1)

The Svecofennian tectonic unit forms a crustal segment (marked P_1 in Figure 4-1) about 1,000 km wide, which can be described both as a single crustal age province and, by and large, as a single structural age province. Large-scale low-angle thrusting only locally affects the province (cf. the thrusts marked 5 in Figure 4-1). The crustal age of the Svecofennian province is ca. 1,925 to 1,800 Ma, and its structural age is about 1,850 to 1,800 Ma. Consequently, the province is characterized by an aboriginal crust, and the uniqueness of many Svecofennian features no doubt stems from this characteristic. A general and detailed review of the Svecofennian geology was recently presented by Lundqvist (1979).

Hietanen (1975), Rickard (1979), Berthelsen (1980), and Wilson (1982) suggest that the accretion of the Svecofennian crust was caused by the rise of calc-alkaline magmas above a descending oceanic slab. The oceanic trench of the northeast-dipping subduction zone, here called the Main Svecofennian subduction zone, must have been situated somewhere west of the present eastern front of the later Sveconorwegian belt in southern Sweden (see Figure 4-9, below). Berthelsen (1982a) suggests that the northwest-southeast trending Main Svecofennian subduction zone may still be traced in the buried Precambrian basement along the Tornquist-Teisseire line in the region of the western Baltic Sea and Poland, and that in post-Permian time this relic subduction suture in the basement controlled the development of the northwest-southeast trending fault systems of the Fennoscandian border zone (see Figure 4-1).

The idea that the crustal accretion was caused by the function of the Main Svecofennian subduction zone implies a horizontal component of subduction of more than 1,000 km. Evidently this could only have been realized if the dip of the descending slab was low (Figure 4-4).

Based on extensive studies of a key area 100 by 125 km around Stockholm and Nyköping, Stålhös (1981) presents an interesting model for the Svecofennian crustal and tectonic evolution (Figure 4-8). He visualizes that the ascending calc-alkaline magmas spread out and solidified as stratiform bodies at a fairly constant level within the sedimentary-volcanic sequence overlying the presumably oceanic, but actually unknown, primordial crust of the Svecofennian province. Repeated emplacement resulted in formation of a thick "sialic layer" within the lower argillites and below the overlying, more competent leptite-limestone formation, probably at depths of 4 to 6 km (leptites are devitrified to fine-grained quartzo-feldspathic metamorphic rocks of probable volcano-clastic origin).

The new "sialic layer," which consists mainly of tonalites and granodiorites, more or less served as a basement during the subsequent Svecofennian evolution. This evolution comprised deposition of an upper sequence several km thick of graywackes, greenstones, impure

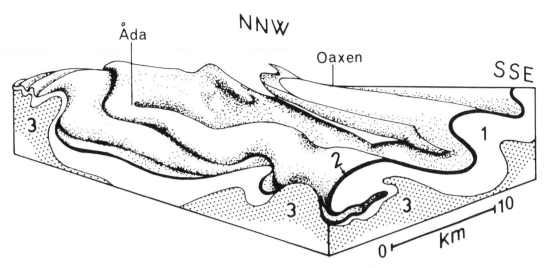

FIGURE 4-8. Tectonogram of the Svecofennian structures northeast of Nyköping, redrawn from Figure 5 in Stålhös (1981). The early granitoid rocks (3) of the pseudo-basement were intruded into a lower sequence of argillites and graywackes (1), which is overlain by the thin leptite-limestone formation (2).

quartzites, and argillites; continued intrusive activity; and a final culmination of the metamorphic, anatectic, and deformational processes.

In the Stockholm-Nyköping area studied by Stålhös (1981) the Svecofennian deformation resulted in formation of a train of west-vergent, north-northwest trending folds as well as east-west cross folds (Figure 4-8).

In contrast to Stephansson's (1975) polydiapiric view on the development of the Svecofennian terrain, Stålhös's (1981) model rests on the assumption that stresses generated at a distant, converging margin can be transmitted more or less through the developing crust because of the basement-like behavior of the earliest stratiform intrusives.

Stålhös's (1981) evolutionary model primarily aims at explaining the structural evolution of the key area. Viewed in relation to the entire Svecofennian development, however, an important problem appears to remain unsolved. The early calc-alkaline intrusives are thought to have intruded a supracrustal sequence of lower argillites and dacitic to rhyodacitic volcanics and limestones. This fails to explain the volcanism that produced the effusive rocks of the leptite-limestone formation. Could it be, as hinted at by Wilson (1982), that this volcanism was merely a submarine to island-arc expression of the magmatic activity simultaneously going on at depth, where new sialic crust was being formed? In this connection the zircon age of 1,862 Ma, presented by Welin et al. (1980) for acid metavolcanics from the upper, mainly metasedimentary sequence of the Grythyttan-Hällefors region of the Bergslagen district, is of particular interest. In Finland intermediate and silicic metavolcanics, with a stratigraphic position similar to that of the Swedish leptite-limestone formation, supply ages between 1,920 and 1,880 when dated by the zircon method (Simonen 1980).

If this idea holds true, the sedimentation and volcanism within the supposed Svecofennian ocean need not have started "well before 2,000 Ma," as is often assumed, but rather after the plate-tectonic shift at ca. 1,950 Ma that caused the formation of the Main Svecofennian subduction zone. No doubt the ca. 2,400 Ma detrital zircons in some Svecofennian meta-sediments bear witness to an old continental source area. This source area should probably be looked for to the east of the present Svecofennian province, where in the basement underlying the Russian platform Archean rocks extend much farther to the south (right to the Ukrainian shield) than do those within the Baltic shield proper (Figure 4-1).

The recent contributions by van der Velden et al. (1982) and Oen et al. (1982) from the Grythyttan field of the Bergslagen ore district point toward a very labile tectonic regime within the leptite belt—with rising and sinking volcanic centers, formation of local rift zones, and downslope spreading of volcanoclastic debris flows. Frietsch (1982) has written a general review of the exhalative-sedimentary quartz-banded iron ores and manganese and sulfide ores of the leptite belt of central Sweden. He suggests that all ore types were formed syngenetically, and that the type of ore precipitation depended on the depth and corresponding volcano-sedimentary facies variations within the basins. The fact that non-manganiferous skarn iron ores, the shallowest type, now occur in anticlines, while quartz-banded iron ores, manganiferous skarn iron ores, and iron-manganese-silicate ores typically are found in synclines, points toward a direct relationship between the primary morphology/bathymetry and the present structural trends. The conclusion of van der Velden et al. (1982) that the present large curvature, shown by both the leptite-limestone formation and the lower part of the upper, largely metasedimentary sequence, was controlled more by the primary position of the volcanic eruption centers than by subsequent folding should also be noted in this connection.

An instructive example of how gravitative forces may have exerted a marked influence on local Svecofennian structures is supplied by Ehlers (1976, 1978) from the Kumlinge-Enklinge area in the archipelago between the Åland islands and the mainland of southwest Finland. A trough-like syncline, filled by leptite and limestone and a thick sequence of pillow lavas, agglomerates, and upper, terrestrial flows, appears to have sunk into a surrounding younger tonalite. A strong flattening is seen in the marginal parts of the syncline, while an elongation-type deformation with an axial ratio of 2/3/75 and a subvertical position of the long axis is observed in its central parts.

These examples illustrate the great diversity in the Svecofennian structural styles. If the dynamic evolution of all these styles were to be characterized in a few words, it would probably best be described as a steadily changing but, in the long run, almost equal competitive interaction between vertically and tangentially operating forces. This characterization would reconcile the opposing views of Stephansson (1975) and Stålhös (1981).

The emplacement of the large late- to post-orogenic granite batholiths in the Bothnian subprovince of Norrland and central southwest Finland, ca. 1,800-1,750 Ma ago, forms so to speak an epilogue to the Svecofennian crustal evolution (Figure 4-5).

However, the Svecofennian magmatic and tectonic activity not only caused the formation of a broad segment with new, aboriginal continental crust, but it also affected neighboring parts in the proto-shield, not least the ca. 2,000 Ma subduction zone, which at that time had already modified the formerly passive margin of the Archean continent. Through this means the original features of the ca. 2,000 Ma subduction zone became blurred and only what may be termed a healed subduction suture was preserved.

The formation of the Kotalahti shear zone in southwest Finland (Figure 4-5) most probably occurred as a sort of ''contrecoup'' in response to the plate movements in the Svecofennian province proper.

THE CRUSTAL EVOLUTION OF THE SOUTHWESTERN PARTS OF THE BALTIC SHIELD

The results of the last two decades of research in the southwestern parts of the Baltic shield serve as an example of how new isotopic age determinations, remapping, and structural studies, can, in effect, turn classic concepts upside down, and how new ideas gradually develop along a zigzag path of progress. This path may be traced through Hjelmqvist (1973), Gorbatschev (1975, 1980), Berthelsen (1977b, 1978, 1980, 1982b), Pedersen et al. (1978), Falkum and Petersen (1980), Skjernaa and Pedersen (1982), Nyström (1982), and Priem and Verschure (1982). Although most authors agree on the predominantly ensialic nature of the Sveconorwegian evolution ca. 1,200 to 850 Ma old, there are considerable variations in opinion as to whether, or which way, the Sveconorwegian tectonic features should be interpreted in plate-tectonic terms. However, the solution of these vigorously debated problems is a necessary prerequisite for obtaining a full understanding of the post-Svecofennian but pre-Sveconorwegian evolution (ca. 1,750 to 1,200 Ma).

Arguing from structural evidence, Berthelsen (1977a and b, 1978, 1980) assumes that the km-broad west-dipping shear zones (marked 6, 7, 8 in Figure 4-1), which separate the eastern sub-units of the Sveconorwegian structural age province, represent deep erosional sections in steepened, originally listric overthrust zones (Figures 4-9 and 4-10) formed in connection with crust-mantle decoupling and crustal peeling or with more shallow duplication of crustal slices. These points of view have been contradicted by Gorbatschev (1980, p. 134) as an ''unsubstantiated extrapolation.''

Oftedahl (1980) took a more positive approach, but unfortunately he misread Figure 8 in Berthelsen (1980) and consequently spoke about ''microcontinents'' instead of ''the now visible part'' of the reconstructed segments (Berthelsen's Figure 9). However, Berthelsen (1982b) has himself changed his original point of view, noting that the strong crustal shortening in the eastern part of the Sveconorwegian unit might just as well have been compensated for by ductile thickening of the underlying lithosphere as by closing of a neighboring late-Jotnian ocean, subduction along its western margin, and ultimate collision between the eastern and western sub-units of the Sveconorwegian belt in the so-called cryptic Oslo suture. It is hoped that the combined efforts to carry out geophysical and geological studies in a swath across the Sveconorwegian unit, within the framework of the European Geotraverse program, will clear up these problems in the near future.

In the meantime it can be stated that the oldest crustal rocks of probable mantle derivation encountered in the eastern sub-units of the Sveconorwegian belt either are Svecofennian, as in the extreme east, or were formed in connection with a period of accretion between ca. 1,750 and 1,600 Ma. Where exactly the geographical limit between the ca. 1,700 Ma crustal age province P_2 and the older Svecofennian crust P_1 should be placed is an open question (Figure 4-9).

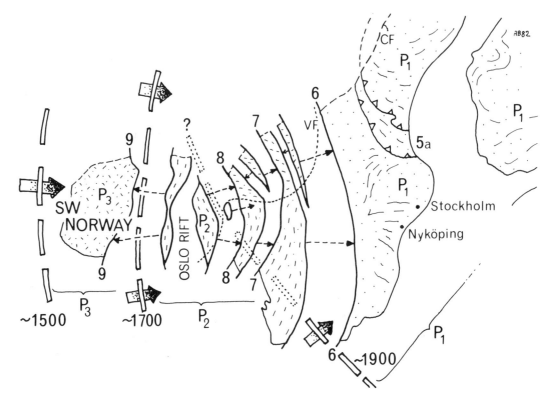

FIGURE 4-9. Tentative palinspastic reconstruction of the southwestern part of the Baltic shield around 1,400 Ma ago, redrawn with some alterations from Berthelsen (1980, Figure 8). The arrows with dashed shafts indicate the amount of crustal shortening that the P_2 and P_3 crustal age provinces are believed to have suffered in connection with the ca. 1,200 to 850 Ma old Sveconorwegian orogeny. The broken double lines and heavy arrows show the supposed sites of the Main Svecofennian subduction zone (ca. 1,900 Ma), the ca. 1,700 Ma subduction zone, and the ca. 1,500 Ma subduction zone. VF stands for the south boundary of the Värmlandian fold train and CF for the Caledonian front.

As emphasized by Priem and Verschure (1982), there cannot be much doubt that the first formation of the continental crust of southwest Norway took place around 1,500 Ma ago. Where exactly the boundary between the ca. 1,700 Ma and the ca. 1,500 Ma crustal age provinces is to be found is difficult to say for sure, because of the subsequent strong Sveconorwegian deformation. A qualified guess is that the boundary or transition zone has been cut out by the east-dipping Kristiansand-Bang shear zone (Hageskov 1980), which forms the present border (marked 9 in Figure 4-9) between the eastern and western sub-units of the Sveconorwegian structural age unit (Berthelsen 1980).

No formal names have yet been given to the ca. 1,700 Ma and ca. 1,500 Ma crustal accretion periods or to the corresponding crustal age provinces P_2 and P_3. Berthelsen (1980) referred to them respectively as the "Ghost-Gothian" and "Western" orogenies. Gorbatschev (1980) spoke of the "South-Western Orogen" in southwest Sweden and adjacent parts of

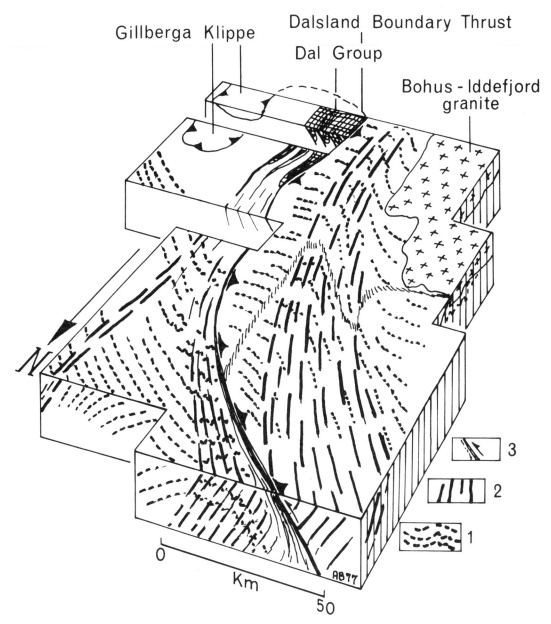

FIGURE 4-10. Block diagram showing the principal Sveconorwegian structures in southeast Norway and parts of Värmland and Dalsland (Sweden), from Berthelsen (1977b). The location of the area is shown in Figure 4-1 (small frame marked by an asterisk). 1: Trend of Värmlandian fold axes. 2: Axial trend of superimposed north-plunging folds. 3: Trend of the Dalsland Boundary thrust/shear zone (marked 8 in Figures 4-1 and 4-9), which in the southeast caused older crustal units to override the Dal Group rocks.

southeast Norway, but he did not discuss the relations in southwest Norway west of the Kristiansand-Bang shear zone.

At the same time as large-scale formation of mantle-derived tonalitic to granodioritic rocks occurred between ca. 1,700 and 1,600 Ma in a belt within Gorbatschev's "South-Western Orogen," the granites of the Småland-Värmland belt (marked SVB in Figure 4-1) were being emplaced in the neighboring Svecofennian crust to the east (Nyström 1982, Lindh and Gorbatschev 1984). Reasoning from the almost north-south trend of the Småland-Värmland granite belt, which cuts earlier Svecofennian trends, Gorbatschev (1980, Figure 1 and p. 131) concluded that the active subduction zones must have had different directions in relation to the Baltic shield during the ca. 1,900 Ma Svecofennian and the ca. 1,700 Ma accretionary periods.

Recent plate-tectonic interpretations dealing with the Sveconorwegian evolution presume the former existence of a major east-dipping subduction zone off the present coast of southwest Norway (Berthelsen 1980, Falkum and Petersen 1980) in order to explain the widespread occurrence of acid plutonics with an age range from ca. 1,200 to 950 Ma in the western Sveconorwegian sub-unit (Pedersen et al. 1978, Priem and Verschure 1982).

In the eastern sub-unit of the Sveconorwegian tectonic belt, no comparable magmatic activity occurred in the 1,200 to 950 Ma interval. This is probably the reason why the crust of the eastern sub-unit reacted in a more rigid way and, toward the later part of the Sveconorwegian tectonic evolution, became divided into north-south trending segments separated by relatively narrow zones with intense ductile (to brittle) deformation.

The Sveconorwegian deformational history is fairly well known (Pedersen et al. 1978, Park et al. 1979, Berthelsen 1980, Falkum and Petersen 1980, Hageskov 1980, Huijsmans et al. 1981, Skjernaa and Pedersen 1982). In the southern part of the western sub-unit the overall structural pattern is controlled by large, tight to isoclinal structures that were formed during a deformational event ca. 1,100 Ma old but were refolded into more open, west-vergent folds with more or less north-south trending axes during an event 1,000 to 990 Ma old.

In the northern and western parts of the eastern sub-units, the oldest large-scale Sveconorwegian structures are represented by a regionally developed foliation, which outlines south-vergent recumbent to overturned folds, the axial-trend pattern of which describes large S-shaped bends (Figure 4-10). Obviously these bends have been controlled by the formation of later open north-plunging structures, and not least by large-scale inhomogeneous deformation of the crustal compartments between the major shear zones that truncate the axial-trend pattern. Berthelsen (1977a) first suggested a pre-Sveconorwegian age for this oldest, now deflected fold train, and he related it to a ? 2,000-1,900 Ma old Värmlandian orogeny, but when new age determinations became available, he corrected this mistake in his 1978 paper, and the distribution and origin of the post-1,200 Ma old Värmlandian structures were discussed in his 1980 contribution. A more detailed study of a part of the region influenced by the Värmlandian deformation has been carried out by Skjernaa and Pedersen (1982).

The absence of Värmlandian fold trends in the southern parts of the segments of the eastern sub-unit is probably an original feature, in which case the Dalslandian supracrustals or Dal Group west of Lake Vänern rest on a sort of foreland to the Värmland fold train (see Figure 4-10).

If Berthelsen's (1980 and this volume) mobilistic view of the Sveconorwegian tectonic evolution is accepted, the growth stage of the Baltic shield can be said to have come to an end before 1,200 Ma B.P. In the now preserved parts of the shield the Sveconorwegian deformation resulted in tectonic reworking and partial destruction of crust formed during the ca. 1,900 Ma, the ca. 1,700 Ma, and the ca. 1,500 Ma accretionary periods, and in this respect the Sveconorwegian orogeny is rather analogous to the Caledonian orogeny in Scandinavia.

SUMMARY OF CONCLUSIONS

1. The Archean evolution of the Baltic proto-shield during the age interval from 3,100 to 2,800 Ma cannot yet be fully traced.

2. As is the case in Sydvaranger, the late Archean tectonic evolution includes a stage with predominantly horizontal tectonics, resulting in large-scale overthrusting of crustal slices. This necessitates large-scale plate movements, and the related magmatic activity points toward contemporaneous subduction or underplating.

3. The Proterozoic tectonic evolution appears to be divisible into a sequence of plate-tectonically controlled accretionary periods and tectonic episodes. Orogenic activity thus culminated around 2,000 to 1,900 Ma ago in the Granulite belt and the adjacent Kola suture belt, 2,050 to 1,900 Ma ago along the southwestern margin of the Karelian tectonic unit, 1,950 to 1,800 Ma ago within the Svecofennian province, and ca. 1,700 Ma and 1,500 Ma ago in the southwesternmost parts of the shield. This sequence of events brought the growth stage of the shield to an end, and reworking and shortening of the shield occurred in connection with the later Sveconorwegian (and Caledonian) tectonic events.

4. In the light of this complex evolution, it seems no longer advisable to speak of a shield-wide Svecokarelian orogen or of one prolonged (2,000 or earlier to 1,500 Ma B.P) Svecokarelian orogeny, or to undertake formal stratigraphic divisions according to corresponding megacycle concepts.

5. It does not appear that tectonic styles have changed systematically with time during the interval ca. 2,600 to 900 Ma. Thus contrasting tectonic regimes should be explained in terms of aboriginal or reworked crust. In aboriginal crustal sections, such as most of the Svecofennian province, primitive magma tectonics (with diapirism and accompanying gravity-induced sliding and downfolding) apparently competed ably with crustal shortening. Tectonic styles reflecting strong horizontal movements, crustal decoupling, peeling, and duplications appear only to have been developed where the preexisting continental crust was tectonically reworked in response to, or as a proximal or distal effect of, subduction (or collision) zones. Where a preexisting crust was "softened" due to subsequent magma rise, composite tectonic styles developed.

6. From these considerations it can be deduced that at least the late Archean and Proterozoic tectonic development was influenced by decoupling of the upper and the lower crust—or of the crust and the underlying mantle.

7. The described, assumed, or postulated tectonic evolution of the Baltic shield, which was arrived at from geological and geochronological data, is not in conflict with available paleomagnetic results (Pesonen and Neuvonen 1981).

8. There is still much to do.

ACKNOWLEDGMENTS

The author is grateful to Ms. Pernille Andersen for typing the manuscript, Max Schierling for assistance with the figures, Ole Bang Berthelsen for photographic work, and Lektor T.C.R. Pulvertaft, B.A., for improving the English.

REFERENCES

Backlund, H. G., 1936: Der "Magmaaufsteig" in Faltengebirgen. Comm. Géol. Finlande Bull. *115*, 293-347.

Backlund, H. G., 1943: Einblicke in das geologische Geschehen des Präkambriums. Geol. Rundschau *34*, 79-148.

Barbey, P., Convert, J., Martin, H., Moreau, B., Capdevila, R., and Hameurt, J., 1980: Relationships between granite-gneiss terrains, greenstone belts and granulite belts in the Archean crust of Lapland (Fennoscandia). Geol. Rundschau *69*, 648-658.

Barbey, P., Convert, J., Moreau, B., Capdevila, R., and Hameurt, J., 1984: Petrogenesis and evolution of an early Proterozoic collisional orogenic belt: The Granulite belt of Lapland and the Belomorides (Fennoscandia). Geol. Soc. Finland Bull. *56*, 161-188.

Bernard-Griffiths, J., Peucat, J. J., Postaire, B., Vidal, P., Convert, J., and Moreau, B., 1984: Isotopic data (U-Pb, Rb-Sr, Pb-Pb and Sm-Nd) of mafic granulite from Finnish Lapland. Precambrian Res. *23*, 325-348.

Berthelsen, A., 1977a: Himalayan tectonics: A key to the understanding of Precambrian shield patterns. C.N.R.S. Coll. Internat. *268*, 2, 61-67.

Berthelsen, A., 1977b: Field guide COMTEC 1977 (I.U.G.S. Comm. on Tectonics). Copenhagen, 11 p.

Berthelsen, A., 1978: Himalayan and Sveconorwegian tectonics—a comparison. *In*: Saklani, P. S. (ed), Tectonic Geology of the Himalaya. Today and Tomorrow's Printers & Publishers, New Delhi, 287-294.

Berthelsen, A., 1980: Towards a palinspastic tectonic analysis of the Baltic shield. *In*: Cogné, J., and Slausky, M. (eds), Geology of Europe. Bur. Rech. Géol. Min. Mém. *108*, 5-21; 26th Internat. Geol. Congr., Paris 1980, Colloque *C6*: Geology of Europe, 5-21.

Berthelsen, A., ms., 1982a: Interrelations between basement and cover tectonics around the Tornquist line. Internal Rept. to ESRC's Working Group on the European Geotraverse. Copenhagen, 10 p.

Berthelsen, A., ms., 1982b: Contributions to section on "Northern Segment." Proposal for a European Geotraverse (EGT), prepared by ESRC's Working Group.

Berthelsen, A., and Marker, M., 1986: Tectonics of the Kola collision suture and adjacent Archaean and Early Proterozoic terrains in the northeastern region of the Baltic Shield. Tectonophysics *126*, 31-55.

Berthelsen, A., and Marker, M., in prep.: 1.9-1.8 Ga old strike-slip megashears in the Baltic shield, and their plate-tectonic implications.

Bogdanoff, A. A., 1963: Sur le terme "étage structural." Rév. Géog. phys. Géol. dyn. *5*, 245-253.

Bugge, J.A.W., 1978a: The Sydvaranger type of quartz-banded iron ore, with a synopsis of Precambrian geology and ore deposits in Finnmark. Geol. Surv. Finland Bull. *307*, 15-24.

Bugge, J.A.W., 1978b: Norway. *In*: Bowie, S.H.U., Kvalheim, A., and Haslam, H. W. (eds), Mineral Deposits of Europe, Vol. 1, Northwest Europe. Adlard & Son Ltd., Bartholomew Press, Dorking, 199-249.

Bugge, J.A.W., and Iversen, E., 1981: Geologisk guide, NORDKALOT Ekskursjon til Sydvaranger, June 1981. Issued by Sydvaranger A/S. Oslo, 13 p.

Ehlers, C., 1976: Homogeneous deformation in Precambrian supracrustal rocks of Kumlinge area, SW Finland. Precambrian Res. *3*, 481-504.

Ehlers, C., 1978: Gravity tectonics and folding around a basic volcanic centre in the Kumlinge area, southwest Finland. Geol. Surv. Finland Bull. *295*, 43 p.

Eskola, P., 1929: Om mineralfacies. Geol. Fören. Stockholm Förh. *51*, 157-172.

Eskola, P., 1948: The problem of mantled gneiss domes. Geol. Soc. London Quart. Jour. *104*, 461-476.

Eskola, P., 1963: The Precambrian of Finland. *In*: Rankama, K. (ed), The Precambrian, *1*, 145-263.

Falkum, T., and Petersen, J. S., 1980: The Sveconorwegian orogenic belt, a case of late-Proterozoic plate-collision. Geol. Rundschau *69*, 622-647.

Frietsch, R., 1982: A model for the formation of the iron, manganese and sulphide ores of central Sweden. Geol. Rundschau *71*, 206-212.

Gaal, G., 1972: Tectonic control of some Ni-Cu deposits in Finland. 24th Internat. Geol. Congr., Montreal 1972, Sect. *4*, 215-224.

Gaal, G., 1982: Proterozoic tectonic evolution and late Svecokarelian plate deformation of the Central Baltic shield. Geol. Rundschau *71*, 158-170.

Gaal, G., Mikkola, A., and Söderholm, B., 1978: Evolution of the Archaean crust in Finland. Precambrian Res. *6*, 199-215.

Gorbatschev, R., 1975: Fundamental subdivisions of Precambrian granitoids in the Åmål meta-unit and the evolution of the south-western Baltic Shield, Sweden. Geol. Fören. Stockholm Förh. *97*, 109-114.

Gorbatschev, R., 1980: The Precambrian development of southern Sweden. Geol. Fören. Stockholm Förh. *102*, 129-136.

Gorbunov, G. I., 1971: The role of structural and tectonic factors in the formation of sulphide copper-nickel deposits. Soc. Mining Geol. Japan, Spec. Issue *3*, 20-22.

Gorbunov, G. I., 1980: Geologicheskaya karta—schema Kol'skogo Polustrova 1:1,000,000 (Geological map of the Kola Peninsula, scale 1:1,000,000). Akademiya Nauk SSSR.

Grip, E., 1978: Sweden. *In*: Bowie, S.H.U., Kvalheim, A., and Haslam, H. W. (eds), Mineral Deposits of Europe, Vol. 1, Northwest Europe. Adlard & Son Ltd., Bartholomew Press, Dorking, 93-198.

Haapala, I., and Front, K. 1985: Petrology of the Nattanen-type granites, northern Finland (abstr.). Helsinki Symposium on the Baltic Shield, 4-6 March 1985.

Hageskov, B., 1980: The Sveconorwegian structures of the Norwegian part of the Kongsberg-Bamble-Østfold segment. Geol. Fören. Stockholm Förh. *102*, 150-155.

Hietanen, A., 1975: Generation of potassium-poor magmas in the northern Sierra Nevada and the Svecofennian of Finland. U.S. Geol. Survey Jour. Res. *3*, 631-645.

Hjelmqvist, S., 1973: An old evolution and a young "model." Sveriges Geol. Unders., Ser. C, *686* (Årsbok 67, 5), 11 p.

Huijsmans, J.P.P., Kabel, A.B.E.T., and Steenstra, S. E., 1981: On the structure of a high-grade metamorphic Precambrian terrain in Rogaland, south Norway. Norsk Geol. Tidsskrift *61*, 183-192.

Isokangas, P., 1978: Finland. *In*: Bowie, S.H.U., Kvalheim, A., and Haslam, H. W. (eds), Mineral Deposits of Europe, Vol. 1, Northwest Europe. Adlard & Son Ltd., Batholomew Press, Dorking, 39-92.

Khain, V. Ye., and Leonov, Yu. G. (chief editors), 1979: Carte tectonique de l'Europe et des régions avoisinantes. Moscow.

Kozlovsky, E. A., 1984: Kol'skaya sverchglubokaya skvazhina. Issledovanie glubinnogo stroeniya kontinental'noy kory s pomosch'u bureniya Kol'skoy sverchglubokoy skvazhiny. Nedra (Ministerstvo geologii SSSR), Moskva, 490 p. (in Russian).

Kröner, A., 1980: New aspects of craton–mobile belt relationships in the Archaean and early Proterozoic: Examples from southern Africa and Finland. *In*: Closs, H., et al. (eds), Mobile Earth, Internat. Geodynamics Project, Final Report of the Federal Republic of Germany. Harald Boldt Verlag, Boppard, 225-234.

Kröner, A., Puustinen, K., and Hickman, M., 1981: Geochronology of an Archaean tonalitic gneiss dome in northern Finland and its relation with an unusual volcanic conglomerate and komatiitic greenstone. Contrib. Mineral. Petrol. *76*, 33-41.

Lindh, A., and Gorbatschev, R. 1984: Chemical variation in a Proterozoic suite of granitoids extending across a mobile belt–craton boundary. Geol. Rundschau *73*, 881-893.

Lundberg, B., 1980: Aspects of the geology of the Skellefte field, northern Sweden. Geol. Fören. Stockholm Förh. *102*, 156-166.

Lundqvist, T., 1979: The Precambrian of Sweden. Sveriges Geol. Unders., Ser. C, *768* (Årsbok 73, 9), 87 p.

Meriläinen, K., 1976: The granulite complex and adjacent rocks in Lapland, northern Finland. Geol. Survey Finland Bull. *281*, 129 p.

Meriläinen, K., 1980: Stratigraphy of the Precambrian in Finland. Geol. Fören. Stockholm Förh. *102*, 177-180.

Mikkola, A. K., and Niini, H., 1968: Structural position of ore-bearing areas in Finland. Geol. Soc. Finland Bull. *40*, 17-33.

Nyström, J. O., 1982: Post-Svecokarelian Andinotype evolution in central Sweden. Geol. Rundschau *71*, 141-157.

Oen, I. S., Helmers, H., Verschure, R. H., and Wiklander, U., 1982: Ore deposition in a Proterozoic incipient rift zone environment: A tentative model for the Filipstad-Grythyttan-Hjulsjö region, Bergslagen, Sweden. Geol. Rundschau *71*, 182-194.

Oftedahl, C., 1980: Geology of Norway. Norges Geol. Unders., Bull. *54*, 114 p.

Park, G., Bailey, A., Craine, A., Creswell, D., and Standley, R., 1979: Structure and geological history of the Stora Le-Marstrand rocks in western Orust, southwestern Sweden. Sveriges Geol. Unders., Ser. C, *763* (Årsbok 73, 4), 36 p.

Pedersen, S., Berthelsen, A., Falkum, T., Graversen, O., Hageskov, B., Maaløe, S., Petersen, J. S., Skjernaa, L., and Wilson, J. R., 1978: Rb-Sr dating of the plutonic and tectonic evolution of the Sveconorwegian province, southern Norway. ICOG Denver, U.S. Geol. Survey, Open-File Rept. *78-701*, 329-331.

Pesonen, L. J., and Neuvonen, K. J., 1981: Palaeomagnetism of the Baltic shield: Implications for Precambrian tectonics. *In*: Kröner, A. (ed), Precambrian Plate Tectonics. Elsevier, Amsterdam, 623-648.

Preston, J., 1954: The geology of the pre-Cambrian rocks of the Kuopio district. Acad. Sci. Fenniae Ann., Ser. A III, *40*, 111 p.

Priem, H.N.A., and Verschure, R. H., 1982: Review of the isotope geochronology of the high-grade metamorphic Precambrian of SW Norway. Geol. Rundschau *71*, 81-84.

Raith, M., Raase, P., and Hörmann, P. K., 1982: The Precambrian of Finnish Lapland: Evolution and regime of metamorphism. Geol. Rundschau *71*, 230-244.

Rickard, D. T., 1979: Scandinavian metallogenesis. Geo-Journal *3*, 235-252.

Rickard, D. T., and Zweifel, H., 1975: Genesis of Precambrian sulfide ores, Skellefte district, Sweden. Econ. Geology *70*, 255-274.

Sederholm, J. J., 1907: On granite and gneiss, their origin, relations and occurrence in the pre-Cambrian complex of Fenno-Scandia. Comm. géol. Finlande Bull. *23*, 91-110.

Sederholm, J. J., 1927: Precambrian of Fennoscandia with special reference to Finland. Geol. Soc. America Bull. *38*, 813-836.

Sederholm, J. J., 1932: On the geology of Fennoscandia with special reference to the pre-Cambrian. Explanatory notes to accompany a general geological map of Fennoscandia. Comm. géol. Finlande Bull. *98*, 30 p.

Silvennoinen, A., 1972: On the stratigraphic and structural geology of the Rukatunturi area, northeastern Finland. Geol. Survey Finland Bull. *257*, 48 p.

Simonen, A., 1980: The Precambrian in Finland. Geol. Survey Finland Bull. *304*, 58 p.

Skjernaa, L., and Pedersen, S., 1982: The effects of penetrative Sveconorwegian deformations on Rb-Sr isotope systems in the Römskog-Aurskog-Höland area, SE Norway. Precambrian Res. *17*, 215-243.

Stålhös, G., 1981: A tectonic model for the Svecokarelian folding in east central Sweden. Geol. Fören. Stockholm Förh. *103*, 33-46.

Stephansson, O., 1975: Polydiapirism of granitic rocks in the Svecofennian of central Sweden. Precambrian Res. *2*, 189-214.

Stockwell, C., 1982: Proposals for time classification and correlation of Precambrian rocks and events in Canada and adjacent areas of the Canadian shield. Part 1, A time classification of Precambrian rocks and events. Geol. Survey Canada Paper *80-19*, 135 p.

van der Velden, W., Baker, J., de Maesschalck, S., and van Meerten, T., 1982: Bimodal early Proterozoic volcanism in the Grythytte field and associated volcano-plutonic complexes, Bergslagen, Central Sweden. Geol. Rundschau *71*, 171-181.

Väyrynen, H., 1931: Über die geologische Struktur des Erzfeldes Kammikivitunturi in Petsamo. Comm. géol. Finlande Bull. *92*, 17-32.

Walser, G., and Einarsson, O., 1982: The geological context of molybdenum occurrences in the southern Norrbotten region, northern Sweden. Geol. Rundschau *71*, 213-229.

Wegmann, C. E., 1928: Über die Tektonik der jüngeren Faltung in Ostfinnland. Fennia *50*, no. 16, 22 p.

Welin, E., Wicklander, U., and Kähr, A-M., 1980: Radiometric dating of a quartz-porphyritic potassium rhyolite at Hällefors, south central Sweden. Geol. Fören. Stockholm Förh. *102*, 269-272.

Wilson, M. R., 1980: Granite types in Sweden. Geol. Fören. Stockholm Förh. *102*, 167-176.

Wilson, M. R., 1982: Magma types and the tectonic evolution of the Swedish Proterozoic. Geol. Rundschau *71*, 120-129.

Witschard, F., 1980: Stratigraphy and geotectonic evolution of northern Norrbotten. Geol. Fören. Stockholm Förh. *102*, 188-190.

Chapter 5

THE LOWER ALLOCHTHON IN SOUTHERN NORWAY: AN EXHUMED ANALOG OF THE SOUTHERN APPALACHIANS DEEP DETACHMENT?

A. G. MILNES

Swiss Federal Institute of Technology, Zürich

The Caledonian orogenic belt in southern Norway can be thought of in terms of three major units (Figure 5-1): (1) a lower unit of Precambrian crystalline rocks, representing the northwestern margin of the Baltic shield (autochthon); (2) an upper complex of far-traveled nappes, of which the main member is a large mass of Precambrian igneous and metamorphic rocks known as the Jotun complex (allochthon); and (3) a separating zone of smaller parautochthonous and allochthonous units, containing lower Paleozoic sediments from different facies belts, slices of Precambrian basement, and some ophiolitic fragments (lower allochthon = detachment zone). The last unit is now exposed over wide areas because of subsequent warping, faulting, tilting, uplift, and erosion, but at the time of the main translational movements (Late Silurian to Early Devonian), the main décollement dipped gently northwestward, suggesting an analogy with the deep detachment identified below the southern Appalachians (Cook et al. 1979, Cook et al. 1983) and similar mega-structures identified below other collisional-type orogens (see, e.g., Price 1981, Seeber et al. 1981, Bally 1983).

The late Caledonian detachment in southern Norway is most clearly illustrated by the Jotunheim and Sognefjord cross sections (Figure 5-2). The lower limit of the detachment is marked by a sole thrust which seems to follow closely the peneplained upper surface of the Baltic shield basement, although some involvement of basement slices has been postulated (e.g., Hossack et al. 1985). The upper limit of the zone is more diffuse, but it can be defined by the appearance above it of complexes whose main tectonic and metamorphic overprinting is of pre-detachment age (early Caledonian or Precambrian). Between these limits, the zone can be subdivided into three tectonic regimes, which from southeast to northwest represent increasing depth in the crust at the time of deformation (Figure 5-1).

FORELAND FOLD-AND-THRUST REGIME

The most southeasterly and, originally, superficial regime in the detachment is represented by a thin-skinned fold-and-thrust belt (Figure 5-2a) comparable to the Valley-and-Ridge province of the Appalachians (e.g., Roeder et al. 1978). In its inner, most highly deformed parts,

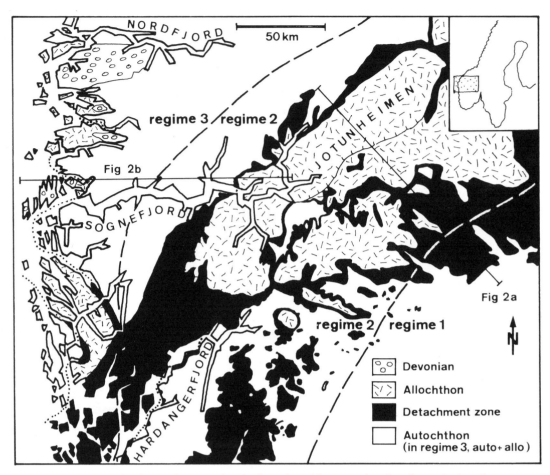

FIGURE 5-1. Generalized tectonic map of part of southern Norway, showing the main detachment zone of the lower allochthon and the approximate boundaries of the three structural regimes recognized.

Vendian to lower Ordovician sediments (the original cover of the Baltic shield) are involved (Bjørlykke 1978, Nystuen 1981, Hossack et al. 1985). In its outermost parts (Oslo graben), the youngest deposits affected by the folding are of upper Silurian age (Ludlow; see Størmer 1967, Bryhni et al. 1981) and are non-marine. The width of the belt is about 150 km and the depth of burial of the inner parts is estimated to be about 10 km (Hossack et al. 1985), giving an average original northwesterly dip of the basal décollement of 4°. The original width of sedimentary cover now telescoped within the belt is such that the pre-orogenic position of the innermost parts would have been at least 240 km northwest of their present position (Hossack et al. 1985), giving a minimum estimate for the original width of the overthrust Baltic shield margin.

a) Jotunheimen cross section

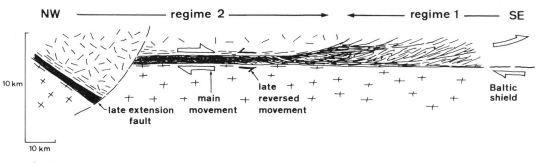

b) Sognefjord cross section

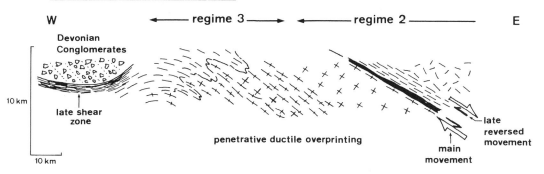

FIGURE 5-2. Schematic cross sections through the Caledonides of southern Norway (see Figure 5-1 for position). Vertical exaggeration, ca. 4:1.

INTRA-BASEMENT CONFINED REGIME

The change from regime 1 to regime 2 is marked by the appearance of undeformed Precambrian crystalline masses or earlier-formed Caledonian complexes above the mainly sedimentary units, or both (Figures 5-2a and 5-2b). The relatively ordered lithostratigraphic units of regime 1 grade into a narrow zone of disordered elements, characterized by complex structural sequences, cleavage and mylonite formation, and low-grade (or retrograde) metamorphism (Milnes and Koestler 1985). This is the equivalent of the prominent seismic reflector which is interpreted as a partially sediment-filled, mylonitized shear zone below the Blue Ridge and Inner Piedmont of the southern Appalachians (Hatcher 1981, Iverson and Smithson 1982, Fountain et al. 1984). In southern Norway, radiometric dating suggests that the main penetrative deformation in this zone took place around 400 Ma ago (spread 415 to 385 Ma; see Andresen et al. 1974, Corfu 1980, Schärer 1980), whereas the Precambrian basement above and below it did not receive a Caledonian overprint. This intra-basement confined regime of the detachment extends about 100 km northwest of regime 1. The upper greenschist facies assemblages typical for the northwesternmost exposures suggest original depths of 15 km, roughly a continuation of the first regime's dip. The main translation of the allochthon was up-dip, southeastward over the shield margin, and it must be measured in hundreds of kilometers (Hossack et al. 1981). However, at a late stage a few kilometers of reversed (north-westward) movement took place in the same zone (Milnes and Koestler 1985).

OVERPRINTED BASEMENT REGIME

A number of changes mark the transition to the deepest identified regime (regime 3, Figure 5-2b). First, recognizable lower Paleozoic cover rocks no longer appear as a continuous separating screen between the shield basement and the allochthonous units. Second, the deformation is no longer confined within a narrow zone, but rather spreads out as heterogeneous shearing and complex polyphase folding of all units on all scales. In southern Norway, relations in this regime—now visible in the western half of the ''Basal Gneiss Complex'' and the Paleozoic and Jotun-like complexes along the west coast—are still incompletely known, controversial, and contradictory (cf. Brynhi 1966, Guezou et al. 1980, Bryhni et al. 1981, Griffin et al. 1981, Cuthbert et al. 1983). However, it is clear that regime 3, had it remained buried, would have yielded complex seismic reflections of the type defining the deep structure of the eastern Piedmont in the southern Appalachians (Iverson and Smithson 1982, Cook et al. 1983). Perhaps studies of the structural relations in the exhumed southern Norwegian equivalents will lead to a better understanding of this buried and inaccessible Appalachian ''root zone.''

Worldwide, deep detachments are known from two main plate-tectonic situations: in compressive, back-arc terrains, in association with Cordilleran-type orogenesis (always antithetic to the contemporaneous subduction; see, e.g., Allmendinger et al. 1983, Jordan et al. 1983, Milnes, this volume), and as intracontinental mega-shears, as a sequel to continent-continent collision (either synthetic or antithetic to the pre-collisional subduction, not at the site of the collision suture; cf. Hatcher and Odom 1980 and Seeber et al. 1981). The detachment zone in southern Norway is of the latter type, intracontinental and post-collisional, but it does contain remnants of oceanic crust (Bergen arcs, Karmøy; see Furnes et al. 1980). By analogy, these remnants do not necessarily indicate that the detachment zone itself is a collision suture; it is more likely that earlier collisional events resulted in ophiolite-bearing units being overthrust onto the continental margin, to be later involved in the intracontinental mega-shear now exposed as the zone of detachment.

REFERENCES

Allmendinger, R. W., Ramos, V. E., Jordan, T. E., Palma, M., and Isacks, B. L., 1983: Paleogeography and Andean structural geometry, northwest Argentina. Tectonics 2, 1-16.

Andresen, A., Heier, K. S., Jorde, R., and Naterstad, J., 1974: A preliminary Rb/Sr geochronological study of the Hardangervidda-Ryfylke nappe system in the Røldal area, south Norway. Norsk. Geol. Tidsskr. 54, 35-47.

Bally, A. W., 1983: Seismic expression of structural styles, Vol. 3, Tectonics of compressional provinces/strike-slip tectonics. Am. Assoc. Petroleum Geologists Studies in Geology 15, 448 p.

Bjørlykke, K., 1978: The eastern marginal zone of the Caledonide orogen in Norway. Geol. Survey Canada Paper 78-13, 49-55.

Bryhni, I., 1966: Reconnaissance studies of gneisses, ultrabasites, eclogites and anorthosites in the outer Nordfjord, western Norway. Norges Geol. Unders. *241*, 68 p.

Bryhni, I., Bockelie, J. F., and Nysteun, J. P., 1981: The Southern Norwegian Caledonides—Oslo to Sognefjord and Ålesund. Field guide to excursion Al, Uppsala Caledonide Symposium.

Cook, F. A., Albaugh, D. S., Brown, L. D., Kaufman, S., Oliver, J. E., and Hatcher, R. D., 1979: Thin-skinned tectonics in the crystalline southern Appalachians; COCORP seismic-reflection profiling of the Blue Ridge and Piedmont. Geology 7, 563-567.

Cook, F. A., Brown, L. D., Kaufman, S., and Oliver, J. E., 1983: The COCORP seismic reflection traverse across the southern Appalachians. Am. Assoc. Petroleum Geologists Studies in Geology *14*, 61 p.

Corfu, F., 1980: U-Pb and Rb-Sr systematics in a polyorogenic segment of the Precambrian shield, central southern Norway. Lithos *13*, 305-323.

Cuthbert, S. J., Harvey, M. A., and Carswell, D. A., 1983: A tectonic model for the metamorphic evolution of the Basal Gneiss Complex, western South Norway. Jour. Metamorphic Geology *1*, 63-90.

Fountain, D. M., Hurich, C. A., and Smithson, S. B., 1984: Seismic reflectivity of mylonite zones in the crust. Geology *12*, 195-198.

Furnes, H., Roberts, D., Sturt, B. A., Thon, A., and Gale, G. H., 1980: Ophiolite fragments in the Scandinavian Caledonides. *In*: Panayiotou, A. (ed), Ophiolites. Cyprus Geological Survey Department, 582-600.

Griffin, W. L., et al., 1981: Eclogites and basal gneisses in western Norway. Field guide to excursion Bl, Uppsala Caledonide Symposium.

Guezou, J.-C., Lécorché, J.-P., and Quenardel, J.-M., 1980: Les Calédonides scandinaves. Réunion extraordinaire Soc. géol. France, 19-28 août, 1979. Soc. géol. France Bull. (7), *22*, 251-295.

Hatcher, R. D., Jr., 1981: Thrusts and nappes in the North American Appalachian orogen. *In*: McClay, K. R., and Price, N. J. (eds), Thrust and Nappe Tectonics. Geol. Soc. London Spec. Publ. *9*, 491-499.

Hatcher, R. D., Jr., and Odom, A. L., 1980: Time of thrusting in the Southern Appalachians, U.S.A.: Model for orogeny? Geol. Soc. London Jour. *137*, 321-327.

Hossack, J. R., Garton, M. R., and Nickelsen, R. P., 1985: The geological section from the foreland up to the Jotun thrust sheet in the Valdres area, south Norway. *In*: Gee, D. G., and Sturt, B. A. (eds), The Caledonide Orogen—Scandinavia and Related Areas. Wiley, New York, 443-456.

Hossack, J. R., Koestler, A. G., Lutro, O., Milnes, A. G., and Nickelsen, R. P., 1981: A traverse from the foreland through the thrust sheet complex of Jotunheimen. Field guide to excursion B3, Uppsala Caledonide Symposium.

Iverson, W. P., and Smithson, S. B., 1982: Master décollement root zone beneath the southern Appalachians and crustal balance. Geology *10*, 241-245.

Jordan, T. E., Isacks, B. L., Ramos, V. A., and Allmendinger, R. W., 1983: Mountain building in the Central Andes. Episodes, *1983/3*, 20-26.

Milnes, A. G., and Koestler, A. G., 1985: Geological structure of Jotunheimen, southern Norway (Sognefjell-Valdres cross-section). *In*: Gee, D. G., and Sturt, B. A. (eds), The Caledonide Orogen—Scandinavia and Related Areas. Wiley, New York, 457-474.

Naterstad, J., Andresen, A., and Jorde, K., 1973: Tectonic succession of the Caledonian nappe front in the Haukelisaeter-Røldal area, southwest Norway. Norges Geol. Unders. *292*, 20 p.

Nystuen, J. P., 1981: The late Precambrian "sparagmites" of southern Norway: A major Caledonian allochthon—the Osen-Røa nappe complex. Am. Jour. Sci. *281*, 69-94.

Price, R. A., 1981: The Cordilleran foreland thrust and fold belt in the southern Canadian Rocky

Mountains. *In*: McClay, K. R., and Price, N. J. (eds), Thrust and Nappe Tectonics. Geol. Soc. London Spec. Publ. *9*, 427-448.

Roeder, D., Gilbert, O. E., Jr., and Witherspoon, W. D., 1978: Evolution and macroscopic structure of Valley and Ridge thrust belt, Tennessee and Virginia. Univ. Tennessee, Studies in Geology *2*, 25 p.

Schärer, U., 1980: U-Pb and Rb-Sr dating of polymetamorphic nappe terrain: The Caledonian Jotun nappe, southern Norway. Earth Planet. Sci. Lett. *49*, 205-218.

Seeber, L., Armbruster, J. G., and Quittmeyer, R. C., 1981: Seismicity and continental subduction in the Himalayan arc. *In*: Gupta, M. K., et al. (eds), Zagros. Hindu Kush. Himalaya. Geodynamic Evolution. Am. Geophys. Union Geodynamics Ser. *3*, 215-242.

Størmer, L., 1967: Some aspects of the Caledonian geosyncline and foreland west of the Baltic shield. Geol. Soc. London Quart. Jour. *123*, 183-214.

Chapter 6

THE STRUCTURE AND EVOLUTION OF THE HERCYNIAN FOLD BELT IN THE IBERIAN PENINSULA

M. JULIVERT
(with the collaboration of F. J. MARTÍNEZ for metamorphism and plutonism)

Universitat Autònoma de Barcelona

The core of the Iberian Peninsula is formed by a Hercynian cratonic block, bounded to the west and northwest by the Atlantic Ocean and the Cantabrian Sea (Bay of Biscay) and to the northeast and southeast by the Pyrenean and Betic segments of the Alpine fold belt. Within this block Paleozoic rocks crop out largely in the west, forming the Iberian massif, while to the east they plunge gently below a Mesozoic-Tertiary platform cover.

The Hercynian structures cross the Iberian massif in a northwest-southeast direction, except in the northern part where they describe a sharp bend cut by the Cantabrian continental margin. More to the north, beyond the Bay of Biscay, the Hercynian structures extend into the Armorican massif (Ibero-Armorican arc). To the south, the Hercynian structures are intersected nearly at a right angle by the Betic front.

A cross section normal to the Hercynian structural trends provides the longest (770 km) and most complete cross section of the European Hercynian fold belt. Nevertheless, in contrast to Central Europe, where the Hercynian fold belt borders a stable area (the Baltic shield), in the Iberian Peninsula it is not known what kind of geotectonic units existed beyond the boundaries of the belt. In the northern border the arc is so tight that no space is left in its core (Figure 6-1), and to the south the belt is at present bounded by a continental margin.

The Iberian massif exhibits a longitudinal zonation. From north to south the following five zones are distinguished (Lotze 1945, Julivert et al. 1972–1974): (1) Cantabrian, (2) West Asturian-Leonese, (3) Central Iberian, (4) Ossa-Morena, and (5) South Portuguese (Figure 6-1).

Upper Paleozoic rocks are concentrated in the two extreme zones of the massif (Cantabrian and South Portuguese), which are also characterized by their thin-skinned tectonics and by an almost complete absence of metamorphism and plutonism. In contrast, in the other zones of the massif, lower Paleozoic and Precambrian rocks predominate; metamorphism, although showing much variation, reaches high-grade amphibolite and even granulite facies; and granitoids are abundant. All these facts, together with the opposite facing of the structures on the two sides of the massif (northeast or east in the Cantabrian and West Asturian-Leonese zones and southwest in the South Portuguese), indicate the existence in the fold belt of two

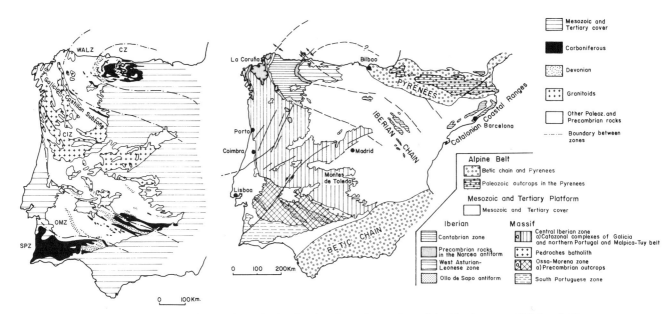

FIGURE 6-1. Structural units of the Iberian Peninsula and zonation of the Iberian massif (after Julivert et al. 1972–1974, based on Lotze 1945) and distribution of the upper Paleozoic rocks and granitoids in the Iberian massif. (Reproduced by permission of the Instituto Geológico y Minero de España.)

branches with opposite polarity. There is not, however, perfect mirror-image symmetry in the belt, since there are many important differences between the two branches (Lotze 1945, Julivert et al. 1972–1974).

THE SEDIMENTARY RECORD

The Paleozoic sequence contains a great deal of information on the geological evolution during Paleozoic time. Differing thicknesses, lateral and vertical facies changes, variations in deposition rate, breaks in the sequence, and angular unconformities record changes in crustal dynamics and permit one to trace the evolution of the different parts of the orogenic belt. Figure 6-2 summarizes the distribution of the most representative facies, sedimentary breaks, angular unconformities, and volcanic events, and Figure 6-3 gives a general idea of the variations in thickness.

Below the fossiliferous Paleozoic sequence, an azoic shale-graywacke succession crops out in several areas, equivalent to the Brioverian of the Armorican massif. In the northern part of the peninsula an angular unconformity intervenes, while in the central part the shale-graywacke succession crops out conformably below the Cambrian. Where the unconformity exists, the Cambrian-Precambrian boundary is drawn somewhat arbitrarily at the unconformity.

In other cases, the deepest rocks exposed are para- and orthogneisses (often augen gneisses) cropping out in the core of late antiformal structures below lower Ordovician or pre-Ordovician (Cambrian and/or Precambrian) sequences, and in general affected by high-grade metamorphism: ''Ollo de Sapo'' antiform, Miranda do Douro antiform, gneisses in the Sierra

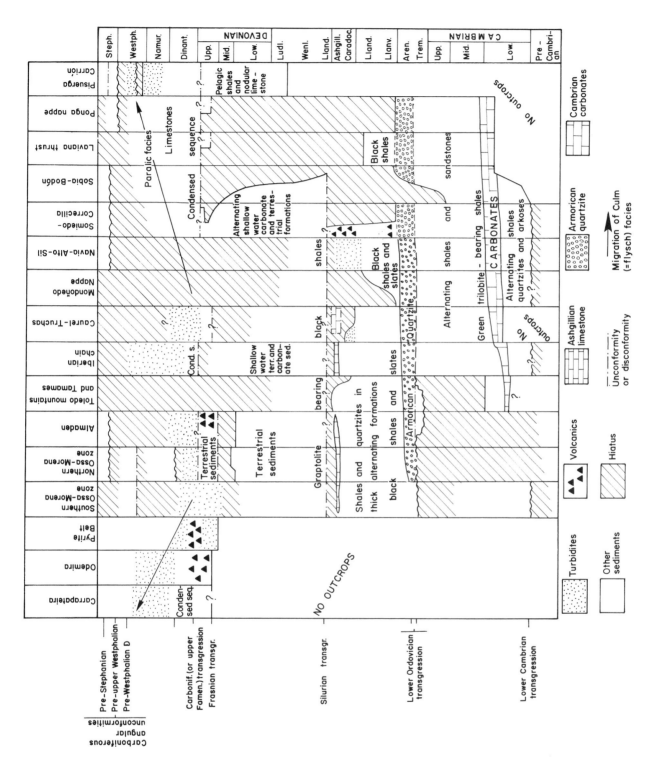

FIGURE 6-2. Main facies and breaks in the stratigraphic sequence throughout the Iberian massif. (According to Julivert 1983b. Reproduced by permission of the Instituto Geológico y Minero de España.)

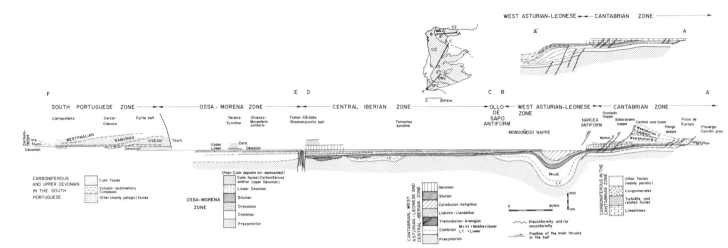

FIGURE 6-3. *Top*: Schematic reconstruction of the Cantabrian zone and eastern part of the West Asturian-Leonese zone, interpreted as a passive continental margin. *Bottom*: Thickness variations throughout the Iberian massif (based mainly on Matte 1968; Schermerhorn 1971; Julivert et al. 1972; Marcos 1973; Julivert 1978, 1981; Oliveira et al. 1979; and Marcos and Pulgar 1982). The horizontal distances correspond to the relative present-day positions of the different localities; no palinspastic reconstruction has been attempted.

de Guadarrama, gneiss massifs in the Pyrenees, and so forth. These rocks are affected by the oldest Hercynian penetrative structures and are consequently pre-Hercynian, but whether they are Precambrian basement and/or early Paleozoic intrusions is still unclear (Parga-Pondal et al. 1964, Floor 1966, Guitard 1970, Martínez 1974a, A. Ribeiro 1974, Bischoff et al. 1978, Gil Ibarguchi 1982b, among others).

All the above rocks have been more or less strongly deformed and metamorphosed during Hercynian time, but no evidence for earlier cleavages or metamorphism has been found. The only areas in which the existence of a crust with a Precambrian metamorphism has been postulated are the katazonal complexes of Galicia and northern Portugal and the Ossa-Morena zone; these areas are discussed below.

THE PRE-OROGENIC EVOLUTION

The Hercynian orogeny deformed a thick pile of sediments not disrupted by an earlier significant orogenic event since the beginning of the Paleozoic or even since the late Proterozoic. However, the pre-Hercynian Paleozoic cannot be regarded simply as a long period of crustal stability and uniform deposition, since thickness and facies variations record a complex sedimentary history. Variations in the thickness of the Cambrian and Ordovician sediments (Julivert et al. 1972, 1980) suggest the existence during that time of several shoals and troughs with different rates of subsidence and deposition. Volcanism was important in some areas, indicating deep faulting, and peralkaline and calc-alkaline anorogenic intrusions also took place (Floor 1966, Priem et al. 1972). All these facts point toward a tensile crustal regime during the early Paleozoic. Nevertheless, during some periods deposition was rather uniform (in both facies and thickness) throughout the Iberian massif, mainly during the deposition of the Cambrian carbonates, the Armorican quartzite (Arenigian), and the Llanvirn-Llandeilian black slates. On the other hand, maximal differences in thickness of sediments correspond to the Upper Middle Cambrian–Upper Cambrian–Tremadocian and to the Caradocian.

There are two main areas of thick early Paleozoic deposition in the Iberian massif: the West Asturian-Leonese zone and the Ossa-Morena zone. In the first a sequence 8,000 m thick was laid down from the Late Middle Cambrian until the end of the Ordovician; this sequence shows successively terrestrial shallow-water, euxinic, and finally turbidite facies. Different depositional rates prevailed at different moments through this time span, but the result was the deposition of a very thick sequence indicating an evolution from shallow-water to deep-water conditions. In the Ossa-Morena zone the rate of deposition seems to have been high during the Cambrian but not during the Ordovician. Thus the evolution of the two areas with thick early Paleozoic sequences was not parallel.

In contrast to the important differences shown by the Cambrian-Ordovician sequence, the Silurian sequence has similar thicknesses throughout the Iberian massif and at least the Llandoverian and Wenlockian exhibit nearly the same facies: black graptolitic slates or shales. The most important differences in the Silurian sequence are due to variations in the volcanic content and are probably a consequence of the distribution of faults throughout the massif.

In many parts of the Iberian massif, the Ordovician succession shows a remarkable non-depositional or erosional break in its upper part, and the basal part of the Silurian is lacking; the first graptolite faunas indicate in general the *convolutus* or perhaps the *gregarius* zone (Truyols and Julivert 1983). Thus a general hiatus may separate the Ordovician and Silurian sequences. In some areas (Cantabrian zone and most of the West Asturian-Leonese), the hiatus is very clear, for Silurian sediments rest on units rather low in the Ordovician, locally even on the Arenigian quartzite. On the other hand, in some areas the hiatus may not exist, for the Ordovician seems to be complete and has yielded Caradocian or even Ashgillian faunas (Iberian chain, southern Central Iberian zone, Ossa-Morena), and the Silurian contains early Llandoverian graptolites in its lower part. Even in these cases, however, there is no reliable biostratigraphic information on the transitional beds, so that a small hiatus could exist between Ordovician and Silurian. Recently Hafenrichter (1980) concluded—based on the detailed paleontological surveying of the transitional beds of some apparently complete successions in the Iberian chain and in the southern part of the Central Iberian zone—that in these sequences a small hiatus represents the Hirnantian.

Hiatus at the Ordovician-Silurian boundary followed by an early Silurian onlap has been claimed to be a worldwide phenomenon (Berry and Boucot 1973, Sheehan 1973, Brenchley and Newall 1980) related to glacio-eustatic effects in connection with the Ordovician glaciations. The question of the Silurian transgression is related to the still puzzling question of the meaning of the Caradocian facies. These are not uniform throughout the Iberian massif, but graywackes are abundant and mud-supported conglomerates and volcanic horizons, together with pebble-rich mudstones, are present in many localities within the Hercynian fold belt. In some localities the sedimentary structures (Bouma cycles, sole marks, etc.) indicate turbidite deposition (northwest Spain; see Crimes et al. 1974), while in other localities isolated pebbles have been interpreted as dropstones related to the Ordovician glaciations (Dangeard and Doré

1971, Doré and Le Gall 1972, Iglesias and Robardet 1980, Robardet 1981). It is difficult to be sure of the influence of the Ordovician glacial events on sedimentation in the Hercynian fold belt, because dating is not precise enough to permit a close correlation with the well-known continental Ordovician glaciation in North Africa (Beuf et al. 1966, 1971) or with environmental changes in stable platform areas. Brenchley and Newall (1980), studying the Upper Ordovician–Lower Silurian in the Oslo region, conclude that glacio-eustatic sea-level changes could influence deep-water deposition; the major effect of sea lowering would be the formation of channels extending across shallow-water areas into the slopes, from which turbidity currents would carry clastics into deep water. On the other hand, the effect of a transgression would be to drown the supply areas. Nevertheless, however important the glacial effects might have been, the great thickness contrasts, the importance of volcanism, and the existence of peralkaline and calc-alkaline anorogenic intrusions point toward crustal instability during Caradocian-Ashgillian(?) time. Upper Ordovician deposition in the Iberian massif probably reflects the interference of diastrophic events and synchronous glacio-eustatic sea-level changes, one of which could be the Silurian transgression.

In conclusion, in spite of the terrigenous character of the Cambrian-Ordovician-Silurian deposits, sequences that could be interpreted either as flysch or as molasse are not found. The Ordovician turbidites are not related to any orogenic event and cannot be interpreted as flysch, and molasse-type sediments recording the erosion of a growing fold chain do not exist in the lower Paleozoic, in sharp contrast to the Carboniferous sequence, which shows a typical pre-, syn-, and post-tectonic facies succession. The lower Paleozoic sequences, with their volcanism and anorogenic intrusions, suggest rifting, although a minor compressive event took place in part of the massif, probably during the Upper Cambrian. This event, affecting both the Central Iberian and Ossa-Morena zones, is very conspicuous in the first, where the Arenigian ''Armorican'' quartzite or some probably Tremadocian sediments rest unconformably on Cambrian and Precambrian rocks. The meaning of this event, which was followed by the Ordovician transgression, is not yet clearly understood, but in any case no differences in cleavages or degree of metamorphism are observed across the unconformity.

THE FIRST RECORDS OF HERCYNIAN OROGENY

Evidence for the beginning of orogeny deduced from the sedimentary record is not completely coincident with that deduced from the metamorphism and plutonism through radiometric dating. The stratigraphic data point to an Upper(?) Carboniferous age of the orogeny. Everywhere in the northern branch of the orogen, Devonian and Carboniferous rocks crop out extensively (Cantabrian and southern Central Iberian zones, Iberian chain, Pyrenees), the oldest angular unconformities are intra-Carboniferous and not older than the Westphalian, and the Culm (= flysch)[1] deposits are not older than the Early Namurian or the Late Viséan.

Below the Carboniferous there is in general a non-depositional and erosional hiatus of variable extent. In general, the following transgression did not coincide exactly with the Car-

---

[1] The term Culm has long been used in the European Hercynian belt to designate shale-graywacke successions of Carboniferous age, formed in large part of turbidites and other deep-sea deposits and representing the Hercynian equivalent of the Alpine flysch.

boniferous-Devonian boundary but was somewhat older. In the northern part of the Iberian massif (Cantabrian zone), however, the Devonian sediments between the disconformity and the base of the Carboniferous are in general quite thin, and in some localities they are absent altogether, the transgressive sequence starting with the Lower Carboniferous. In central Spain (southern Central Iberian zone) and in the Ossa-Morena zone there is a hiatus embracing the Middle Devonian. In these areas, the Upper Devonian forms a thick sequence above which rests the Carboniferous, perhaps in continuity or perhaps separated by a minor hiatus. In the southern Central Iberian zone, the Carboniferous is practically limited to a Culm sequence of Dinantian or at most Namurian age, but in the Ossa-Morena, where it is better represented, the Carboniferous sequence is broken by several unconformities and disconformities. More to the south, in the South Portuguese zone, the succession extends from the Late Devonian into the Early Westphalian without any unconformity but is affected by several tectonic phases, which indicate that deformation was Late Carboniferous (intra-Westphalian) in the area.

The pre-Carboniferous and intra-Devonian sedimentary breaks have often been quoted as evidence of early Hercynian tectonic phases; in the northern and central parts of the Iberian massif, however, the hiatuses are associated with disconformities, never with angular unconformities. In contrast, more to the south, in the Barrancos synclinorium (Ossa-Morena zone), the Terena graywacke, assigned to the Upper Devonian, has been reported to rest unconformably on pre-Devonian rocks (Chacón et al. 1983). If the age assignment is correct, this unconformity could indicate that deformation began earlier in the southern than in the northern branch of the orogen.

Thus, the unconformities indicate Late Carboniferous ages for the orogeny, but Culm facies are older. As these facies indicate the beginning of instability in the basin and the rising of emergent areas inside the fold belt due to orogeny, the beginning of the orogeny would be Early Carboniferous or even perhaps very Late Devonian. An exception might be the southern Ossa-Morena zone, where the Terena graywacke could be still older, resting unconformably on pre-Devonian rocks, although more data and more reliable age determinations are still needed.

A final point to consider is the nature of the clasts in the Culm sequence. In general terms, the beginning of Culm deposition marks a reversal in sediment supply; major sources of supply appeared inside the orogenic belt (Schermerhorn 1971, Julivert 1978, Oliveira et al. 1979), indicating that deformation, producing uplifted areas, was in progress during Culm deposition. Conglomerates in the Culm sequence often contain pebbles of granite, gneiss, and other metamorphic rocks; granite and gneiss pebbles have been reported from the Catalonian Coastal Ranges (Fontboté and Julivert 1954, Julivert and Martínez 1980) and the Eastern Pyrenees, and pebbles of metamorphic rocks from the San Vitero and San Clodio sequences (Early? Carboniferous) in northeastern Portugal (Ribeiro and Ribeiro 1974) and neighboring areas in Spain. The existence of large and rather abundant pebbles of igneous and metamorphic rocks in the Culm sequences implies either the existence of an ancient basement, nowhere observed in outcrop inside the fold belt, or a pre-Carboniferous (or at least pre-Viséan) age for the first metamorphism and igneous intrusions. This problem cannot yet be resolved with the available evidence.

Syn- and Post-Orogenic Deposition

As Carboniferous deposition and Hercynian orogeny were essentially synchronous, much information on the progress of the orogeny can be obtained by studying the Carboniferous stratigraphy and facies distribution.

Carboniferous rocks are concentrated in the outer part of each branch of the Hercynian fold belt. This distribution is found along the whole belt throughout central and western Europe, but it is particularly clear in the Iberian massif. Here the two outer zones of the belt (Cantabrian and South Portuguese) were frontal zones of active deposition during the Carboniferous, as the Hercynian chain was being built. Carboniferous rocks are also found in more internal positions within the fold belt, but in general as minor and more or less scattered outcrops.

The Carboniferous facies succession records pre-, syn-, and post-tectonic conditions. The change from one facies to another is not synchronous in the different zones of the belt, although the facies replace one another everywhere in the following order:

1. Condensed or starved facies, formed by cherts, more or less siliceous shales, and limestones, frequently nodular. These facies are peculiar to the Dinantian and are found in a large part of the fold belt, although they are lacking in some areas, such as the southern branch of the Iberian massif.

2. Culm facies, beginning in the Viséan (or earlier in the southern branch) or in the Namurian (Cantabrian zone). The source of terrigenous material for the Culm sequences lay inside the fold belt, and the Culm deposits must be considered as (early) syntectonic.

3. Paralic facies, beginning in the Namurian (Ossa-Morena) or in the Westphalian and ending in the Westphalian (Ossa-Morena) or in the Stephanian. These facies partially coexisted in time with the turbidite (Culm) sedimentation and can be found in two different positions in the fold belt: (a) in a more internal position than contemporaneous Culm deposits, and consequently well within the belt proper, as in the case of the Spanish paralic basins; and (b) in foredeeps along the border of the fold belt and overlapping the continental stable areas in front of it, even extending far onto the platform, as in the case of the paralic basins of northern Europe, which have no equivalent in the Iberian massif. The paralic facies have in many respects the characteristics of molasse, but they coexist in time with Culm deposition. Their relationships with deformation show that orogeny was still active. The paralic deposits inside the fold belt can consequently be regarded as (late) syntectonic, although paralic deposition outside the belt could have taken place later.

4. Fluviatile and limnic facies, beginning in the Westphalian in some areas (Ossa-Morena), but especially characteristic of the Stephanian. These facies are found in intramontane basins related to normal or strike-slip faults or both, and essentially they can be considered post-tectonic, although the Stephanian was affected by the latest deformation, which is locally severe, mainly along strike-slip fault zones.

Figure 6-4 summarizes the evolution in time of the Carboniferous sedimentation in the two branches of the orogen. In the northern branch, in the Cantabrian zone the Lower Carboniferous is very condensed (25-50 m of black and red cherts and shales, and also red nodular limestones), while the Upper Carboniferous exhibits a mosaic of facies. The first record of instability found in the Cantabrian zone is provided by Namurian turbidites and olistostromes

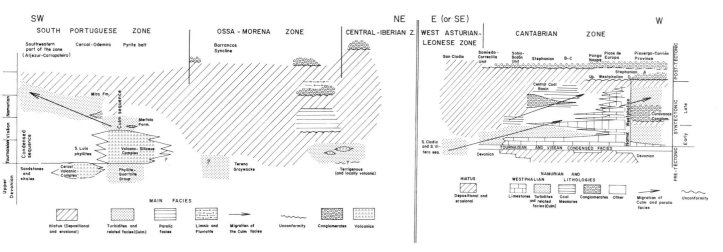

FIGURE 6-4. Migration of Culm and paralic facies in the northern and southern branches of the Hercynian belt in the Iberian Peninsula.

(Culm facies), although in more internal parts older Culm sequences may exist (San Glorio and San Vitero sequences). The beginning of the Namurian Culm deposition marks a reversal in the direction of the terrigenous supply. During early Paleozoic and Devonian time, clastic material reached the Cantabrian zone from the core of the arc, suggesting the existence outside the fold belt of some emergent land mass that is no longer visible at present. During late Carboniferous time the terrigenous supply came from the west or from the south, that is from within the fold belt. Namurian turbidites are widespread throughout the Cantabrian zone, but the distribution of the Namurian facies shows a complicated pattern. As deformation progressed from the Namurian into the Westphalian, important changes in the sedimentary basin took place. The Westphalian sequence shows that active erosion of the chain was already in progress (first thick and coarse conglomerates, first angular unconformity); turbidites are restricted to the most frontal part of the chain, while paralic conditions extended over large areas (Asturian coalfields) (Julivert 1978). Finally, upper Westphalian D and Stephanian sediments form essentially post-tectonic deposits of molasse character.

In the southern branch of the orogen, as in the northern one, Carboniferous sedimentation took place essentially in the frontal part of the belt. The South Portuguese zone was a basin with turbidite deposition, whereas in more internal (northern) parts of the chain deformation was already in progress, as shown by the intra-Carboniferous unconformities in the Ossa-Morena zone and by the appearance of paralic facies as early as the upper Viséan. Culm deposition began very early in the northern part of the South Portuguese zone and moved progressively southward (Pfefferkorn 1968, Schermerhorn 1971, Oliveira et al. 1979). Thus, to the north, the upper Devonian "Phyllite-Quartzite Group," with its sole marks, graded bedding, cross lamination, and trace fossils such as *Nereites*, represents the beginning of Culm deposition, while to the south the Culm sequence is of Namurian-Westphalian age and rests on an Early Carboniferous condensed sequence.

A comparison of the two branches of the orogen reveals some similarities but also some important differences. Following are the most important similarities:

1. The predominance and great thickness of Carboniferous rocks in both outermost zones (Cantabrian and South Portuguese), in contrast with the inner parts of the chain.

2. The migration of the Culm facies toward the frontal part in each branch of the belt, probably registering a similar migration of deformation.

3. The Westphalian age of deformation in both outer zones, Cantabrian and South Portuguese.

The most striking differences are:

1. The greater importance of turbidites in the South Portuguese zone than in the Cantabrian zone.

2. The earlier beginning of active sedimentation in the South Portuguese zone. At the same time that a Tournaisian-Viséan condensed sequence was being laid down in the Cantabrian zone, the thick volcano-sedimentary complex and Viséan Culm sequence was actively being deposited in most of the South Portuguese zone.

3. The great importance of Early Carboniferous volcanism in the South Portuguese zone, in contrast to its almost complete absence in the Cantabrian zone.

4. The earlier appearance of Culm, paralic, and fluviatile-limnic facies in the southern branch than in the northern branch, probably because deformation began earlier in the former.

THE STRUCTURE OF THE NORTHERN BRANCH OF THE OROGEN

The best section across the northern branch of the orogen can be obtained in the northwestern part of the Iberian Peninsula, in an east-west direction parallel to the Cantabrian coast.

The structure of this cross section exhibits the typical polarity of fold chains, from an outer non-metamorphic area with thin-skinned tectonics (Cantabrian zone, to the east) into an inner high-grade metamorphic area with abundant granitoids and with structures involving the Precambrian as well as its Paleozoic cover (central and western Galicia, to the west). The changes in structure, metamorphism, and plutonism have been used to define the three classical zones distinguished in northwestern Spain: Cantabrian, West Asturian-Leonese, and Galician (northwestern part of the Central Iberian) (Figure 6-5).

THIN-SKINNED TECTONICS IN THE FRONTAL PART OF THE BELT

The frontal part of the belt is formed by the Cantabrian zone, characterized (Julivert 1979, 1981) by its varied lithology (limestones, sandstones, shales) giving an alternation at different scales of competent and incompetent units, as is normal in sequences laid down on shallow-water platforms. Carboniferous rocks form more than 65% of the surface of the zone. Cleavage is absent or only locally developed, and the zone practically lacks metamorphism and plutonism; the first is limited to some localities close to its western boundary and to a small area in the very core of the arc (chloritoid), and the second is represented only by a few tiny bodies near the western border of the zone.

The structure is the result of polyphase deformation. In general, deformation started with the generation of décollement nappes and slices and associated folds, the latter replacing the

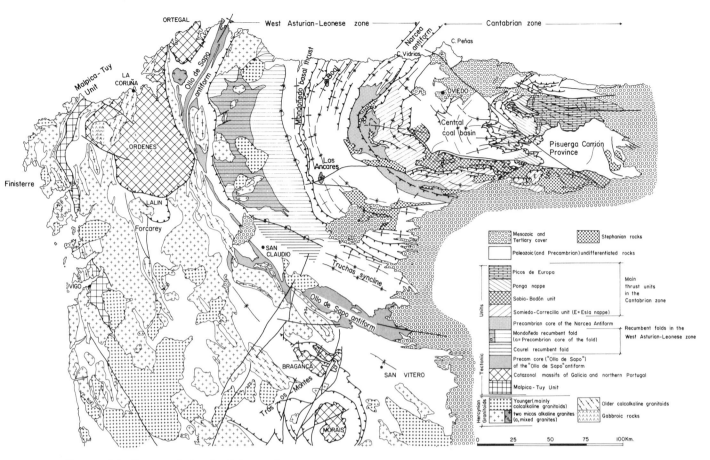

FIGURE 6-5. Structural sketch of the northern part of the Iberian massif.

nappes and slices laterally and representing the structures into which they die out along strike. The trace of these structures describes an arc, now curving through 180°. As deformation progressed, the nappes and slices began to fold flexurally. At the beginning they were folded more or less longitudinally, so that the folds generated describe the same arc as the first structures. Later, a radial set of folds was formed (Julivert and Marcos 1973); they die out toward the *convex* part of the arc. This radial set shows that the arc was being tightened, at least during the last stages of its evolution (Julivert 1971b).

The most striking tectonic feature of the Cantabrian zone is the existence of a general detachment of the Paleozoic sequence, giving an imbricate structure formed by an array of thrust faults, flattening and merging downdip into a single bedding fault. This bedding fault or décollement surface coincides or nearly coincides with the base of the Láncara Formation (limestones and dolostones of Early to Middle Cambrian age) (Julivert 1971a). The décollement plates moved without being internally folded, so that the folds affecting the décollement structures are younger than their emplacement and different décollement units were folded together along with the surfaces that separated them. Also, the nappe emplacement took place without significant strain in the rocks, as shown by undeformed fossils, oolites, and so forth, which are found unstrained even very close to the thrust surface.

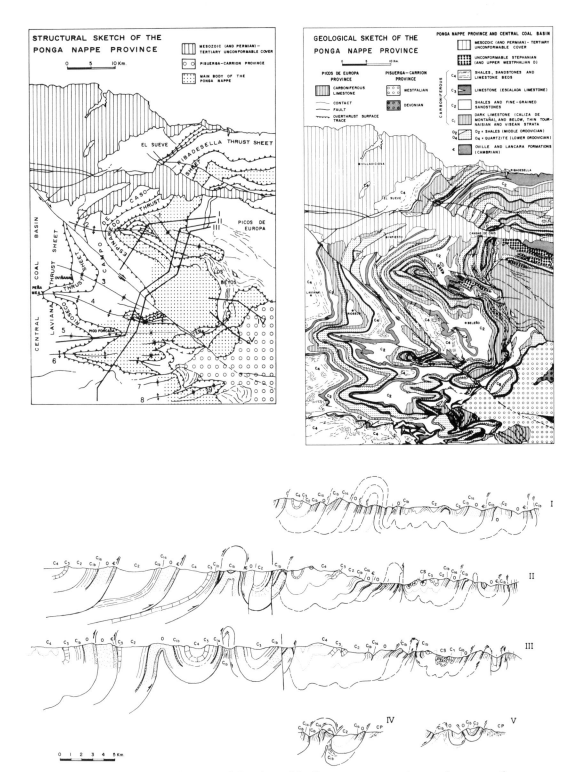

FIGURE 6-6. Geological and structural sketches of the Ponga nappe province and cross sections through the Ponga nappe (after Julivert 1971a). $\mathbb{C}$: Cambrian. O: Lower Ordovician. C_{1a}: Dinantian. C_{1b}: Lower Namurian. C_2: Upper Namurian-Lower Westphalian. C_3 and C_4: Westphalian. CS: Stephanian. CP: Pisuerga-Carrión province. (Reproduced by permission of the *American Journal of Science*.)

In some localities (Esla nappe; see Arboleya 1981), there is a thin layer (about 1 m) of strongly deformed rocks at the base of the nappes, probably formed during the advance of the thrust sheets on their substratum. The deformed zone extends for a short distance above and below the thrust plane and consists of different kinds of cataclastic rocks, including some ultracataclasites (Arboleya 1981). Beyond this narrow layer the rocks are unstrained, both in the thrust sheet and in its relative autochthon.

In the Cantabrian zone there are three important décollement units: the Somiedo-Correcilla unit, the La Sobia-Bodón unit, and the Ponga nappe unit. The Picos de Europa are also a detached thrust unit, but with some peculiarities. Each of these major units consists of a main body and several minor slices more or less directly related to it. The stratigraphy changes from one unit to another (Julivert 1971b, 1981, Zamarreño 1972, 1975; Mendez-Bedia 1976), and on the basis of the trace of the main thrust units, the stratigraphic characteristics, and the structural features, the Cantabrian zone has been subdivided into five provinces (Julivert 1981): (1) fold and nappe province, (2) central coal basin, (3) Ponga nappe province (Figure 6-6), (4) Picos de Europa, and (5) Pisuerga-Carrión province. The western boundary of the zone is an antiformal structure (Narcea antiform), in whose core a Precambrian shale/graywacke succession crops out.

The thrust units die out laterally into folds. This change is particularly clear in the fold and nappe province, where both thrust units (Somiedo-Correcilla and La Sobia-Bodón) die out northward into a fold system, approximately at the parallel of Ovideo (Julivert and Arboleya 1984).

In the pre-Carboniferous part of the sequence, the sedimentary prism was wedge-shaped, thinning toward the concave side of the arc and thickening radially in the opposite direction. This shape is due mainly to the wedging out of the different Devonian and Silurian formations toward the concave side of the arc. With the progress of deformation, the wedge was distorted by telescoping due to thrusting and also because Carboniferous sedimentation in subsiding areas was superimposed on the pre-Carboniferous sedimentary wedge. However, the original shape was more or less preserved, and the sedimentary wedge turned into a wedge-shaped deforming region, thicker at the back end, from which the thrusts came. This evolution has recently been described by Julivert and Arboleya (1986).

Figure 6-7 shows the position of the thrust surfaces for the main thrust units. In the most forward slices (Los Beyos, Picos de Europa), only the Carboniferous is involved. In the Ponga nappe province, as in most of the Cantabrian zone, the Paleozoic sequence has been detached at the base of the Láncara Formation. More toward the convex part of the arc, in the La Sobia-Bodón unit, the Herreria Formation (the unit below the Láncara Formation) has been incorporated into the thrust sheet. Finally, at the western boundary of the Cantabrian zone, in the Narcea antiform, there are two tectonic windows (Narcea and Villabandín) where wedges of Precambrian rocks override the basal part of the Paleozoic sequence. The structure can be interpreted as the root zone of the Somiedo-Correcilla thrust unit, where the thrust cuts down into the Precambrian.

In summary, all along the Narcea antiform there is a zone of deep faults, which flatten upon reaching the Paleozoic sequence and merge with the bedding faults of the Cantabrian zone. This fault zone forms the boundary between the area with a thin-skinned tectonics, where the Paleozoic sequence was stripped off the Precambrian (Cantabrian zone), and the

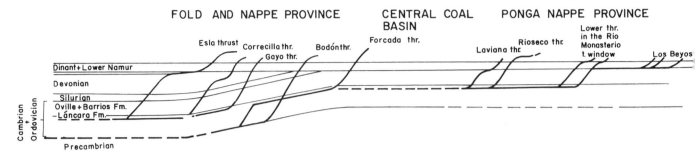

FIGURE 6-7. Position of the thrust surfaces (flats and ramps) within the wedge of pre-Carboniferous sediments, for the main décollement units of the Cantabrian zone. (Reproduced by permission of the *Journal of Structural Geology*.)

area where the fold structure involves the Precambrian and is associated with the development of cleavages and metamorphism (West Asturian-Leonese zone).

THE STRUCTURE OF THE INTERMEDIATE PART OF THE BELT

Between the Cantabrian zone with its thin-skinned tectonics and the inner part of the belt with its high-grade metamorphism over large areas and with abundant granitoids (central and western Galicia), there is a wide zone with intermediate characteristics. This zone, the West Asturian-Leonese zone, is bounded to the east and west by antiformal structures: the Narcea antiform, with a core formed by the late Precambrian shale/graywacke sequence, and the "Ollo de Sapo" antiform, with a core formed by a porphyroid sequence known by the local name of "Ollo de Sapo."

This zone has the following characteristics (Matte 1968, Capdevila 1969, Marcos 1973, Julivert 1983a): (1) the rocks are mainly schists, and to a lesser extent quartzites; (2) Cambrian and Ordovician rocks form more than 70% of the surface of the zone; (3) the structure is the result of polyphase deformation and consists of early asymmetrical and recumbent folds, facing toward the concavity of the arc, and later superimposed, more open folds with steeply dipping axial planes; (4) the largest first-phase folds involve the Precambrian (shales and graywackes); (5) a generalized slaty cleavage, related to the early structures, is found throughout the zone; (6) crenulation cleavages, related to later deformation phases, are also more or less widespread; (7) there is a generalized regional metamorphism throughout the zone; (8) greenschist facies prevail in most of the zone, but in its western part the sillimanite isograd is reached; and (9) plutonism becomes important in the westernmost part of the zone, where different kinds of granitoids show different relationships with deformation.

The overall structure of the West Asturian-Leonese zone consists of a pile of recumbent folds that form as a whole a large fold-nappe (Mondoñedo nappe) in the western part of the zone and a series of asymmetrical folds in the eastern part. These structures, generated during the first deformation phase, are the most conspicuous structures, although they have been modified to a greater or lesser degree by later deformations, such as thrusting (second phase) and a slight back-folding (third phase). The main structures in the zone are shown in a map and in cross section in Figures 6-5 and 6-8.

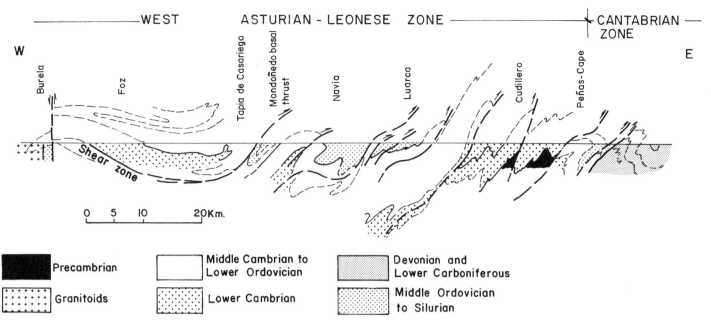

FIGURE 6-8. Cross section through the West Asturian-Leonese zone. (Based, with slight modification, on Pulgar 1980. Reproduced by permission of the author.)

The structure of the Mondoñedo nappe has been described mainly by Matte (1968), Bastida and Pulgar (1978), and Martínez-Catalán (1981). The nappe is a large composite structure consisting of a pile of second-order recumbent folds (Figure 6-9) forming a first-order recumbent anticline, with a core of Precambrian schists and graywackes, followed to the east by a recumbent syncline. At the base of the complex a shear zone (second phase) forms its true sole (Mondoñedo basal thrust). Superimposed on the nappe structure is a very gentle anticline-syncline couple (P_3) with nearly vertical axial planes.

The reverse limbs of the second-order recumbent folds have a length of about 10 to 20 km (Bastida 1980), and according to Martínez-Catalán (1981) the tectonic overlap of the Mondoñedo nappe is about 70 km. The nappe plunges to the south, making it possible to view the shear zone forming the sole below the body of the nappe at the surface along the Cantabrian coast.

Individual folds can be followed southward and southeastward for quite long distances, but the overlap becomes progressively less important and the nappe as a first-order structure dies out in that direction, while a new recumbent fold (El Caurel fold nappe) appears (Matte 1968). The El Caurel nappe is smaller than the Mondoñedo nappe and to some degree could be considered simply a second-order fold on its normal limb. Toward the southeast, the El Caurel nappe itself loses importance and the structure changes to a series of asymmetrical folds facing northeast (Pérez-Estaún 1978).

Taking into account only the stratigraphic units involved in the folds, the total thickness of the Mondoñedo recumbent fold complex above the basal shear zone is about 15 km, but there must have been an important additional overburden provided by younger Paleozoic strata and structures at higher levels.

Variations in angle between limbs and variations in shape of the minor folds through the

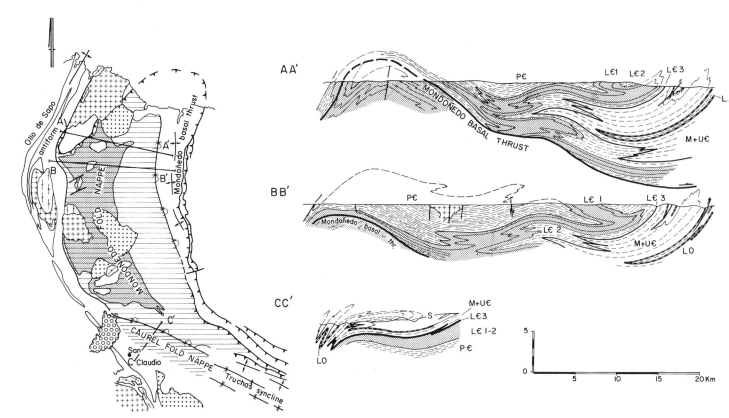

FIGURE 6-9. Cross section through the Mondoñedo and El Caurel fold nappes (after Martínez-Catalán 1981). P€: Precambrian. L€1, L€2, and L€3: Lower Cambrian. M + U€: Middle and Upper Cambrian. LO: Lower Ordovician. S: Silurian. (Reproduced by permission of the Barrie de la Maza Foundation.)

nappe pile, together with some strain measurements, show an increase of deformation with depth. The axial planes of second-order folds are curved, flattening and converging downdip to become parallel to the shear zone at the base of the complex (Martínez-Catalán 1981). There is a well-developed axial-plane slaty cleavage (S_1) throughout the Mondoñedo nappe, grading with depth into a schistosity. The stretching lineation is nearly at right angles to the fold axes, so that it fans some 40° around the arc as the folds curve (Martínez-Catalán 1981). More to the south in the El Caurel nappe, however, the stretching lineation is nearly parallel to the fold axes (Matte 1968), which seems to indicate that it remains in a more or less east-west direction regardless of the curvature of the structures describing the arc (Figure 6-9). More measurements are still needed to establish the overall strain pattern throughout the West Asturian-Leonese zone.

The trace of the shear zone forming the sole of the nappe complex can be followed along the frontal part of the nappe (Marcos 1973), and also more internally, below its main body (Bastida and Pulgar 1978), because of the axial plunge of the complex. Below the nappe, it is a wide ductile shear zone, but it narrows and is transformed into a brittle-ductile zone toward the frontal part, as it reaches higher structural levels. In its ductile part, below the nappe body, it shows many minor folds with horizontal axial planes and curved axes, and with a schistosity (S_2) whose origin by crenulation of an earlier (S_1) cleavage can still be recognized in many places (Bastida and Pulgar 1978). In its brittle-ductile part, the zone has an associated crenulation cleavage and also some folds with curved axes (Marcos 1973, Pérez-Estaún 1978).

The strain variation in the Mondoñedo nappe complex and the change of attitude of S_1 and of axial planes of second-order folds with depth suggest simple shear due to relative transport of the upper part of the complex, with a displacement gradient increasing toward the base. The conspicuous P_2 ductile shear zone forming the sole of the complex might simply be a consequence of progressive shearing, with a deformation path able to enter the foliation in the shortening field, as observed in other shear zones (Carreras et al. 1977), and therefore permitting the folding of S_1 inside the shear zone. The first and second deformation phases can therefore be interpreted as the consequence of progressive deformation rather than as two true tectonic phases widely separated in time.

The Mondoñedo nappe is affected by an important metamorphism, increasing with depth and reaching the sillimanite isograd. Migmatization occurs at several points, and granitoids are quite abundant. The isograds cut through the nappe body and are centered in the superimposed third-phase dome structure. The general characteristics of metamorphism and plutonism are discussed below.

East of the frontal trace of the Mondoñedo basal thrust, the P_1 folds show, in general, horizontal axes and a rather cylindrical character, as in the Mondoñedo area, but they are asymmetrical, show reverse limbs 2 to 4 km long, and are less flattened than the second-order folds of the Mondoñedo area, or at least not so flattened as the deepest folds of the nappe complex. The dip of their axial planes can be evaluated at 20 to 30°, discounting the effects of the third deformation phase (Pulgar 1980). The metamorphism is, in general, in greenschist facies, and there is a well-developed axial-plane slaty cleavage, at least in the incompetent levels. Beside the folds, and (slightly?) younger, there are several important thrusts facing east (or northeast). These thrusts can be regarded as more superficial equivalents—with brittle characteristics—of shear zones such as the one described in the Mondoñedo area.

The structures of the third phase consist of folds of different sizes, nearly homoaxial to the first folds, with a crenulation cleavage, and with vertical or east-dipping (back-folding) axial planes. The crenulation cleavage (S_3) is not homogeneously distributed but is concentrated in belts of different magnitudes (Marcos 1973, Pulgar 1980) trending parallel to the major structures and, therefore, describing the arc. The attitude of the third folds depends largely on the structures on which they are superimposed. In the Mondoñedo area, where the first folds are tight recumbent folds, the third ones have vertical axial planes and the resulting structures correspond perfectly to Ramsay's type 3 interference pattern (Ramsay 1962). In contrast, to the east, where the first folds are asymmetrical, the superimposition of the third deformation gives two-hinged folds (Figure 6-10) because of the different dip of the two limbs (5-25° for the normal limb and about 40° for the reverse one) and the deformation by buckling of the competent beds. Thus, assuming a more or less horizontal compression, the normal limbs buckle and fold while the reverse ones just rotate, the result being a two-hinged fold (Pulgar 1980). Also in the eastern part of the zone, where the axial planes and the first cleavage (S_1) dip west, the new folds have eastward-dipping axial planes and S_1 is deformed in a step-like fashion. The crenulation cleavage (S_3) develops in the flats and is absent where S_1 was vertical (Figure 6-11); the result is the distribution of S_3 in belts. In the belts with vertical S_1 a horizontal set of kink bands (KB on Figure 6-11) develops; less frequently the kink bands

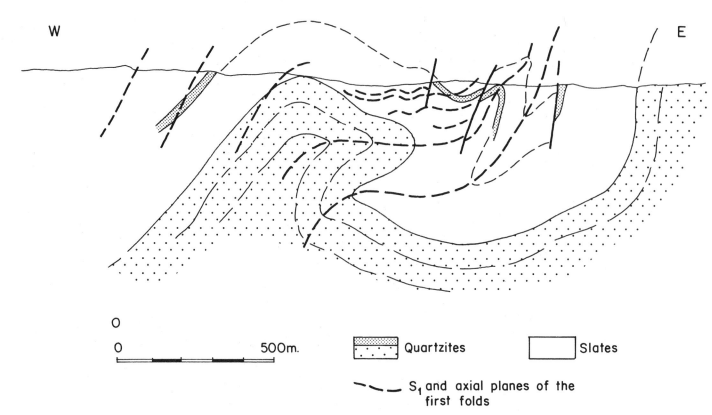

FIGURE 6-10. First-phase fold deformed during the third phase in Portizuelo, east of Luarca. (After Pulgar 1980. Reproduced by permission of the author.)

are also found in the S_3 belts, kinking S_3. The structures can be interpreted as being due to lithostatic load (Matte 1969) during a horizontal distension or simply during the release of P_3 stresses.

THE STRUCTURE OF THE INTERNAL PART OF THE BELT

The more internal part of the belt in the cross section described is found in central and western Galicia and in northern Portugal. The structure of this area is largely masked by the intensity of metamorphism and the large amount of granitoids. The last regionally important deformations produced domes (in general elongated domes) separated by synformal areas, forming in some instances narrow synclines and in others broad basin-like structures. Structurally the domes show many differences. In some cases they are clearly late structures in which the schistosity that developed during older phases is domed. In other cases they are the result of the reshaping of former folds.

The structure discussed above is closely related to the distribution of metamorphism. In general metamorphism increases from outer to inner parts of the fold belt, that is, from east to west in the cross section considered, although in detail this variation is not regular. The zones showing increased metamorphism form several belts parallel to the main structural trends and separated by areas with a lower degree. The metamorphic belts essentially coincide with antiformal or domal structures, in which the isograds are more or less centered. These

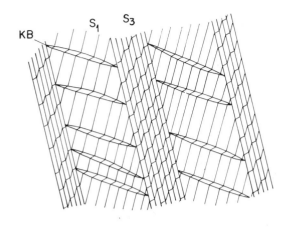

FIGURE 6-11. Alternating S_3 belts with a flatter-lying S_1, and belts with a vertical and kinked S_1 and without development of S_3.

belts and antiformal structures are also outlined by granitic plutons of alkaline tendency (two-mica granites) (Martínez and Gil Ibarguchi 1983).

The synformal areas separating the domes show a lower grade metamorphism, generally in greenschist facies. The synforms are sometimes narrow, cuspate, and more or less anasto-mosing, but they can also be broad basin-like structures (Gil Ibarguchi et al. 1983a). In the first case, the synformal core consists of Silurian and occasionally (Lower?) Devonian rocks, which are the youngest Paleozoic rocks in the area. In the second case, rock complexes affected by high-pressure-temperature (P-T) metamorphism and including mafic-ultramafic materials are found in the cores of the basins. These complexes are the most striking feature of Galicia and northern Portugal and can be grouped as follows: (1) the Ortegal, Ordenes, Bragança, and Morais complexes, (2) the Malpica-Tuy unit, and (3) the Lalín and Forcarey complexes.

The complexes of the first group (Ferreira 1965, Vogel 1967, van Zuuren 1969, Anthonioz 1970, Ribeiro 1974, Kuijper 1979, Kuijper and Arps 1983, among others) consist of mafic and felsic (partially metasedimentary) rocks in amphibolite and granulite high-pressure facies, although lower grade rocks are also found in the cores of the Ordenes and Morais complexes. In any case, a tectonic contact coincides with a sharp contrast in metamorphism between the complexes and the surrounding underlying rocks. The Malpica-Tuy unit ("Blastomylonitic graben" of the Dutch authors—den Tex and Floor 1967, among others) is characterized by its high P and low (to intermediate) T metamorphism (Ortega and Gil Ibarguchi 1983), which contrasts with the lower-grade metamorphism of the rocks surrounding it, as in the first group of complexes. The origin and emplacement of all these complexes has been controversial for many years and is still puzzling in many respects. They have been interpreted as remnants of a thrust sheet, as mushroom-shaped domes, as blocks bounded by tension faults reactivated during the Hercynian orogeny, as the result of the alternation between phases of elevation and tension and phases of burial and compression (Blastomylonitic graben), and as the result of the evolution of a mantle plume (Ribeiro et al. 1964, den Tex 1966, Matte and Ribeiro 1967, Anthonioz 1970, Ries and Shackleton 1971, Keasberry et al. 1976, van Calsteren 1977, Bard et al. 1980, den Tex 1981). In any case, however, they reached their present-day position tectonically, and most probably they represent remnants of one or several (Anthonioz 1970, Iglesias et al. 1983) thrust sheets.

The Lalín and Forcarey complexes, like those described above, also show a synformal structure, but a metamorphic discontinuity with the surrounding rocks is not clearly observed, suggesting that these complexes have perhaps a somewhat different significance.

The domes and the rock complexes described above have very different characteristics. The domes record the mesocrustal evolution of the autochthon, while the complexes record events in the lower crust (some perhaps during the Proterozoic) previous to their tectonic emplacement from a more westerly source into their present-day position.

The domes show the superimposition of three foliations (S_1, S_2, S_3) with the following characteristics (Minnigh 1975, Noronha et al. 1979, Julivert et al. 1980, among others): (1) S_1 is a slaty cleavage in the low-grade areas and the main cleavage observed there, but in the high-grade areas, that is, in the core of the domes, S_1 is recorded only by some relict structures such as oriented inclusions inside metamorphic mineral grains or polygonal arcs inside S_2; (2) S_2 has the characteristics of a schistosity with alternating mica and quartz layers in the high-grade areas, where S_2 is the main cleavage, nearly obliterating S_1, but it dies out or is only very poorly developed in low-grade areas; (3) S_3 is a typical crenulation cleavage.

S_3 in this part of the belt can be perfectly correlated to S_3 in the West Asturian-Leonese zone and is generally vertical or steeply inclined. S_2 is in general gently inclined and very often is flat-lying on top of the antiforms and dips away in their limbs, conforming to the antiform or dome structure. Finally, the attitude of S_1 cannot be known in the high-grade areas, that is, in most of central and western Galicia.

The fixing of the time of overthrusting of the complexes in relation to the tectonic phases recorded by foliations in the antiformal areas, the correlation of events inside and outside the complexes, and even the separation between Hercynian and pre-Hercynian events inside the complexes are still puzzling. However, it can be stated that the overall structure of the area is the result of the superimposition of domes and narrow anastomosing or open basin-like synclines on an older nappe structure.

The Hercynian metamorphism is plurifacial; the older mineral associations (M_1) indicate in general an intermediate gradient of Barrovian type (with garnet, staurolite, and sometimes kyanite), and the younger mineral associations (M_2) indicate a low-pressure metamorphism (andalusite, cordierite, sillimanite). During the metamorphic climax (M_2), a large amount of melt was generated, giving rise to widespread granitic areas. In this way a granite series with alkaline tendency was generated; rocks of this series are found in bodies ranging from autochthonous and suballochthonous anatectic products to clearly intrusive plutons. These plutons intruded in a continuous process of telescoping (Martínez 1974b), cutting across the metamorphic zones. The thermal and tectonic structure of the high-grade metamorphic belts is the result of a complex evolution in time of the metamorphic isograds and accompanying granitic intrusions (plutono-metamorphism of Oen Ing Soen 1970), and also of the tectonic evolution of the domes, generated or accentuated during the phase that generated S_3 (Arboleya et al. 1983).

Apart from the granites mentioned above, there are also calc-alkaline granitoids; thus, traditionally, two main series of granitoids have been distinguished in the Iberian massif (Capdevila and Floor 1970, Oen Ing Soen 1970, Capdevila et al. 1973): the series of granites of alkaline tendency and mesocrustal origin, described above, and a calc-alkaline series of deeper origin. The calc-alkaline granitoids have been subdivided in turn into two groups according to their age: a group of older granitoids (granodiorites and adamellites with abundant tonalitic, dioritic, and even gabbroic inclusions), deformed to a variable degree, and a group of younger post-tectonic granitoids (essentially coarse-grained biotite and hornblende grano-diorites). In addition to these groups, and with petrographic characteristics intermediate between them, there is still a third group of granitoids called "of mixed type."

Figure 6-12 shows the relationships between deformation phases, metamorphism, and plutonism in different parts of the belt. Chlorite and biotite began to grow during P_1 and define S_1, although in low-grade areas there is evidence of a very early growth of chlorite. The metamorphism of intermediate gradient (Barrovian) is older (syn-P_1 to syn-P_2) than that of low-pressure type. The first granitoids were intruded before, during, or after P_2 (older calc-alkaline granitoids and muscovite or two-mica leucogranites). The intrusion of the muscovite or two-mica leucogranites (series with alkaline tendency) reached its climax, coinciding with the metamorphic climax, after P_2 and during P_3, and it continued even later. At the end of and after P_3, the younger calc-alkaline granitoids and the granitoids of "mixed type" were intruded, their relationships with P_3 ranging from late syn-P_3 to clearly post-tectonic.

THE OVERALL STRUCTURE OF THE NORTHERN BRANCH

The cross section represented in Figure 6-13 gives a good idea of the overall structure of the northern branch of the Hercynian fold belt. The three zones to be distinguished appear clearly

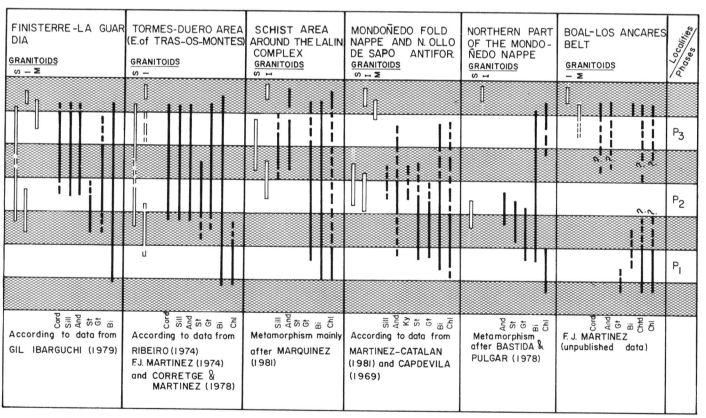

FIGURE 6-12. Relations between deformation, metamorphism, and plutonism (after Gil Ibarguchi et al. 1983b, compiled from several sources). S and I for granitoids are used in the sense of White and Chappell (1977). M: granitoids of the mixed group. (Reproduced by permission of the Instituto Geológico y Minero de España.)

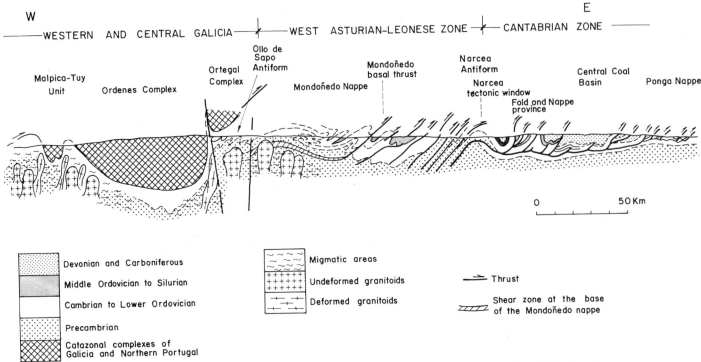

FIGURE 6-13. General cross section through the northern branch of the Hercynian fold belt in the Iberian Peninsula (based mainly on Julivert 1971a, 1981; Ries and Shackleton 1971; and Pulgar 1980).

differentiated in the cross section; from east to west they represent (1) a wedge of shallow-water platform sediments stripped off their basement and broken into many slices that override one another forming a wedge-shaped deformed region (Cantabrian zone); (2) a metamorphic area with a fold and fold-nappe structure involving the upper Proterozoic, separated from the first zone by a structural front representing a zone of deep faulting and showing a generalized cleavage and metamorphism (West Asturian-Leonese zone); and (3) a more internal area with a large amount of granitoids, overridden by a large thrust sheet of materials from the lower crust.

The cross section also shows progressively deeper crustal levels involved in the structure from east to west (from outer to inner zones of the belt) and an important overall shortening, calculable in the Cantabrian zone and affecting also the internal zones, although impossible to evaluate for the moment. This shortening must have produced a very important crustal thickening in central and western Galicia.

The basement below the sedimentary sequences is not seen except perhaps in the Galician complexes, for where the Precambrian is observed it consists of rocks that show only the effects of Hercynian metamorphism and penetrative deformations. Nevertheless, the Paleozoic sequences in the Cantabrian zone are typical shallow-water platform sequences, indicating that a continental crust floors the zone. More to the west, from the eastern to the western limb of the Narcea antiform, the Cambrian and Lower Ordovician increase very much in thickness, suggesting proximity to a shelf margin. The existence more to the west of more distal facies in the Cambrian–Lower Ordovician sequence and of turbidites in the Caradocian may indicate a passive continental margin, with the Cantabrian zone representing the shelf and the eastern part of the West Asturian-Leonese zone the shelf margin, slope, and continental rise. Still more to the west, however, instead of an oceanic basin, the Paleozoic facies and the existence of some erosional breaks in the sequence seem to indicate shallower conditions. West of the "Ollo de Sapo" antiform the Paleozoic sediments are more distal, but there is still no evidence of oceanic conditions. Thus, analysis of the stratigraphic sequence suggests the existence of continental crust throughout the whole area, although with thinned sections in the West Asturian-Leonese zone (eastern part) and in central and western Galicia.

From isotopic data, Vidal (1977) and Jaeger (1977) concluded some years ago that, except very locally, there is no old continental crust below the Paleozoic and upper Proterozoic sedimentary sequences, although this conclusion was questioned by Zwart and Dornsiepen (1978). According to Vidal (1977), the upper Proterozoic sediments were laid down on an oceanic crust bordering an ancient shield, and the present-day continental crust of Europe was generated during the Cadomian and Hercynian orogenies, but according to some other authors (Zwart and Dornsiepen 1978, Vitrac et al. 1981, Priem and den Tex 1982, Lancelot et al. 1985), an old continental crust is to be found underlying Europe.

More work is needed if we are to solve all the questions concerning the generation and evolution of the crust in the Hercynian belt, but putting aside the question of when the crust was generated, the existence over large areas of platform Paleozoic sequences indicates a crust isostatically compatible with the existence of epicontinental seas. Therefore, we can assume that these sequences were laid down on a continental-type crust.

According to the above analysis, if some remnants of oceanic crust or oceanic sediments exist, they must lie west of the present-day autochthonous and parautochthonous materials of Galicia and are to be found only in the overthrust complexes or at their base.

Below some of the Galician and northern Portugal complexes there is a more or less wide zone formed by blocks of different sizes embedded in a volcanosedimentary matrix. These materials are best observed around the Ortegal complex, especially along its eastern border, where they have been described as a mélange by Dutch authors (den Tex 1981). The blocks inside the mélange are up to 2 to 3 km across and include limestones and serpentinites. Other rocks found are mafic volcanics, keratophyres, flaser gabbros, phyllites, cherts, and fragments of rocks exposed in the complex (den Tex 1981). Recently the existence of pillow lavas has also been reported (Arenas and Peinado 1981).

According to Iglesias et al. (1983), these materials represent an ophiolitic mélange. From this point of view, the complexes could represent the thinned continental crust, intruded by mafic and ultramafic rocks, of the continent on the other side of the oceanic basin, and the chain can be interpreted as a collisional one.

In summary, all the observations discussed above seem to suggest the following model:

1. During the early Paleozoic a tensile crustal regime prevailed, causing thinning of the crust and the differentiation of a platform (Cantabrian zone) with a slope to the west connecting the platform with a deeper water area, bounded in turn to the west by a marginal plateau; still more to the west an oceanic area may have existed. Cambrian, Ordovician, and Silurian volcanism and also the pre-Hercynian acid and basic intrusions can be interpreted as being related to this period of crustal thinning (Lancelot et al. 1982, Priem and den Tex 1982).

2. Probably during the Devonian (Julivert 1983b), as suggested by radiometric data (Gil Ibarguchi et al. 1983b), the tectonic regime changed to compressional, marking the beginning

of the orogeny. The oceanic area closed in until the continents collided, probably in the Carboniferous (Iglesias et al. 1983). The existence of a high-pressure metamorphism, related to the thrust sheet forming the complexes of Galicia and northern Portugal and also found in the southern part of the Armorican massif (Triboulet 1974, van der Wegen 1978, Gil Ibarguchi and Ortega 1982), supports a plate-tectonic model. Radiometric data from the internal part of the chain point toward a (Middle or Late) Devonian age for the beginning of the orogeny, but in the external parts there are no unconformities older than the Westphalian. Thus orogeny seems to have migrated from inner to outer parts of the belt, an inference that is also suggested by the migration of Culm and paralic facies.

THE STRUCTURE OF THE SOUTHERN BRANCH OF THE OROGEN

The southern branch of the orogen is formed by the Ossa-Morena and South Portuguese zones (Figure 6-14). This branch is 300 km wide and thus narrower than the northern branch, whose width reaches some 470 km. Its polarity is symmetrical with respect to the northern branch: folds and thrusts facing south, thin-skinned tectonics to the south versus structures involving the Precambrian to the north, and decrease of metamorphism and plutonism from north to south.

THE STRUCTURE OF THE SOUTH PORTUGUESE ZONE

In this zone only Carboniferous and Upper Devonian rocks are found in outcrop. Stratigraphically, the most striking feature is the predominance of Culm facies and the existence of important Lower Carboniferous volcanism, related to sulfide deposits (Pyrite belt).

The regional metamorphism is weak and decreases from northeast to southwest. To the northeast it is in greenschist facies (chlorite zone); in the Pyrite belt parageneses of the prehnite-pumpellyite facies are well defined in the basic and intermediate volcanic rocks (Schermerhorn 1975, Routhier et al. 1978); finally, more to the southwest, anchimetamorphic conditions prevail.

The structure consists of southwest-facing folds and overthrusts, some of them of nappe size, with complicated mutual relationships (Ribeiro et al. 1983). The zone is affected by a slaty cleavage syngenetic with the folds and a younger crenulation cleavage. The general structure of the zone corresponds to a thrust belt, formed probably by décollement of the upper Paleozoic sequence, which would explain the lack of rocks older than the upper Devonian (Ribeiro et al. 1983).

THE STRUCTURE OF THE OSSA-MORENA ZONE

The structure of this zone is the result of two major folding phases. The first produced folds and overthrusts, some of them with important displacements (Juromenha overthrust), and the second refolded the first structures.

The zone shows several belts formed by Precambrian (Coimbra-Badajoz-Córdoba and

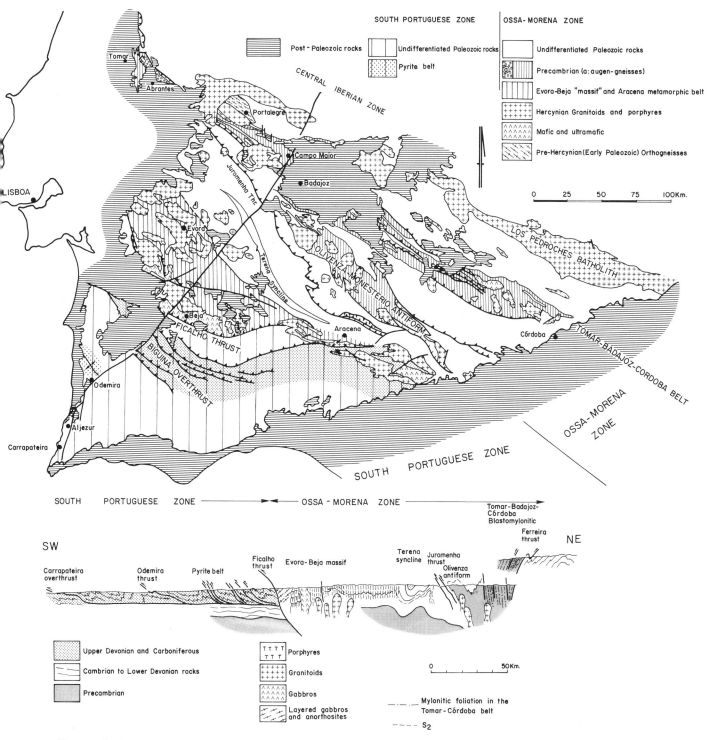

FIGURE 6-14. *Top*: Geological sketch of the Ossa-Morena and South Portuguese zones. *Bottom*: Cross section through the Ossa-Morena and South Portuguese zones (based on Chacón et al. 1983 and Ribeiro et al. 1983).

Olivenza-Monesterio belts) and Paleozoic (Evora-Beja massif and Aracena belt) rocks—demonstrating a higher metamorphism than the surrounding rocks and a rather important plutonism—separated by belts with a lower degree of metamorphism without plutonism, and formed by Paleozoic materials only.

The overall structure of the zone displays a fan-like pattern centered in the Badajoz-Córdoba belt; a narrow strip to the north is thrust to the northeast while the rest of the zone shows important overthrusts to the south. Along its southern boundary, the Ossa-Morena zone overrides the South Portuguese zone.

An outstanding structure in the zone is a blastomylonitic belt (Bladier 1974, Burg et al. 1981, Chacón et al. 1983) trending southeast-northwest, from Córdoba to Abrantes, where it is cut off by the north-south strike-slip fault of Porto-Tomar. This belt coincides with one of the Precambrian cores of the Ossa-Morena zone and is bounded by steep shear surfaces. It shows a metamorphism of Barrovian type reaching the sillimanite isograd and a later general retrogression, and it is intruded by calc-alkaline and peralkaline granites and syenites (Priem et al. 1970, Gonçalves 1978). The fabric and minor structures indicate a left-lateral slip motion along the belt with a displacement of at least 72 km, according to Burg et al. (1981). Inside the belt there is a steeply dipping foliation oblique to the shear direction and a horizontal lineation. Intrafolial folds are scarce and are variously oriented; sheath folds, where present, have their noses parallel to the lineation (Burg et al. 1981). The delimitation between Hercynian and possible Precambrian events has been controversial in recent years. The blastomylonitic foliation is younger than the first, Barrovian metamorphism and the pre-Hercynian early Paleozoic calc-alkaline and peralkaline granites but older than the youngest calc-alkaline Hercynian granitoids (316 Ma; see Deloche et al. 1979), which cut across the belt (Burg et al. 1981). The foliation and consequently the strike-slip movement along the belt are, therefore, Hercynian. Concerning the metamorphism, there is no evidence for dating the Barrovian one, which could be early Hercynian or Cadomian (Chacón et al. 1983); the second one, of low-pressure type, is Hercynian. However, it has to be kept in mind that change of gradient with time is a common characteristic of Hercynian metamorphism in the northern branch of the orogen, and that not far from the blastomylonitic belt, in the Porto-Viseu area, the existence of relict kyanite of Hercynian age is well established.

South of the blastomylonitic belt the Precambrian crops out again in the core of the Olivenza-Monesterio antiform. This is, broadly speaking, a large antiform with Precambrian rocks in its core, affected by an important metamorphism, with a thick Cambrian cover forming, to the north, a synclinorial area between the Olivenza-Monesterio and the Córdoba-Badajoz Precambrian belts and, to the south, a large overturned limb with many second-order folds (Vauchez 1976, Apalategui 1980), so that the Cambrian beds override to the south. South of the antiform is a wide synclinorial area, its core (Terena syncline) formed by the Upper Devonian or Lower Carboniferous Terena graywacke.

Farther to the south there is a metamorphic belt, consisting of a series of thermal domes cutting through the Hercynian structures but showing an overall southeast-northwest trend. The metamorphism is of low pressure–high temperature type, and where the sillimanite zone is reached migmatization is important. This area is very wide to the northwest, where it forms the so-called Evora-Beja "massif," but it narrows to the southeast, extending along the Aracena metamorphic belt.

The boundary with the South Portuguese zone is a major overthrust dipping northeast.

Near the overthrust are synmetamorphic intrusions of flaser gabbros and serpentinites (Batista et al. 1976), which suggest that it is a very deep accident cutting across the whole crust.

The synorogenic magmatism has peculiar characteristics in the Ossa-Morena zone. To the northeast there is a domain with granitoids that fit into the model generally accepted for the granitoids of the northern branch of the orogen. To the south, in the Evora-Beja massif, there are still two-mica granites similar to those in the central and northern part of the Iberian Peninsula, but basic intrusions increase from northeast to southwest, reaching a maximum in the Beja massif.

THE OVERALL STRUCTURE OF THE SOUTHERN BRANCH

The structure of the southern branch (Figure 6-14) consists of a series of thrusts and asymmetrical folds facing southwest and involving progressively deeper levels from the external to the internal part of the branch. Also, from the external to the internal part metamorphism and plutonism increase, although not regularly but along belts coinciding with antiformal structures. These characteristics coincide with those of the northern branch, but there are also many differences, which have been described above. Thus there is not a perfect symmetry in the whole fold belt, and the two branches probably have a different significance.

In the southern branch, as in the northern branch, it can be assumed that continental crust underlies the sedimentary sequence. This crust has been supposed by some authors to crop out in the Ossa-Morena zone, but actually there is no evidence of tectono-metamorphic events older than the Cadomian. In the South Portuguese zone, geophysical data indicate a present-day crustal thickness ranging from 30 (close to the Atlantic coast, in the northwest) to 35 km (in the south) (Mueller et al. 1973, Prodehl et al. 1976).

Also as in the northern branch, plate-tectonic models have been proposed to explain the structure of the southern branch (Bard 1971, Ribeiro et al. 1983). These models propose subduction of an oceanic plate toward the north, but while Bard (1971) proposes a Cordilleran-type model, Ribeiro et al. (1983) support a collisional model, the suture lying along the boundary between the Ossa-Morena and South Portuguese zones.

Finally, we might point out that the boundary between the two branches of the orogen is the Córdoba-Badajoz blastomylonitic belt, interpreted as a left-lateral shear zone. This very important structural feature has been regarded as a cryptic suture or a transform fault active during the closing of an oceanic area placed west of Galicia (Lefort and Ribeiro 1980, Burg et al. 1981) or both.

THE IBERO-ARMORICAN ARC

The Ibero-Armorican arc is a first-order structure of the Hercynian fold belt in western Europe and has attracted attention for a long time; whether it is primary or secondary has been a matter of controversy.

The extension of the Iberian structures into the Armorican massif, on the other side of the Bay of Biscay, and thence to the Massif Central and central Europe is at present well established. Several geological lines of different kinds can be traced from one side of the Bay of Biscay to the other (Figure 6-15):

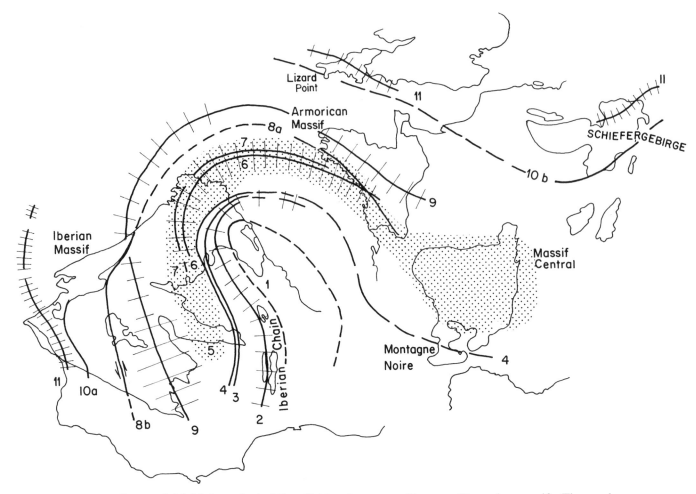

FIGURE 6-15. Main geological lines linking the western European Hercynian massifs. The numbers mark the lines described in the text.

Line 1 marks the western boundary of décollement in the Cantabrian zone.

Line 2 marks the belt of thick Cambrian-Ordovician deposits.

Line 3 marks the eastern limit of Hercynian granitoids in the Iberian Peninsula.

Line 4 separates the area (the concave side of the arc) where the Cambrian and Ordovician sequences are conformable from the area where the Lower Ordovician is transgressive and rests on different Cambrian and Precambrian formations.

Line 5 is the high-grade metamorphic belt of the northern branch (southern in France and central Europe) of the orogen.

Line 6 links the complexes of Galicia and northern Portugal with those in the southern part of the Armorican massif.

Line 7 links the points where a high-pressure metamorphism has been reported. This line and also line 6 are only indicative, for they link fragments of wide thrust sheets instead of narrow linear structures.

Lines 8a and 8b trace the "zone broyée sud-armoricaine" and the Tomar-Badajoz-Córdoba belt, which have dextral and sinistral strike-slip motion, respectively, and are supposed by some authors (Lefort and Ribeiro 1980, Burg et al. 1981) to be connected by a suture zone west of Galicia.

Line 9 marks a belt with similar Ordovician facies and faunas: Tremadocian(?) red beds, disconformable Armorican quartzite, quartzite levels in the Caradocian, and so forth.

Line 10a is the internal boundary of the décollement in the South Portuguese zone (southern branch of the Iberian segment of the belt) and line 10b is a similar boundary, between the Rhenoherzinikum and the Saxothuringikum. The two lines have a similar meaning and can be linked.

Line 11 marks the belt of synorogenic Culm facies along one side of the fold belt; it is continuous regionally, although it has interruptions in detail.

These lines are roughly parallel, and as some are early Paleozoic paleogeographic trends and others are Hercynian structures of Carboniferous age, their parallelism indicates the persistence of the same trends throughout the Paleozoic. Figure 6-15 also shows that most of the lines describe perfectly the Ibero-Armorican arc, but some are interrupted, for example, the Galicia-northern Portugal complexes and the high-pressure metamorphism, whose extension is not found in central Spain. Another peculiar trace is represented by the line linking the "zone broyée sud-armoricaine" and the Tomar-Badajoz-Córdoba belt. This line crosses the Ordovician belt of sediments showing similar facies and a similar stratigraphic succession; that belt extends from the Central Iberian zone to the Central Armorican zone, indicating that the line must be younger than Late Ordovician.

Many problems have arisen about the origin of the arc and its evolution, such as whether it is of secondary or primary origin, how it has evolved to its present shape, and whether it is related to an ancient arcuate, consuming plate boundary. Although these problems have not been satisfactorily resolved as yet, the following models have been proposed to explain the arc: (1) the arc has evolved by progressive increase in curvature, due to the (relative) westward motion of its core between a pair of shear zones with opposite strike slip and parallel to the arc flanks (Matte and Ribeiro 1975, Julivert et al. 1977); (2) the arc has been formed (Ries and Shackleton 1976) or has been tightened (Julivert 1971b) by bending of the whole belt, the flanks of the arc moving toward one another; and (3) the arc is a "corner effect" at the merging point of a collision belt and a transcurrent shear zone (Brun and Burg 1982).

None of these models explains completely the structure of the arc. The first model emphasizes the importance of the strike-slip faults and shear zones that are more or less parallel to the arc flanks, as observed in several areas ("zone broyée sud-armoricaine," Tomar-Badajoz-Córdoba belt, faults bounding the core of the Cantabrian zone). The second model is based on the finite strain patterns; thus Ries and Shackleton (1976) pointed out that the high-grade metamorphic belt (Galicia and northern Portugal, and its extension to the southern part of the Armorican massif), placed in the convex part of the arc, shows a stretching lineation parallel to the main fold axes and turning with them along the arc, while toward the core of the arc (West Asturian-Leonese zone) the stretching lineation plunges downdip and appears to fan around the arc, so that it remains perpendicular to the fold axes. The third model is based on the strain pattern in the southern part of the Armorican massif and on some evidence south of that massif of a plate-consuming margin, which does not seem to extend along the arc beyond Galicia.

These models do not necessarily exclude one another; on the other hand, they can be used either to explain the arc as purely secondary or to explain it as a primary structure, tightened during orogeny. Recently paleomagnetic data have been used to quantify the closing of the arc since the time of magnetization. From NRM directions measured along the fold and nappe province (Cantabrian zone), Ries et al. (1980) found that from Cape Peñas, in the northern flank of the arc, to the Luna region in the south, the NRM shows a rotation of 110°. For the same area Perroud and Bonhommet (1981) found an angle of 80°. According to Ries et al. (1980) the structures were bent 110° after the magnetization, which is supposed to be Carboniferous; there remains therefore an arcuation of about 55°, which must be older. According to Perroud and Bonhommet (1981), 80° of curvature of the arc is secondary and 70° is primary. Most of the NRM data quoted above are difficult to use in precise geometrical analysis. Detailed geometrical analysis has been done in some localities (Bonhommet et al. 1981), but in some cases unfolding the structures has not brought the magnetic vectors into alignment, and in most localities the only correction has been a tilt correction rotating about a horizontal axis. Consequently, the NRM data can only be used in a very general way to explain the evolution of the arc.

Because strain in the Cantabrian zone is not significant, unfolding of the structures can be attempted. The radial set of folds gives 40% of shortening in the core of the arc, normal to its trace. If the above paleomagnetic data are accepted as a rough indication of the arc tightening and the effect of the radial folds is discounted, an angle of some 40 to 60° remains between paleomagnetic vectors measured on the three sides of the arc (Julivert and Arboleya 1984). This angle must be a consequence of rotation during longitudinal folding and nappe emplacement.

In regard to the décollement nappes, we must remember that commonly they die out along strike into a converging anticline-syncline couple. This model implies that their emplacement included a rotational motion around a center placed at the converging point of anticline and syncline. If a primarily curved trace is accepted, the resulting model is a crescent-shaped thrust sheet, ending at one end in a couple of folds and at the other in a strike-slip fault. Such a model conforms with the field data (Julivert and Arboleya 1984). As each main thrust unit consists of several second-order slices, each slice moved with a certain independence, suffering an additional rotation around a center placed at its own end, and the result is that the plate, instead of moving as a rigid body, was being curved as it was being emplaced (Julivert and Arboleya 1984). In such a model, a curved thrust belt could have formed with an apparent movement toward the center of the arc and without space problems. As deformation progressed, the thrust sheets began to fold longitudinally. This folding probably tightened the arc somewhat more, but only to a lesser degree. Finally, the generation of the radial folds gave the arc its present-day shape.

As the facies boundaries in the Cantabrian zone are in their broad lines parallel to the traces of tectonic units, presumably the paleogeographic trends were also curved. Thus, data from the Cantabrian zone supports the idea of a primary gently arcuate belt, progressively closed during the orogeny. This model is consistent with the interpretation that (relative) westward motion of the core of the arc relative to its flanks was an important mechanism, giving way in turn to progressive tightening of the arc, but it cannot be considered sufficient to explain the overall structure of the belt around the arc. We must remember that deformation in the Cantabrian zone began only during the Late Carboniferous, although in inner zones of the belt it probably started earlier, and that changes in plate motion are common in later times, where there are sufficient data to show their presence.

THE OVERALL STRUCTURE OF THE BELT: CONCLUDING REMARKS

The Hercynian orogen can be defined as a linear bilateral fold belt, with two well-differentiated, although asymmetrical, branches. For many years the Hercynian orogen was considered to be very different from the Mesozoic-Tertiary fold belts. The main peculiarities were thought to be (1) the absence of nappe structures; (2) the great width, versus the narrower and more linear character of younger fold belts; (3) the low pressure-high temperature metamorphism; and (4) the abundance of granitoids. As knowledge of the belt advances, however, these differences seem to be progressively less significant and do not support the idea of different behavior or important variations in the nature of the geological processes.

The most prominent characteristic of the Hercynian fold belt is its linearity. The belt extends for many thousands of kilometers, from the Urals through Europe into the Iberian Peninsula, thence to Morocco and probably to the Mauritanides, and beyond the Atlantic to the southern Appalachians; it is a truly linear structure. Moreover, the belt is not significantly wider than other Phanerozoic fold belts (Julivert 1979); compare the following widths: 320 to 460 km for the northern branch of the Hercynian fold belt in the Iberian Peninsula, 300 km for its southern branch, 330 km for the Appalachians in the cross section from Tennessee to the coastal plain, 250 km for the Alpine belt from the Jura front to the ophiolite belt, 1,000 km for the North American Cordilleran system, 200 to 600 km for the Andean Cordillera in Colombia.

The sedimentary evolution of the Hercynian fold belt is also similar to that of the Mesozoic-Tertiary fold belts. It consists of a long pre-orogenic period, a Culm or flysch stage, and a molasse stage, as described in the stratigraphic part of this paper. Both Culm and molasse deposits are comparable—in their facies, relationships with orogeny, position within the belt, and even volume—to the flysch and molasse deposits in younger fold belts.

The structure is the result of polyphase deformation. The oldest structures have a tangential character: décollement-, fold-, and basement nappes of different magnitudes. Later, folds with steeply dipping axial planes or backfolds were formed. The youngest structures are faults and indicate an evolution to brittle behavior.

The plurifacial character of metamorphism is now well defined. The metamorphism during the oldest deformation phases (tangential phases) was at least of intermediate pressure, while later it evolved to high T–low P conditions, suggesting a decompression related to the last stages of deformation (doming) (Arboleya et al. 1983). Blueschist facies were supposed to be absent in the Hercynian fold belt, but during recent years they have been found at several points; nevertheless, a continuous and well-developed high P–low T belt like that in younger fold belts has not been observed.

Synorogenic plutonism is very important. From geochemical and isotopic data, many authors have concluded that these granitoids came mainly (if not entirely) from recycling of a preexisting continental crust rather than from mixing of mantle and crust material.

Many different models have been proposed to explain the structure and evolution of the Hercynian fold belt. Some authors have supported an ensialic evolution (Zwart and Dornsiepen 1978), and among them some have proposed an evolution ultimately due to the diapiric vertical rise of magmas (Krebs and Wachendorf 1973), which produced basement uplifts and gravitational sliding at the rims of the diapirs.

According to the characteristics of the Iberian segment described here, such domed structure, with granitoid intrusions associated with thermal domes, is only the last stage of the Hercynian orogenic evolution, superposed on older and very important tectonic and metamorphic events. The older tectonic events are compressional and must have given rise to an important crustal shortening. This conclusion, together with the similarities between Hercynian and younger fold belts, supports plate-tectonic interpretations.

The plate-tectonic models that have been proposed have varied. Some authors have interpreted the Hercynian fold belt as a Cordilleran-type orogen (Nicolas 1972), while others have supported different collisional models (Burrett 1972, Laurent 1972, Dewey and Burke 1973, Bard et al. 1980). The importance of basement reactivation favors the collisional models; Dewey and Burke (1973) have compared the Hercynian fold belt to the Tibetan plateau and proposed a basement reactivation model for such belts. Nevertheless, the tracing of the sutures and consequently of former oceanic areas, the importance of such areas, and the subduction direction, are still subject to controversy. Two suture traces can tentatively be drawn along the belt. One is represented by the Lizard Point complex in southwestern Great Britain, interpreted by Thayer (1969) and by Kirby (1979) as an ophiolitic suite. This complex supports the idea of a mid-European ocean, as accepted by many collisional models of the chain, but its trace through Germany has been questioned (Krebs and Wachendorf 1973, Behr and Weber 1980). The other possible suture is based upon the complexes of northern Portugal, Galicia, and the southern part of the Armorican massif and the associated blueschist facies metamorphism, but the trace of this suture is also difficult to follow. In this respect, it has been suggested that the suture could end southward, against the Tomar-Badajoz-Córdoba blastomylonitic belt, which would play the role of a transform shear zone. Many of the problems that arise when one attempts to construct a detailed model to fit all the known facts seem to suggest a complicated mosaic of microplates and small oceanic or intermediate-type areas—compare the evolution of the Mediterranean realm during later time.

REFERENCES

Anthonioz, P. M., 1970: Etude des complexes polymétamorphiques précambriens de Morais et Bragança (NE du Portugal). Sci. de la Terre *15*, 145-166.

Apalategui, O., 1980: Consideraciones estratigráficas y tectónicas en Sierra Morena occidental. Inst. Geol. Min. Esp., Temas Geol. Min. (Primera Reunión sobre la Geología de Ossa-Morena), 23-41.

Arboleya, M. L., 1981: La estructura del manto del Esla (Cordillera Cantábrica, León). Bol. Geol. Min. *92*, 19-40.

Arboleya, M. L., Gil Ibarguchi, I., Julivert, M., and Martínez, F. J., 1983: Thermal domes and late Hercynian structures in the NW of the Iberian Massif (abstr.). Terra Cognita *3*, 193.

Arenas, R., and Peinado, M., 1981: Presencia de pilow-lavas en las metavolcanitas submarinas de las proximidades de Espasante, Cabo Ortegal, NW de España. Cuadernos de Geología Ibérica *7*, 105-119.

Bard, J. P., 1971: Sur l'alternance des zones métamorphiques et granitiques dans le segment hercynien sud-ibérique; comparaison de la variabilité des caractères géotectoniques de ces zones avec les orogènes ''orthotectoniques.'' Bol. Geol. Min. *82*, 324-345.

Bard, J. P., Burg, J. P., Matte, P., and Ribeiro, A., 1980: La chaîne hercynienne d'Europe occidentale en termes de tectonique des plaques. Bur. Rech. Géol. Min. Mém. *108*, 233-246; 26th Congr. Géol. Internat., Paris 1980, Colloque *C6*: Géologie de l'Europe, du Précambrien aux bassins sédimentaires post-hercyniens, 233-246.

Bastida, F., 1980: Las estructuras de la primera fase de deformación hercínica en la zona asturoccidental-leonesa (Costa cantábrica, NW de España). Tesis doctoral, Univ. Oviedo, 276 p.

Bastida, F., Marcos, A., Arboleya, M. L., and Mendez, I., 1976: La unidad de Peña Corada y su relación con el manto del Esla (Zona Cantábrica, NW de España). Breviora Geol. Asturica *20*, 49-55.

Bastida, F., and Pulgar, J. A., 1978: La estructura del manto de Mondoñedo entre Burela y Tapia de Casariego (Costa cantábrica, NW de España. Univ. Oviedo Trabajos Geología *10*, 75-159.

Batista, J. A., Munha, J., Oliveira, V., and Ribeiro, L., 1976: Alguns aspectos geológico-petrográficos da bordadura Sul do complexo eruptivo de Beja. Serv. Geol. Portugal Com. *60*, 203-213.

Behr, H. J., and Weber, K., 1980: Subduktion oder Subfluenz in mitteleuropäischen Variszikum? Berliner Geow. Abh., Reihe A, *19* (Internat. A. Wegener-Symposium, Berlin), 22-23.

Berry, W.B.N., and Boucot, A. J., 1973: Glacio-eustatic control of Late Ordovician-Early Silurian platform sedimentation and faunal changes. Geol. Soc. America Bull. *84*, 275-283.

Beuf, S., Biju-Duval, B., de Charpal, O., and Gariel, O., 1971: Les grès du Paléozoique inférieur au Sahara. Paris, Technip (Inst. Fr. Pétrole, Coll. Sci. Tech. Pétrole *18*), 464 p.

Beuf, S., Biju-Duval, B., Stevaux, J., and Kulbicki, G., 1966: Ampleur des glaciations ''Siluriennes'' au Sahara: Leurs influences et leurs conséquences sur la sédimentation. Inst. Fr. Pétrole Rév. *21*, 363-381.

Bischoff, L., Lenz, H., Müller, P., and Schmidt, K., 1978: Geochemische und geochronologische Untersuchungen an Metavulkaniten und Orthogneisen der östlichen Sierra de Guadarrama (Spanien). N. Jb. Geol. Paläont., Abh. *155*, 275-299.

Bladier, Y., 1974: Structure et pétrologie de la bande blastomylonitique de Badajoz-Cordoue. Les roches cataclastiques. Classification-interprétation. Thèse 3e cycle, Univ. Montpellier, 105 p.

Bonhommet, N., Cobbold, P. R., Perroud, H., and Richardson, A., 1981: Paleomagnetism and cross-folding in a key area of the Asturian arc (Spain). Jour. Geophys. Res. *86*, 1873-1887.

Brenchley, P. J., and Newall, G., 1980: A facies analysis of Upper Ordovician regressive sequences in the Oslo region, Norway; a record of glacio-eustatic changes. Palaeogeogr. Palaeoclimat. Palaeoecol. *31*, 1-38.

Brun, J. P., and Burg, J. P., 1982: Combined thrusting and wrenching in the Ibero-Armorican arc: A corner effect during continental collision. Earth Planet. Sci. Lett. *61*, 319-332.

Burg, J. P., Iglesias, M., Laurent, P., Matte, P., and Ribeiro, A., 1981: Variscan intracontinental deformation. The Coimbra-Cordoba shear zone (SW Iberian Peninsula). Tectonophysics *78*, 161-177.

Burrett, C. F., 1972: Plate tectonics and the Hercynian orogeny. Nature *239*, 155-157.

Calsteren, P.W.C. van, 1977: Geochronological, geochemical and geophysical investigations in the high-grade mafic-ultramafic complex at Cabo Ortegal and other pre-existing elements in the Hercynian basement of Galicia (NW Spain). Z.W.O. Labor. voor Isotopen Geologie, Amsterdam, Verh. *2*, 74 p. (see also Leidse Geol. Meded. *51*, 57-60).

Capdevila, R., 1969: Le métamorphisme régional progressif et les granites dans le segment hercynien de Galice Nord Orientale (NW de l'Espagne). Thèse, Univ. Montpellier, 430 p.

Capdevila, R., Corretgé, G., and Floor, P., 1973: Les granitoïdes varisques de la Méséta Ibérique. Soc. Géol. France Bull. *15*, 209-228.

Capdevila, R., and Floor, P., 1970: Les différents types de granites hercyniens et leur distribution dans le nordouest de l'Espagne. Bol. Geol. Min. *81*, 215-225.

Carreras, J., Estrada, A., and White, S., 1977: The effects of folding on the c-axis fabrics of a quartz mylonite. Tectonophysics *39*, 3-24.

Chacón, J., Oliveira, V., Ribeiro, A., and Tomas Oliveira, J., 1983: La estructura de la zona de Ossa-Morena. Libro Jubilar J. M. Rios (Geología de España) *1*, 490-540.

Corretgé, L. G., and Martínez, F. J., 1978: Problemas sobre estructura y emplazamiento de los granitoïdes: Aplicación a los batolitos hercínicos del Centro-Oeste de la Meseta Ibérica. *In*: Geol. Parte Norte Macizo Ibérico (Ed. homenaje I. Parga Pondal). Ed. del Castro, La Coruña, 113-134.

Crimes, T. P., Marcos, A., and Pérez-Estaún, A., 1974: Upper Ordovician turbidites in western Asturias: A facies analysis with particular reference to vertical and lateral variations. Palaeogeogr. Palaeoclimat. Palaeoecol. *15*, 169-184.

Dangeard, L., and Doré, F., 1971: Faciès glaciaires de l'Ordovicien supérieur en Normandie. Bur. Rech. Géol. Min. Mém. (Colloque Ordovicien-Silurien, Brest 1971), *73*, 119-128.

Deloche, C., Simon, D., and Tamain, G., 1979: Le charriage majeur de type himalayen du Cerro-Muriano (Cordoue) dans les Cadomides du Sud-Est hespérique. Acad. Sci. Paris C. R. (D) *289*, 253-256.

Dewey, J. F., and Burke, K.C.A., 1973: Tibetan, Variscan, and Precambrian basement reactivation: Products of continental collision. Jour. Geology *81*, 683-692.

Doré, F., and Le Gall, J., 1972: Sédimentologie de la ''Tillite de Feuguerolles'' (Ordovicien supérieur de Normandie). Soc. Géol. France Bull. *14*, 199-211.

Ferreira, M.P.V., 1965: Geologia e petrologia de região de Rebordelo-Vinhais. Univ. Coimbra Rev. Fac. Ciênc. *36*, 287 p.

Floor, P., 1966: Petrology of an aegirine-riebeckite gneiss-bearing part of the Hesperian Massif; the Galiñeiro and surrounding areas, Vigo, Spain. Leidse Geol. Meded. *36*, 1-203.

Fontboté, J. M., and Julivert, M., 1954: Algunas precisiones sobre la cronología de los plegamientos hercinianos en Cataluña. 19th Congr. Géol. Internat., Alger 1952, C. R. Sect. 13, Fasc. *15*, 575-591.

Gil Ibarguchi, I., 1982a: The metamorphic evolution of the Muxia-Finisterre region (Galice, NW Spain) during the Hercynian orogenesis. Geol. Rundschau *71*, 657-686.

Gil Ibarguchi, I., 1982b: Metamorfismo y plutonismo en la región de Muxia-Finisterre (NW de España). Corpus Geol. Gallaeciae *1*, 250 p.

Gil Ibarguchi, I., Julivert, M., and Martínez, F. J., 1983a: Los rasgos estructurales generales de la parte Noroeste de la zona centroibérica. Libro Jubilar J. M. Rios (Geología de España), *1*, 420-422.

Gil Ibarguchi, I., Julivert, M., and Martínez, F. J., 1983b: La evolución de la Cordillera Herciniana en el tiempo. Libro Jubilar J. M. Rios (Geología de España), *1*, 607-612.

Gil Ibarguchi, I., and Ortega, E., 1982: Petrology, structure and geotectonic implications of glaucophane-bearing eclogites and related rocks from the Malpica-Tuy unit, Galicia, NW Spain (abstr.). Terra Cognita *2*, 311.

Gonçalves, F., 1978: Estado actual do conhecimiento geológico do nordeste alentejano. Univ. Lisboa Fac. Ciênc., Ciênc. Geol. IV Curso Ext., 1-23.

Guitard, G., 1970: Le métamorphisme hercynien mésozonal et les gneiss oeillés du massif du Canigou (Pyrénées orientales). Bur. Rech. Géol. Min. Mém. *63*, 303 p.

Hafenrichter, M., 1980: The lower and upper boundary of the Ordovician System of some selected regions (Celtiberia, Eastern Sierra Morena) in Spain. Part II, The Ordovician/Silurian boundary in Spain. N. Jb. Geol. Paläont., Abh. *160*, 138-148.

Iglesias, M., Ribeiro, M. L., and Ribeiro, A., 1983: La intrerpretación aloctonista de la estructura del noroeste peninsular. Libro Jubilar J. M. Rios (Geología de España), *1*, 459-467.

Iglesias, M., and Robardet, M., 1980: El Silúrico de Galicia media (central): Su importancia en la paleografía varisca. Cuad. Lab. Xeol. Laxe *1*, 99-115.

Jaeger, E., 1977: The evolution of the Central and West European Continent. *In*: La Chaîne varisque d'Europe moyenne et occidentale. C.N.R.S. Coll. Internat. (Rennes 1974) *243*, 227-239.

Julivert, M., 1971a: Décollement tectonics in the Hercynian Cordillera of northwest Spain. Am. Jour. Sci. *270*, 1-29.

Julivert, M., 1971b: L'évolution structurale de l'arc asturien. *In*: Histoire structurale du Golfe de Gascogne, 1. Paris, Technip (Inst. Fr. Pétrole, Collect. Coll. Sem. *22*) I.2-1/I.2-28.

Julivert, M., 1978: Hercynian orogeny and Carboniferous palaeogeography in northwestern Spain: A model of deformation-sedimentation relationships. Zeits. deutscher Geol. Gesell. *129*, 565-593.

Julivert, M., 1979: A cross-section through the northern part of the Iberian massif: Its position within the Hercynian fold belt. Krystalinikum *14*, 51-67.

Julivert, M., 1981: A cross-section through the northern part of the Iberian Massif. Geol. en Mijnbouw *60*, 107-128.

Julivert, M., 1983a: La estructura de la Zona Asturoccidental-Leonesa. Libro Jubilar J. M. Rios (Geología de España), *1*, 381-408.

Julivert, M., 1983b: La evolución sedimentaria durante el Paleozoico y el registro de la deformación en la columna estratigráfica paleozoica. Libro Jubilar J. M. Rios (Geología de España), *1*, 593-601.

Julivert, M., and Arboleya, M. L., 1984: A geometrical and kinematical approach to the nappe structure in an arcuate fold belt: The Cantabrian nappes (Hercynian chain, NW Spain). Jour. Struct. Geology *6*, 499-519.

Julivert, M., and Arboleya, M. L., 1986: Areal balancing and estimate of areal reduction in a thin-skinned fold-and-thrust belt (Cantabrian Zone, Northwest Spain): Constraints to its emplacement mechanism. Jour. Struct. Geol. *8*, 407-414.

Julivert, M., Fontboté, J. M., Ribeiro, A., and Nabais Conde, L. E., 1972–1974: Mapa Tectónico de la Península Ibérica y Baleares, 1:1,000,000. Memoria explicativa (Inst. Geol. Min. España), 113 p.

Julivert, M., and Marcos, A., 1973: Superimposed folding under flexural conditions in the Cantabrian Zone (Hercynian Cordillera, Northwest Spain). Am. Jour. Sci. *273*, 353-375.

Julivert, M., Marcos, A., and Pérez-Estaún, A., 1977: La structure de la chaîne hercynienne dans le secteur ibérique et l'arc ibéro-armoricain. *In*: La Chaîne varisque d'Europe moyenne et occidentale. C.N.R.S. Coll. Internat. (Rennes 1974) *243*, 429-440.

Julivert, M., Marcos, A., and Truyols, J., 1972: L'évolution paleogéographique du nord-ouest de l'Espagne pendant l'Ordovicien-Silurien. Soc. Géol. Minéral. Bretagne Bull. *4*, 1-7.

Julivert, M., and Martínez, F. J., 1980: The Paleozoic of the Catalonian Coastal Ranges (northwestern Mediterranean). IGCP 5, Newsletter (Sassi, F. P., ed.) *2*, 124-128.

Julivert, M., Martínez, F. J., and Ribeiro, A., 1980: The Iberian segment of the European Hercynian fold belt. Bur. Rech. Géol. Min. Mém. *108*, 132-158; 26th Congr. Géol. Internat., Paris 1980, Colloque *C6*, Géologie de l'Europe, du Précambrien aux bassins sédimentaires post-hercyniens, 132-158.

Keasberry, E. J., van Calsteren, P.W.C., and Kuijper, R. P., 1976: Early Paleozoic mantle diapirism in Galicia. Tectonophysics *31*, T61-T65.

Kirby, G. A., 1979: The Lizard complex as an ophiolite. Nature *282*, 58-61.

Krebs, W., and Wachendorf, H., 1973: Proterozoic-Paleozoic geosynclinal and orogenic evolution of central Europe. Geol. Soc. America Bull. *84*, 2611-2629.

Kuijper, R. P., 1979: U-Pb systematics and the petrogenetic evolution of infracrustal rocks in the Paleozoic basement of Western Galicia, NW Spain. Z.W.O. Labor. voor Isotopen Geologie, Amsterdam, Verh. *5*, 101 p.

Kuijper, R. P., and Arps, G.E.S., 1983: Los complejos de Ordenes, Lalín y Forcarey. Libro Jubilar J. M. Rios (Geología de España), *1*, 422-430.

Lancelot, J. R., Allegret, A., and Iglesias, M., 1985: Outline of Upper Precambrian and Lower Paleozoic evolution of the Iberian Peninsula according to U-Pb dating of zircons. Earth Planet. Sci. Lett. *74*, 325-337.

Lancelot, J. R., Ducrot, J., and Allegret, A., 1982: Signification géodynamique et âge U/Pb sur zircons des orthogeneiss et leptynites alcalins du socle antéhercynien d'Europe Occidentale. Coll. Internat. Géoch. Petrol. Granit., Clermont-Ferrand.

Laurent, R., 1972: The Hercynides of South Europe—a model. 24th Internat. Geol. Congr., Montreal 1972, Proc., Sect. *3*, 363-370.

Lefort, J. P., and Ribeiro, A., 1980: La faille Porto-Badajoz-Cordoue a-t-elle controlé l'évolution de l'océan paléozoïque sudarmoricain? Soc. Géol. France Bull. *22*, 455-462.

Lotze, F., 1945: Zur Gliederung der Varisziden der Iberischen Meseta. Geotekt. Forsch. *6*, 78-92.

Marcos, A., 1973: Las series del Paleozoico inferior y la estructura herciniana del occidente de Asturias (NW de España). Univ. Oviedo Trabajos Geología *6*, 113 p.

Marcos, A., and Pulgar, J. A., 1982: An approach to the tectonostratigraphic evolution of the Cantabrian foreland thrust and fold belt, Hercynian Cordillera of NW Spain. N. Jb. Geol. Paläont., Abh. *163*, 256-260.

Marquínez, J., 1981: Estudio geológico del área esquistosa de Galicia central (zona de Lalín-Forcarey-Beariz). Cuad. Lab. Xeol. Laxe *2*, 135-154.

Martínez, F. J., 1974a: Estudio del área metamórfica del NW de Salamanca (Cordillera Herciniana, España). Univ. Oviedo Trabajos Geología *7*, 3-59.

Martínez, F. J., 1974b: Petrografía, estructura y geoquímica de los diferentes tipos de granitoides del NW de Salamanca (Cordillera Herciniana, España). Univ. Oviedo Trabajos Geología *7*, 61-141.

Martínez, F. J., and Gil Ibarguchi, I., 1983: El metamorfismo en el macizo ibérico. Libro Jubilar J. M. Rios (Geología de España), *1*, 555-569.

Martínez-Catalán, J. R., 1981: Estratigrafía y estructura del domo de Lugo (sector Oeste de la zona Asturoccidental-leonesa). Tesis, Univ. Salamanca, 317 p. The Barrie de la Maza Foundation.

Matte, P., 1968: La structure de la virgation hercynienne de Galice (Espagne). Géologie Alpine (Fac. Sci. Grenoble, Lab. Géol., Trav.) *44*, 157-280.

Matte, P., 1969: Les kink-bands; exemple de déformation tardive dans l'hercynien du nord-ouest de l'Espagne. Tectonophysics *7*, 309-322.

Matte, P., and Ribeiro, A., 1967: Les rapports tectoniques entre le Précambrien ancien et le Paléozoique dans le nord-ouest de la Peninsule Ibérique: Grandes nappes ou extrusions? Acad. Sci. Paris C. R. *264*, 2268-2271.

Matte, P., and Ribeiro, A., 1975: Forme et orientation de l'ellipsoïde de déformation dans la virgation hercynienne de Galice; relations avec le plissement et hypothèses sur la genèse de l'arc ibéro-armoricain. Acad. Sci. Paris C. R. (D) *280*, 2825-2828.

Mendez-Bedia, I., 1976: Biofacies y litofacies de la Formación Moniello-Santa Lucía (Devónico de la Cordillera Cantábrica, NW de España). Univ. Oviedo Trabajos Geología *9*, 93 p.

Minnigh, L. D., 1975: Tectonic and petrographic aspects of an area SW of the Lalín unit (Prov. Orense and Pontevedra, NW Spain). Leidse Geol. Meded. *49*, 499-504.

Mueller, S., Prodehl, C., Mendes, A. S., and Sousa Moreira, V., 1973: Crustal structure in the southwestern part of the Iberian Peninsula. Tectonophysics *20*, 307-318.

Nicolas, A., 1972: Was the Hercynian orogenic belt of Europe of the Andean type? Nature *236*, 221-223.

Noronha, F., Ramos, J.M.F., Rebelo, J. A., Ribeiro, A., and Ribeiro, M. L., 1979: Essai de corrélation des phases de déformation hercynienne dans le Nord-Ouest péninsulaire. Soc. Geol. Portugal Bol. *2*, 227-246.

Oen Ing Soen, 1970: Granite intrusion, folding and metamorphism in central northern Portugal. Bol. Geol. Min. *81*, 271-298.

Oliveira, J. T., Horn, M., and Paproth, E., 1979: Preliminary note on the stratigraphy of the Baixo Alentejo Flysch Group, Carboniferous of Southern Portugal and on the palaeogeographic development, compared to corresponding units in Northwest Germany. Serv. Geol. Portugal Com. *65*, 151-168.

Ortega, E., and Gil Ibarguchi, I., 1983: La Unidad de Malpica-Tuy ("Complejo antiguo"—"fosa blastomilonítica"). Libro Jubilar J. M. Rios (Geología de España), *1*, 430-440.

Parga-Pondal, I., Matte, P., and Capdevila, R., 1964: Introduction à la géologie de "l'Ollo de Sapo," formation porphyroïde antésilurienne du nord ouest de l'Espagne. Inst. Geol. Min. España Not. Com. *76*, 119-154.

Pérez-Estaún, A., 1978: Estratigrafía y estructura de la rama S de la zona Asturoccidental-Leonesa. Inst. Geol. Min. España Mem. *92*, 151 p.

Perroud, H., and Bonhommet, N., 1981: Paleomagnetism of the Ibero-Armorican arc and the Hercynian orogeny in Western Europe. Nature *292*, 445-448.

Pfefferkorn, H. W., 1968: Geologie des Gebietes zwischen Serpa und Mértola (Baixo Alentejo, Portugal). Münster Forsch. Geol. Paläont *9*, 143 p.

Priem, H.N.A., Boelrijk, N.A.I.M., Hebeda, E. H., Verdurmen, E.A.T., and Verschure, R. H., 1972: Upper Ordovician/Lower Silurian acidic magmatism in the pre-Hercynian basement of western Galicia, NW Spain. Z.W.O. Labor. voor Isotopen Geologie, Amsterdam, Reports Invs. 1870/1972, 123-127.

Priem, H.N.A., Boelrijk, N.A.I.M., Verschure, R. H., Hebeda, E. H., and Verdurmen, E.A.T., 1970: Dating events of acid plutonism through the Paleozoic of the western Iberian peninsula. Eclogae Geol. Helvetiae *63*, 255-274.

Priem, H.N.A., and den Tex, E., 1982: Tracing crustal evolution in the northwestern Iberian Peninsula through the Rb-Sr and U-Pb systematics of granitoids. Coll. Internat. Géoch. Petrol. Granit., Clermont-Ferrand.

Prodehl, C., Sousa Moreira, V., Mueller, S., and Mendes, A. S., 1976: Deep seismic sounding experiments in central and southern Portugal. 14th General Assembly Europ. Seismol. Com., Berlin, 261-266.

Pulgar, J. A., 1980: Análisis e interpretación de las estructuras originadas durante la fase de replegamiento en la Zona Asturoccidental-Leonesa (Cordillera Herciniana, NW de España). Tesis, Univ. Oviedo, 334 p.

Ramsay, J., 1962: Interference patterns produced by the superimposition of folds of similar type. Jour. Geology 70, 466-481.

Ribeiro, A., 1974: Contribution à l'étude tectonique de Trás-os-Montes oriental. Serv. Geol. Portugal Mem. 24, 168 p.

Ribeiro, A., Cramez, C., and Rebelo, J. A., 1964: Sur la structure de Trás-os-Montes (nord-est du Portugal). Acad. Sci. Paris C. R. 258, 263-265.

Ribeiro, A., Oliveira, J. T., and Brandão Silva, J., 1983: La estructura de la zona Surportuguesa. Libro Jubilar J. M. Rios (Geología de España), 1, 504-511.

Ribeiro, M. L., and Ribeiro, A., 1974: Signification paléogeographique et tectonique de la présence de galets de roches métamorphiques dans un flysch d'âge dévonien supérieur du Trás-os-Montes oriental (Nord-Est du Portugal). Acad. Sci. Paris C. R. 278, 3161-3163.

Ries, A. C., Richardson, A., and Shackleton, R. M., 1980: Rotation of the Iberian arc: Palaeomagnetic results from North Spain. Earth Planet. Sci. Lett. 50, 301-310.

Ries, A. C., and Shackleton, R. M., 1971: Catazonal complexes of North-West Spain and North Portugal, remnants of a Hercynian thrust plate. Nature Phys. Sci. 234, 65-68.

Ries, A. C., and Shackleton, R. M., 1976: Patterns of strain variation in arcuate fold belts. Royal Soc. London Phil. Trans. (A) 283, 281-288.

Robardet, M., 1981: Late Ordovician tillites in the Iberian Peninsula. In: Hambrey, M. J., and Harland, W. B. (eds), Earth's Pre-Pleistocene Glacial Record. Cambridge Univ. Press, 585-589.

Routhier, P., Aye, F., Boyer, C., Lecolle, M., Molière, P., Picot, P., and Roger, G., 1978: La ceinture sud-ibérique à amas sulfurés dans sa partie espagnole médiane. Tableau géologique et metallogénique. Synthèse sur le type amas sulfurés volcano-sédimentaires. Bur. Rech. Géol. Min. Mém. 94, 265 p.

Schermerhorn, L.J.G., 1971: An outline stratigraphy of the Iberian pyrite belt. Bol. Geol. Min. 82, 239-268.

Schermerhorn, L.J.G., 1975: Pumpellyite-facies metamorphism in the Spanish pyrite belt. Pétrologie 1, 71-86.

Sheehan, P. M., 1973: The relation of Late Ordovician glaciation to the Ordovician-Silurian changeover in North American brachiopod faunas. Lethaia 6, 147-154.

Tex, E. den, 1966: Aperçu pétrologique et structural de la Galice cristalline. Leidse Geol. Meded. 36, 211-222.

Tex, E. den, 1981: A geological section across the Hesperian massif in western and central Galicia. Geol. en Mijnbouw 60, 33-40.

Tex, E. den, and Floor, P., 1967: A blastomylonitic and polymetamorphic ''graben'' in western Galicia (NW Spain). In: Etages Tectoniques. Univ. Neuchâtel Inst. Géol., 169-178.

Thayer, T. P., 1969: Peridotite-gabbro complexes as keys to petrology of mid-oceanic ridges. Geol. Soc. America Bull. 80, 1515-1522.

Triboulet, C., 1974: Les glaucophanites et roches associées de l'île de Groix (Morbihan, France): Étude minéralogique et petrogénétique. Contr. Mineral. Petrol. 45, 65-90.

Truyols, J., and Julivert, M., 1983: El Silúrico en el Macizo Ibérico. Libro Jubilar J. M. Rios (Geología de España), *1*, 246-265.

Vauchez, A., 1976: Les structures hercyniennes dans la région de Fregenal-Oliva de la Frontera (Badajoz, Espagne); un exemple de tectoniques tangentielles superposées. Serv. Geol. Portugal Com. *60*, 261-265.

Vidal, P., 1977: Limitations isotopiques à l'âge et à l'evolution de la croûte continental en Europe moyenne et occidentale. C.N.R.S. Coll. Internat. (Rennes 1974) *243*, 129-141.

Vitrac, A. M., Albarède, F., and Allègre, C. J., 1981: Lead isotopic composition of Hercynian granitic K-feldspars constrains continental genesis. Nature *291*, 460-464.

Vogel, D. E., 1967: Petrology of an eclogite- and pyrigarnite-bearing polymetamorphic rock complex at Cabo Ortegal, NW Spain. Leidse Geol. Meded. *40*, 121-213.

Wegen, G., van der, 1978: Garnet-bearing metabasites from the blastomylonitic graben, western Galicia, Spain. Scripta Geol. *45*, 95 p.

White, A.J.R., and Chappell, B. W., 1977: Ultrametamorphism and granitoid genesis. Tectonophysics *43*, 7-22.

Zamarreño, I., 1972: Las litofacies carbonatadas del Cámbrico de la zone cantábrica (NW España) y su distribución paleogeográfica. Univ. Oviedo Trabajos Geología *5*, 118 p.

Zamarreño, I., 1975: Peritidal origin of Cambrian carbonates in northwest Spain. *In*: Ginsburg, R. N. (ed), Tidal Deposits: A Casebook of Recent Examples and Fossil Counterparts. Springer-Verlag, New York, 289-298.

Zuuren, A. van, 1969: Structural petrology of an area near Santiago de Compostela (NW Spain). Leidse Geol. Meded. *45*, 71 p.

Zwart, H. J., and Dornsiepen, U. F., 1978: The tectonic framework of central and western Europe. Geol. en Mijnbouw *57*, 627-654.

Part III

AFRICA

Chapter 7

EVOLUTION AND STRUCTURE OF THE HIGH ATLAS OF MOROCCO

JEAN-PAUL SCHAER

Université de Neuchâtel

THE ATLAS IN GENERAL

Located between the African craton in the south and the Rif-Tell thrust belts in the north, bordering the Mediterranean from Morocco to Tunisia, the Atlas can be considered an advanced component of the Alpine orogeny in North Africa (Figure 7-1). Whereas in the Rif and Tell thrusts from north to south are widely developed and imply important transports of the upper units southward, the Atlas appears first and foremost as an autochthonous land where essentially Mesozoic strata are folded over an ill-known Paleozoic and Precambrian basement. The Atlas range extends over more than 2,000 km from Atlantic Morocco to Tunisia, but it is rarely more than 100 km across, except in its eastern end where folded structures spread widely between Gabès in the south and Tunis in the north (Figure 7-1).

During recent years, the post-Triassic evolution of the Moroccan Atlas province, or High Atlas, and of the Atlas Mountains as a whole has often been considered in relation to the opening of the Atlantic Ocean, which offers the advantage of being geographically nearby and rather well known. This viewpoint may not prove very fruitful, however, since there are few opportunities for correlating continental and oceanic structures; it is known moreover that Atlantic waters transgressing from the west covered only a very small surface of the future range. All modern global paleogeographic reconstructions of this part of North Africa (Busson 1970, Wildi 1983) show that the marine connection came mostly from the Tethys and led to the emplacement of a succession of paleogeographic domains trending mainly east-west but with some northeast-southwest segments, probably reflecting ancient Hercynian structures (Michard 1976).

In analyzing the evolution of the Atlas Mountains from an overall view of this part of North Africa, one is easily tempted to include in the palinspastic picture all the territories that spread between the stable African continent in the south and the instable Mediterranean margin in the north, from which the Tell and Rif ranges originated. Problems arise, however, for it is admitted that important lateral offsets occurred in the old Tethys domains during the Mesozoic and Tertiary, along probable faults that are thought on valid grounds to have been oriented east-west (Wildi 1983). Similar faults, similarly oriented, were probably active as far

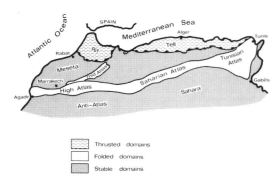

FIGURE 7-1. The main thrusted and folded Alpine structures of Northwest Africa.

south as the margin of the craton. Several recent studies have viewed those within the present Atlas as essential parts of Atlas structure (Mattauer et al. 1977, Jenny et al. 1981, Laville and Harmand 1982, and so on). Identifying them and precisely analyzing their development remain difficult tasks. A few facts and interpretations aimed at elucidating these phenomena are set out below. We also try to illustrate the complexity of the Atlas range through an examination of the segment with which field studies have made us personally familiar, that is, the western and Moroccan part, from the Atlantic to near the Algerian border (Figure 7-2). A rather distinctive morphological unit, this part of the Atlas nevertheless presents, in different areas, evolutions so different that it becomes almost impossible to describe a section that can be used as a model for the structure of the whole.

In describing the Saharian Atlas, we may note that it includes an important, fairly homogeneous Mesozoic series, 4 to 6 km thick, which has been deformed into box folds, often thrusted on its northern border, where major faults can be seen. Folding probably occurred between the middle Eocene and Burdigalian, with a décollement at the level of the Triassic gypsiferous-marly series. Diapirism appears in eastern Algeria and becomes very important in Tunisia. Throughout the Algerian and Tunisian Atlas, en echelon folds are frequent; they arise and die rapidly and are often oriented at a slight angle to the range as a whole (more northeast than the east-northeast orientation of the range). In the Algerian and Tunisian Atlas, the basement can never be seen under the folded cover. The structural and morphological limits of the Saharian Atlas are often clear enough (faults, flexures); they cut facies limits of the Triassic and Jurassic sedimentary basins, thus differing appreciably from what can be seen in the Moroccan Atlas.

THE MOROCCAN OR HIGH ATLAS

Morphologically speaking, the Moroccan or High Atlas (Figure 7-2) is quite clearly marked over the 900 km of its extent, which is almost east-west. It can be divided into four sectors, from west to east (see Figures 7-2 and 7-3):

1. Western High Atlas: a folded range of Mesozoic sediments, formed at the expense of a subsiding basin of the Atlantic continental margin.

2. Paleozoic High Atlas: a horst of Paleozoic terranes intruded by Carboniferous gran-

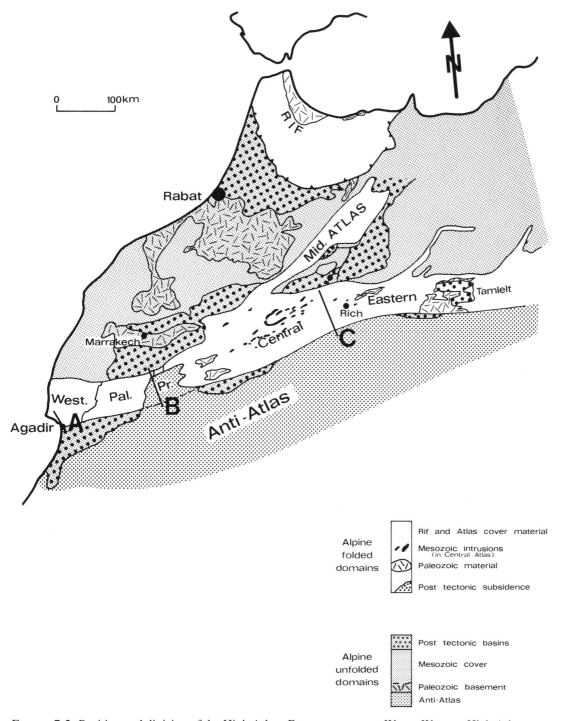

FIGURE 7-2. Position and division of the High Atlas. *From west to east*: West.: Western High Atlas. Pal.: Paleozoic High Atlas. Pr.: Precambrian High Atlas or Ouzellarh Massif. Central: Central High Atlas. Eastern: Eastern High Atlas. Mid.: Middle Atlas. A B C: Traces of profiles shown in Figure 7-5.

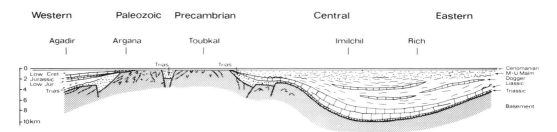

FIGURE 7-3. Schematic longitudinal profile through the High Atlas of Morocco from Agadir to Tamlelt during the Upper Cretaceous, illustrating the differential subsidence during the Mesozoic over a heterogeneous basement (shaded). (Different vertical and horizontal scales.)

ites, only thinly covered by Mesozoic terranes that are often still subhorizontal apart from narrow folds accompanying fault zones.

3. Precambrian High Atlas: a horst made up mostly of little deformed infra-Cambrian terranes and Precambrian rocks, with a very thin Mesozoic cover locally tilted and folded near faults with large vertical offsets.

4. Central and Eastern High Atlas: a range formed of carbonate rocks, laid down in a basin that subsided strongly, especially through the Lower and Middle Jurassic. The structure (folds and faults) was partly formed during the sedimentary evolution. Its main features probably formed during the Middle Jurassic, locally through the influence of magmatic intrusions, but Tertiary uplift gave the whole range its present character.

New deep seismic studies (Makris et al. 1985) show that the crust of the Western Atlas has a normal thickness of 30 km, similar to that of the Meseta to the north and the Anti-Atlas to the south. Farther east, in the western part of the Central High Atlas, these studies constrained by gravity models suggest a thickness of up to 34 or 38 km, depending upon the density assumption.

TRIASSIC PALEOGEOGRAPHY

The Triassic paleogeography is shown in Figure 7-4A. Apart from the central sector south of Marrakech, outcrops in the Moroccan Atlas are mostly Mesozoic sediments, deposited above a basement that had generally been structured and eroded at the end of the Paleozoic. Beginning in the Permian, but mostly during the Triassic, coarse and fine red clastics, often overlain by basalts, accumulated in subsiding basins. The whole scene is broadly homogeneous, but details show fairly great variations in thickness (from about 100 m to more than 5 km) and grain size. Researchers agree about the formation and development of restricted, partially independent basins (Van Houten and Brown 1977), which would be the structural equivalent to what is known from the same time in western Europe (Ziegler 1982b) and eastern North America (Manspeizer 1982). The basin geometry in North America, less affected by subsequent tectonics, permits the recognition of reactivated old faults. In Morocco, by analogy, similar structures may be postulated, but the extent of younger cover masking many Triassic deposits leads to caution. Caution might also be necessary in considering regional stress orientations, as proposed by Laville (1981) for Triassic time; many of the important structures

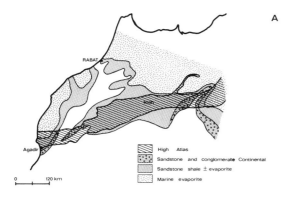

A

High Atlas

Sandstone and conglomerate Continental

Sandstone shale ± evaporite

Marine evaporite

0   120 km

B

Atlas

Mesozoic subsiding zones

Reef Limestone Low Jurassic

Limit of | low Jurassic / middle Jurassic | Transgression

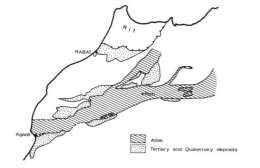

C

Atlas

Tertiary and Quaternary deposits

FIGURE 7-4. Schematic paleogeographic maps of Morocco. A: Triassic time (simplified and with some modifications from Salvan 1974). B: Jurassic time (simplified from du Dresnay 1971). C: Tertiary time.

he used may have been produced during post-Triassic time, whereas small structures may be of dubious value for an appreciation of the general evolution.

The southern limit of the Moroccan Triassic basins followed a general east-northeast line, roughly parallel to the future Atlas range but shifted slightly southward (Figure 7-4A). In that direction, sediments are mostly detrital and probably continental, while marine influences increase toward the Atlantic-Essaouira Basin and especially toward the east and northeast, where major saliferous series, locally over 1000 m, have been found by drilling (Salvan 1974).

In the Moroccan Atlas province, except near the Atlantic coast, gypsum and halite occur in small amounts; their role in tectonics is discussed below. In Algeria and Tunisia, synthetical maps (Busson 1982) show clearly that the saliferous facies extend well beyond the Atlas range, although, in Tunisia especially, they play a determinant part in the evolution of the range.

If locally certain paleogeographic limits of Triassic age follow some lineaments of the actual range, in many places they cut through them. In the Western Atlas sector, along the Argana Corridor (see below), it is not possible to postulate a symmetrical east-west graben during the Triassic, which would prefigure the Tertiary horst (Stets and Wurster 1981); rather, one must consider a complex flexure, oriented roughly parallel to the present shore, that remained active throughout the Mesozoic.

THE WESTERN HIGH ATLAS

The Western High Atlas extends 50 to 70 km inland from the coast to a wide strip of Triassic terranes, which can be as thick as 5 km (Argana Corridor) (Tixeront 1973, Brown 1980). Essentially made up of detrital and locally conglomeratic materials, toward the west the formations include evaporitic sequences producing diapirs (Salvan 1974).

Beginning in the Jurassic, the Mesozoic sedimentation of the Western Atlas basin formed a prism whose thinner eastern part is marked by deposits with frequent continental influences, while the thicker part, toward the Atlantic margin, has clearly marine facies. The southern limit of the basin follows closely enough the southern limit of the Atlas, but northward the basin extends well beyond the limits of the range, though with restricted thickness (Stets and Wurster 1981).

Apart from differential subsidence, no pre-Tertiary deformation has been recognized in this sector. Jurassic and Cretaceous sediments, calcareous and marly, locally evaporitic toward the east, suggest that the hinterland had low relief, except for the Middle Jurassic where conglomerates and sandstones are reported. The Tertiary structure of this part of the range is now marked by rather narrow yet clearly defined anticlines with wide synclines (Figure 7-5A). The axial strikes of these folds are not very constant; though often parallel to the range, they can be oblique and sometimes even transverse. Frequently, Triassic diapirism outlines the major décollement along argillites and evaporites.

THE PALEOZOIC HIGH ATLAS

The Paleozoic High Atlas is a vaguely diamond-shaped horst, about 100 km long east-west and extending almost 70 km north-south. It is made up of lower Paleozoic terranes (mostly Cambrian and Ordovician), with some Carboniferous. Those rocks were folded and peneplained before Triassic time. The very thick sequence of continental red beds and conglomerates of the Argana Corridor did not reach much farther east than its actual outcrops. In that direction, Triassic deposits are first rather thin or absent, but south of Marrakech the remains of other thick basins cut obliquely through the range (Beauchamp and Petit 1983). On the southern and northern edges of the present horst, as well as in the few rare points where the cover remains, the Jurassic is either very thin or absent. The Mesozoic sediments, in which continental and marine facies alternate, belong almost exclusively to the Cretaceous and have

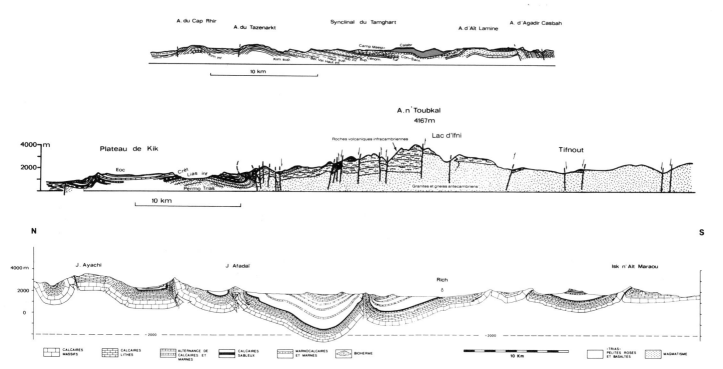

FIGURE 7-5. Profiles through the High Atlas of Morocco (for traces, see Figure 7-2). A: In the Western part, north of Agadir (modified from Ambroggi 1963). B: In the Precambrian part, south of Marrakech (modified from Proust 1973). C: In the Central part, near Rich (Brechbühler 1984).

a thickness of 200 to 500 m (much thinner and less marine than those on the Atlantic margin). The often red continental sediments may be Triassic, Jurassic, or Cretaceous. These terranes do not seem to exhibit any major unconformities that would suggest important movements. In this restricted cover, differential subsidence is expressed by stratigraphic wedging-out within and on the edge of the range.

In this sector of the Atlas, apart from the major reliefs still present during the Triassic, most of the Mesozoic sediments suggest a low topography that lasted until the end of the Cretaceous. Crystalline pebbles above an angular unconformity of that age (Froitzheim 1984) prove the first modification in morphology and structure. But the main change took place during the Miocene when major conglomeratic units were laid down outside the range; they are already turned up to vertical on the edge of the horst. Under the Triassic or Cretaceous cover, the Paleozoic basement, made up of schists, graywackes, conglomerates, granites, and some limestones, behaved rigidly during the Tertiary tectonics. Differential movements along subvertical faults, some already active before Triassic time, produced offsets of 1000 m and more. Where the Triassic or Jurassic is made up of sandstones and conglomerates it remains attached to the basement, and only tilting and faults can be observed. Higher in the stratigraphic series, pelites (sometimes in direct contact with the basement) and gypsum (belonging locally to the Jurassic but mostly to different parts of the Cretaceous) permitted décollements producing folds along the fault zones. On both sides of the central horst (40 km wide), shoulders up to 15 km can be recognized where the basement has been less raised; there, subhorizontal Cretaceous sediments, along with leveled surfaces freshly freed from their cover, outline the extent of these zones, which moved as rigid blocks. Locally, along marginal faults, thrust folds have been induced by gravity gliding. North of Amizmiz (south of Marrakech), folds and flexures in the cover exhibit orthogonal trends, reflecting the influence of faults that strike at right angles with various offsets (Schaer 1967).

In recent years, many articles from the school of Montpellier (Mattauer et al. 1972, Proust and Tapponnier 1973, Petit 1976, Proust et al. 1977) have given a new impetus to Atlas tectonics by insisting on the importance of great strike-slip faults. For example, they have proposed that the complex, north-northeast trending Tizi n'Test fault acted during the Hercynian as a right-lateral strike-slip fault with a very important offset (at least 40 km). After the Triassic, perhaps during the Jurassic, an inversion of movement along the same fault produced a left-lateral slip of a few kilometers. It has been suggested that this major fracture might also be found on the other side of the Atlantic, in the Appalachians. First developed as a hypothesis, this idea frequently has been referred to as fact, especially since Petit (1976) undertook a detailed and quite interesting analysis of the subject. Nevertheless, the arguments developed from the observation of subhorizontal striae, mostly in the Triassic formations, and from analysis of the deformation of Paleozoic terranes do not seem convincing enough to justify raising the proposal above the rank of a hypothesis, especially as the extent of the Hercynian slip is based upon a wrong appraisal of the meaning of Paleozoic cleavage. In the northeastern part of the western Paleozoic Atlas, south of Imi n'Tanoute, precise mapping (Schaer 1974) has enabled me to relate the modest (3-5 km), probably Pre-Triassic, right-lateral slip of a Hercynian syncline to a subvertical fault trending east-southeast.

THE PRECAMBRIAN HIGH ATLAS

In Morocco, south of the High Atlas and separated from it by a succession of small basins, lies the Anti-Atlas (Figure 7-1), a vast province composed of large culminations where a thick sedimentary cover (Paleozoic and Infra-Cambrian) envelops cores of Proterozoic granite, gneiss, and volcanics; the updoming and folding of these structures took place during late Paleozoic time. South-southeast of Marrakech, that province reaches well into the High Atlas, extending over 70 km east-west and reaching almost to the northern border. In this sector, the Precambrian High Atlas (or Ouzellarh Massif), which includes the highest summits in North Africa (Toubkal, 4,167 m, among others), granites and migmatites are overlain by thick late Precambrian lava (Precambrian 3), which has remained subhorizontal to this day. Several Infracambrian and Lower Cambrian sequences wedge out against these formations (Figure 7-5B).

During the Middle Cambrian, this zone, which had acted early on as an upthrown block, was once more submerged but received less sediment than did the adjacent Anti-Atlas. During the Triassic, possibly from the Permian, a thick sequence of continental sandstone covered a great part of the landscape. It is preserved in elongated, collapsed, and tilted zones trending northeast-southwest (former graben?). The cover is completed by a thin, mainly continental but locally marine Cretaceous, which may lie directly over the Precambrian granites. Apart from Triassic normal faults, this part of the High Atlas has remained stable from the end of the Precambrian to the late Cenozoic; the old faults were probably reactivated during the Mio-

Pliocene, with a mainly northeast strike, to form the present horst structure. The Cretaceous cover has remained mostly horizontal, but it can be folded near faults where it suffered local gravity gliding and thus was detached from its substratum. Triassic sandstones more often behave rigidly and follow the basement; locally, however, as a result of bedding-plane glide, they exhibit obvious flexures. The large vertical displacement and the effect of gravity probably produce slight thrusts of Precambrian over Mesozoic rocks (Figure 7-5B).

The Ouzellarh Massif and part of the Paleozoic Atlas have acted as a hinge zone (Figure 7-3). They were the high zone separating the two subsiding Mesozoic marine basins, the Atlantic on the west and the Tethys on the east (Michard 1976). They also appear as a hinge zone when seen in map view, because the east-west trend of the Atlas range seems here to be bent northward. This change in trend can be correlated with a series of faults, and perhaps also with the important development of Triassic basins southeast of Marrakech.

THE CENTRAL AND EASTERN OR CALCAREOUS HIGH ATLAS

The Central and Eastern or Calcareous High Atlas includes everything between the Ouzellarh Massif and the Algerian border. The following distinctive characters can be recognized:

1. Thick marly-calcareous sedimentation during the Jurassic.

2. Early evolution of the structural framework. It was established as early as the lower Liassic, while the internal organization took shape from the Toarcian.

3. In the middle area mostly, intrusions of basic and alkali material during the Middle Jurassic into an edifice whose internal structure was not much modified later.

4. A long period of relative rest extending from the Upper Jurassic to the Miocene.

5. Between the late Tertiary and the Recent, rejuvenation of the range as a horst.

In this part of the range, the pre-Triassic basement appears only locally. Outcrops in the western and central part show that the Paleozoic material was folded into important Paleozoic nappes in the west (Jenny and Le Marrec 1980), and into hectometric recumbent folds in the Mougueur inlier northeast of Rich. At the east end of the Moroccan range, the vast Bou Anane-Tamlelt inlier once more shows slightly deformed lower Paleozoic lying on a Precambrian basement. All along its strike, the Paleozoic basement of the High Atlas is in general much more deformed (recumbent folds, nappes, schistosity) than the equivalent rocks of the Anti-Atlas but there is no sharp boundary between the two tectonic provinces; the presence or absence of schistosity cannot be explained by a set of strike-slip faults.

In the Central and Eastern High Atlas, the Triassic is fairly well exposed. Made of sandstones and pelites, it very locally includes evaporites; in numerous localities, as indeed all over Morocco, the series is capped by upper Triassic basalts, often very altered. Generally and especially toward the east, the detrital Triassic is much thinner than it is directly south or southeast of Marrakech. Ovtracht (1980) thinks that at first the evaporites were very thick in the Central and Eastern Atlas (many hundreds, if not thousands of meters); they purportedly disappeared mainly by diapiric extrusions during various long lasting continental episodes (Jurassic-Cretaceous, Cretaceous-Tertiary, Tertiary-Recent). The Tertiary deformation of the Atlas would result from this halokinesis. For want of arguments and justification, it seems difficult to support such an extreme proposal. If Triassic evaporites were present under the Moroccan Calcareous High Atlas, we think that their volume was modest and that migration was restricted to hydrothermal circulation induced by metamorphism (regional, and local associated with intrusions) during the Jurassic deformation (Brechbühler 1984).

From the beginning of Jurassic time, the eastern part of Morocco was open to the Tethys. At the level of the future High Atlas, a sinking trough established itself, extending westward up to the Ouzellarh Massif (Choubert and Faure-Muret 1962; Dubar 1962; du Dresnay 1971, 1979) (Figure 7-4B). In the center of the trough, the sediments were thick yet do not reflect deep water; on the edges, carbonate reef, platform, and shore facies were built up as series of decreasing thicknesses. The future Atlas range coincides rather closely with this sinking trough and incorporates, on both the south and the north, the edge-facies zones. Mostly carbonate at first, the sedimentation became more argillaceous, modestly during the Domerian, but important from the Toarcian and continuing to the Bathonian. In the center of the trough, Studer (1980) has measured more than 7 km of sediments dating from the lowest Jurassic up into the Dogger. In the western part of the gulf, where terrigenous red sediments become more and more dominant, the total thickness of Jurassic series decreases, from 3000 m at the level of Beni Mellal to almost 1000 m at the level of Telouet (Jenny 1984). In the central part, east of the greatest measured thickness, important erosion has truncated the series. To get some idea of the importance of sedimentary accumulations by determining the depth of burial, Ber-

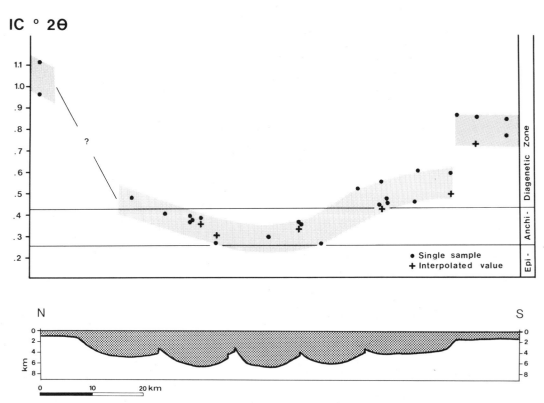

FIGURE 7-6. *Top*: Evolution of illite crystallinity, measured across the Central High Atlas (along the profile in Figure 7-5C) in the Toarcian of the Lower Jurassic (Bernasconi 1983, Brechbühler 1984). *Bottom*: Reconstruction of the post-Toarcian overburden (stippled) along the same profile (Brechbühler 1984).

nasconi (1983) and Brechbühler (1984) have studied the evolution of the illite crystallinity in preserved sediments (Figure 7-6). At the level of Rich, they propose almost 7 km of sediments on top of the lower Toarcian, which would bring the base of the Jurassic to more than 8 km deep. Similar studies have not yet disclosed much about the evolution of subsidence farther east, but at the level of the Tamlelt inlier, du Dresnay's field studies have led him to postulate an emergent sill during a large part of the Jurassic. There also, Jurassic structural evolution individualized a zone that has retained a distinct structural behavior to this day. Thus Quaternary deposits are important.

In this gulf of Tethys, Jurassic subsidence was not uniform. During the lowest part of the Lower Jurassic (lower Liassic), the carbonate facies are rather uniform. At the level of Rich, the thickness doubles from the southern to the northern part of the trough (Brechbühler 1984). From the Toarcian, more important disturbances appeared. Stratigraphic wedge-outs, first brought to the fore by Dubar (1938), occurred on ridges prefiguring future anticlines. Later studies have confirmed the importance of these phenomena (du Dresnay 1979, Studer 1980, Brechbühler 1984), which show that the development of the individual basins and ridges of that part of the High Atlas had already begun by the Toarcian and continued during the Middle Jurassic (Monbaron 1982, Jenny 1984). Beside some anticlines, synsedimentary faults during the Middle Jurassic led to very spectacular wedge-outs, accompanied by breccia and conglomerates which, as at Tizi n'Irhil (Dogger), include lower Liassic and even Triassic basalt pebbles. Erosion of nearly 2 km must be considered (Studer and du Dresnay 1980). Northeast of Rich, Bernasconi (1983) has described a small sedimentary basin of the pull-apart kind (Figure 7-7), whose geometry evokes quite perfectly the global picture of Atlas deformation proposed for that period by Mattauer et al. (1977). The importance of Toarcian-filled tension gashes cutting the reef strata at Bou Dahar must also be mentioned (Agard and du Dresnay 1965).

MAGMATISM

In the Central High Atlas, magmatic intrusions are quite common. Some are subcircular (8-6 km) but more are elongated, mainly parallel to the range (in extreme cases up to 30-40 km long and only 0.5-1.0 km wide). They are associated with both thin and thick dikes, again parallel to the range and mostly present in the synclines (Brechbühler 1984). Monbaron (1980) has illustrated the relationship between gabbroic intrusions, basic dikes and sills, and some lava flows that have been preserved from erosion in the western part of the Central High Atlas. Dikes and flows are essentially basic while the intrusive plutons range from gabbro through diorite to syenite. All these rocks are most abundant where subsidence seems to have reached its maximum.

The age of the magmatism is still much debated. Relying on radiometric data from several laboratories that used the K/Ar method only on isolated minerals (plagioclases, biotites, amphiboles) or on total rocks of basic material (intrusive and extrusive), Laville and Harmand (1982) believe that the magmatism lasted from the Aalenian to the Barremian. They propose that this spread-out igneous activity is the cause of the synsedimentary deformations during the development of the Atlas trough. Unfortunately, no precise stratigraphic argument has

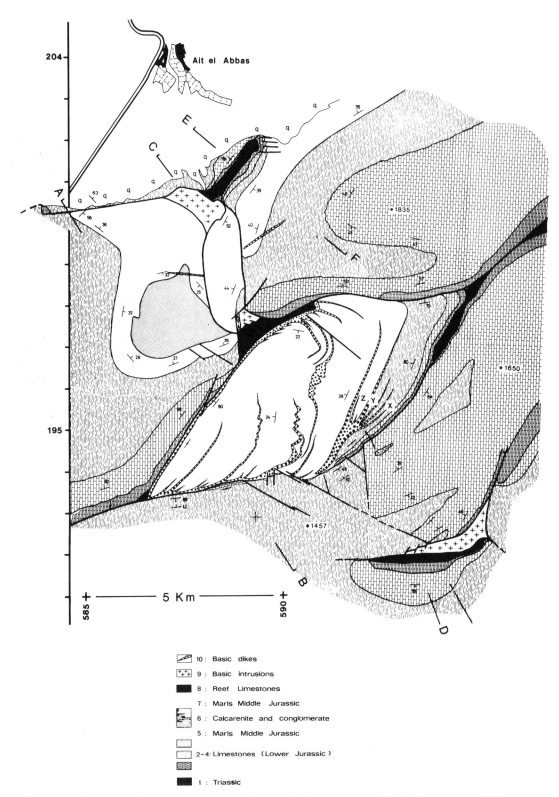

FIGURE 7-7A. Geological map of a small pull-apart basin northeast of Rich (Bernasconi 1983).

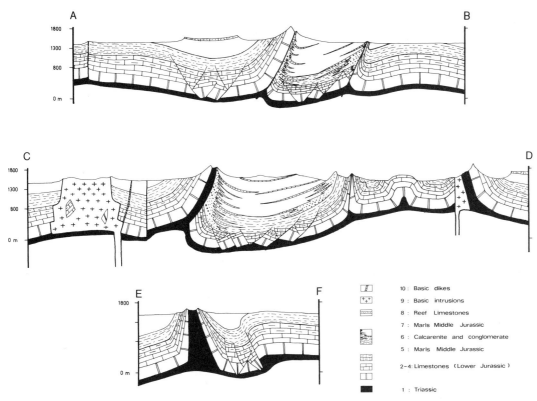

FIGURE 7-7B. Profiles through the small pull-apart basin shown in Figure 7-7A. (Bernasconi 1983).

been put forward to support these proposals. Despite careful regional studies, no researcher has yet found any indication of magmatic activity in the Lias or lower Dogger.

Also, we must question the meaning of radiometric ages obtained from often altered rocks by a method that is not always suited to determining the time of crystallization (cf. Sutter and Smith 1979). We prefer the conclusions of Jenny et al. (1981), who show that in the western part of the Central Atlas the basalts spatially related to dikes and gabbroic bodies are sandwiched between red detrital formations that, through dinosaurs found at the base and pollen found on top, put the igneous event there in the Middle Jurassic. Without asserting that this result definitively solves the problem of the age of Atlas magmatism, we must stress that it agrees perfectly with field data while not requiring that magmatism apparently so homogeneous lasted so long. The nepheline-syenites and associated carbonatites of the northern edge of the Atlas at the level of Midelt (Agard 1960) form a separate group, as confirmed by dates of 33 ± 4 Ma by the Rb/Sr method and 42 ± 3 Ma by the K/Ar method (Tisserant et al. 1976).

A few years ago, we put forward the hypothesis that most of the structural evolution of this part of the Atlas could be related to intrusions (Schaer and Persoz 1976). Recent studies by our collaborators (Studer 1980, Bernasconi 1983, Brechbühler 1984) and field trips with them have convinced us that our proposals were too extreme. They have, however, been retained and extended by Laville and Harmand (1982) on a theoretical basis supported by few precise facts and using a chronology of magmatic events that, as noted above, we cannot accept. Along the contacts of intrusive rocks, the deformation present in limestones and marls varies greatly in both range (from only a few meters to many hundreds of meters) and intensity. Folds, which may be almost isoclinal, and cleavage may or may not be present. Locally, metamorphism has produced marbles which have been deformed by dynamic recrystallization (Brechbühler 1984). Metamorphic acicular amphiboles present in the cleavage planes show a well-oriented fabric; during a last stage of deformation they have been stretched and broken, forming boudins separated by fibrous calcite similar to that present in the pressure shadows around quartz and feldspars. These late crystallizations, developing under the same stress regime as the metamorphic minerals (Brechbühler 1984), are in our opinion related to the last effects of the emplacement of the igneous masses and not to a later (Tertiary) phase of deformation. Moreover, some of the most spectacular deformed limestones are present as inclusions in the plutonic rocks (Studer 1980), confirming the relationship between deformation and intrusion.

The basic dikes larding this part of the High Atlas are very useful chronological and geometrical landmarks. They are spatially, genetically, and temporally related to other Jurassic magmatic events (Monbaron 1980, Studer 1980). Most often subvertical, they locally cut tilted layers and folds (Studer 1980). Post-intrusive deformations have been observed, but globally they appear relatively unimportant and are linked to an increased tilt in the layers cut by the dikes. Brechbühler (1984), relying on a local paleomagmatic analysis he deems still incomplete, concludes that half the tilting of calcareous layers occurred before the intrusion and half after.

The occurrence of cleavage in the Central High Atlas has been studied for several years (Schaer and Persoz 1976, Studer 1980, Bernasconi 1983). It is a pressure-solution cleavage whose presence, absence, intensity, and orientation do not seem to obey simple parameters. Studies on the subject are still restricted regionally to the central part of the High Atlas south of Midelt, and it is therefore not possible to give a clear synthetical picture. It can, however, be stressed that the cleavage is most often subvertical and oriented east-west, parallel to the range. Markedly more visible in marly formations, it may not appear in formations whose position and composition would seem to favor its development. It is not easily observed in massive limestones in the field; its presence, however, is revealed by thin-section studies. Often, but not always, the cleavage is more intense near the intrusions, where its orientation can be variable, sometimes radial (Schaer and Persoz 1976). Strongly marked along some anticlinal faults, it is almost always oblique to these structures and can even be orthogonal. In our view, the cleavage in the Central High Atlas reflects the heterogeneous state of stress in a cover cut into independent slabs by faults and intrusions (Figure 7-8).

FOLDS AND THE POSITION OF THE BASEMENT

The major structures of the Central High Atlas include folds with tight or faulted isoclinal anticlines and wide flat-bottomed synclines. The section shown in Figure 7-5C, drawn by Brechbühler and Bernasconi, gives a good picture of the folding in the area where intrusions, though present, are not very important. Folds most often trend east-northeast; the same trend,

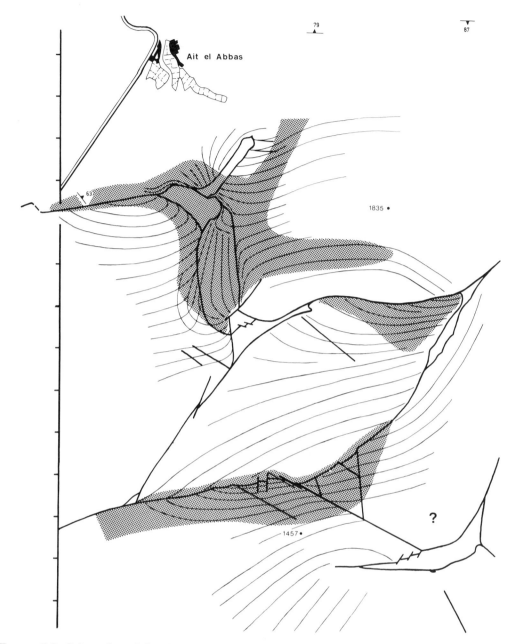

FIGURE 7-8. Orientation of dissolution cleavage northeast of Rich (for geology, see Figure 7-7A). Shading represents areas where a well-marked schistosity is developed (Bernasconi 1983).

slightly oblique to the range, is marked in the Hercynian basement by faults to the south (Sarho), west (Ouzellarh Massif), and northwest (central Morocco). Some major folds in the Atlas show important axial flexures and may trend north-northeast or almost northeast for many kilometers. In the western part of the Central High Atlas, the progressive eastward structural plunge of the Ouzellarh basement beneath the Mesozoic sediments allows one to conclude that faulting of the basement extending under the cover gives rise to folds (Michard

1976). In this area, where anticlines are not as marked as farther east, the geometry of the Cretaceous-filled basins seems controlled mostly by the action of faults, some systems of which were already active during the Middle Jurassic (Jenny 1984). To the east, as the thickness of the cover increases, the geometry of the top of the basement becomes more uncertain, but probably it is never flat over large areas, as it is under some classical foreland ranges like the Jura. In the Central Atlas, southeast of Midelt, the Mouguer inlier brings the basement to an altitude of more than 2000 m, whereas under the bottom of several nearby Mesozoic synclines it must be as low as -2000 m (Figure 7-5C). We know from what is visible outside the High Atlas that the basement is quite heterogeneous. For a body of rock which has been drawn to a depth of nearly 10 km during the Jurassic subsidence and in which epimetamorphic schists are likely to be frequent, overall rigid behavior must not be postulated. The Mesozoic cover lies on a basement cut by faults and divided into independent compartments whose upper surface has been brought to various altitudes, bent, and folded; basement schists may be injected along some major faults.

The Triassic formations that transgress over the basement can stick on it or be detached. These rocks, consisting of conglomerates, sandstones, and silts, accompanied by rare evaporites (a few salty springs, local secondary recrystallization of gypsum), are rather widespread, and their compositions and thicknesses vary greatly. Often, red pelites and some sandstones underline the main anticline faults; concordant or discordant stripes of similar material are present around and among intrusive rocks, where they may separate plutonic masses of different compositions. In some synclines, Studer (1980) has reported pelitic Triassic material interjected as sedimentary dikes into Middle Jurassic rocks that normally overlie the Triassic by more than 5 km. In the Mouguer inlier, according to Bernasconi (1983), a thick mass of red silts without any sign of stratification overlies the well-bedded basal Triassic sandstones. He suggests that these rocks may have formed a subcompacted zone, perhaps partly protected by evaporites. During the folding and intrusions (?), these muds were injected and their mineralogy may have been partially rebalanced according to the fluids they were carrying. These observations and reasonings lead us to accept the idea of a partial décollement of the cover, especially important where subsidence and intrusions have been active.

The vital role of strike-slip faults in the development of the Atlas structure was first expounded by Mattauer et al. (1977). This fruitful idea has been taken up by numerous authors who have endeavored to recognize the stress regime or regimes that conditioned all or part of the structural evolution, including sedimentary and magmatic evolution up to the development of cleavage. Their results, often based on careful local studies, cannot as yet be integrated into a simple picture. For some (Mattauer et al. 1977, Laville and Harmand 1982, Bernasconi 1983, Brechbühler 1984), the major evolution proceeded from a wide zone of left-lateral shear trending east or N 70° E (Laville favors a trend of N 120° E for the major deformation). For others (Studer 1980, Monbaron 1981, Jenny 1984), the major Atlas structures reflect right-lateral strike-slip faults trending N 70° E. While preferring left-lateral shear, we cannot ignore some arguments developed by the other group of researchers. We hope that new studies will take up the problem again and try to confirm not only movements on old Hercynian faults, but also movements in opposite directions occurring at different times. We must also hope that concrete proposals are soon put forth about Mesozoic and Cenozoic differential movement between the territories located north and south of the Atlas.

It does not seem possible that the almost 10 km of subsidence (0.5 mm/year!) that accompanied Jurassic sedimentation in the Calcareous High Atlas could have occurred without an important thinning of the ancient continental crust on which it rests. Listric faults related to the formation of pull-apart basins could have played a leading part in this evolution. Since the Middle Jurassic subsidence, 3 to 5 km of sediments have been eroded in a large part of the Central High Atlas. Perhaps this lightening of the crust has been compensated by a new thickening caused by tightening during the magmatic and cleavage episode.

CONCLUSIONS

The Moroccan High Atlas, as it appears now with its well individualized morphology, mostly bordered on the north and south by Tertiary and Quaternary basins, was born from the juxtaposition of structural components that underwent fairly different Mesozoic evolutions. In the west, active subsidence took place, mostly during the Jurassic and the Cretaceous, in a province whose southern limit almost coincided with the southern limit of the future range (Stets and Wurster 1981). Northward, the limit of the mobile basin extended beyond, into domains that were not later integrated into the range. At the level of Marrakech (eastern part of the Paleozoic High Atlas, Precambrian Ouzellarh Massif), the Mesozoic evolution is not well known for lack of sediments apart from those of the Triassic. There, the Atlas was never marked by active subsidence; it appears as a simple horst bordered by subvertical faults, sometimes developing into overthrusts in which gravity may have taken an active part (Proust 1973). In this area, where Tertiary basins and faults are lacking, the southern Atlas limit is not always clearly marked; numerous right-lateral strike-slip faults, slightly oblique to the range and probably active before the end of the Cretaceous, cut through the range (Proust et al. 1977). They are superimposed on ancient Hercynian faults.

In the eastern part of Morocco, the present Atlas is rather closely superimposed on a relatively narrow (50-80 km) subsiding zone extending over 500 km to the Algerian border and beyond. In its length and width this basin resembles the classical graben of continental domains. In the High Atlas, however, no elevated marginal escarpments can be observed or presumed. A comparison with the development of graben in the North Sea and the North Atlantic margin (Ziegler 1982a) offers some analogies (Celtic Sea), but the important igneous inputs in the High Atlas where subsidence seems to have reached its maximum make it a special case.

Despite its morphological unity, the Moroccan High Atlas is thus clearly heterogeneous. Its external geometry reflects the uplift of a horst during the Tertiary and Quaternary, but it has also been influenced by the specific evolution of each partially independent segment during the Mesozoic. Moreover, Hercynian faults and structures also imposed divisions, some of which came to expression only during Triassic time. As for the folding of the cover, notwithstanding the fairly simple structure made mostly of upright folds and subvertical faults combining into a quite simple unitary picture, the Moroccan High Atlas is complex enough. Inherited influences are often concealed and not easily detected; this may explain the diverse interpretations proposed for the evolution of the stress regime in the range. In a simple evolution regime such as that of the North Sea graben (Ziegler 1982a), oil exploration has shown the complex geometries that were formed. There is no reason why structures as complex as these should not arise in a continental range such as the High Atlas.

In the western part of North Africa, the Atlas can be presented as a foreland fold belt of the Alpine orogeny (Figure 7-1). This picture is particularly attractive in Tunisia and Algeria, where the thrust belt of the Tell and the Atlas range are placed edge to edge or separated only by a single undeformed domain. In Morocco, the situation is more complex; from the Rif in the north to the High Atlas, the following domains are present (Figure 7-2): (1) thrusted Rif, (2) active Tertiary and Quaternary basin, (3) stable domain of the Moroccan Meseta, (4) small Tertiary basin, (5) small horst of the Jbilet, (6) Tertiary and Quaternary basin of Marrakech, and (7) High Atlas. These structures and their relations are not easily explained by a simple convergence of the two rigid plates of Africa and Europe. Many factors perhaps contribute to the observed complexity, including:

1. The shape and position of the Arc of Gibraltar.

2. The boundary conditions along the contact between the African continental crust and the Atlantic oceanic crust.

3. The possible obliquity between some basement lineaments and the convergence vector.

4. Heterogeneity within the African crust.

These last two conditions, imposed by the Hercynian structure and the Mesozoic evolution, seem fundamental. During Jurassic time, the probable thinning of the crust in the Central High Atlas nearly led to the development of an oceanic domain. The growth of an oceanic ridge at that time between Spain and North Africa (Dercourt et al. 1985) was probably enough to release the tensions in that part of North Africa and to stop a nearly similar evolution farther to the south that was already well advanced. In our view, the Central and Eastern High Atlas was folded in the Jurassic. During the Tertiary, the range was shaped as a unit by the formation of a horst; some existing folds were reactivated, while others, especially to the west, were newly formed in the cover above basement faults. Today, the High Atlas of Morocco, a foreland range of the Alpine orogeny, does not present any important thrusts; in the geological future this situation may change if the convergence between Africa and Spain should continue.

ACKNOWLEDGMENTS

This article utilizes ideas and discussions from many sources, but they are presented here with our own interpretations. Particularly helpful were the contributions of Bernasconi, Brechbühler, Heitzmann, Monbaron, and Studer. Many thanks to John Rodgers for corrections and helpful criticism of early drafts of this work. The Moroccan authorities, especially the Geological Survey of Morocco, facilitated our field investigations. The Swiss National Foundation supported our research (Grant 2.206-081).

REFERENCES

Agard, J., 1960: Les carbonatites et les roches à silicates et carbonates associés du massif de roches alcalines du Tamazert (Haut Atlas de Midelt, Maroc) et les problèmes de leur genèse. 21st Internat. Geol. Congr., Copenhagen 1960, Rept. *13*, 293-303.

Agard, J., and du Dresnay, R., 1965: La région minéralisée du jbel Bou-Dahar près de Beni-Tajjite (Haut Atlas oriental). Etude géologique et métallogénique. *In*: Colloque sur des gisements stratiformes de plomb, zinc et manganese du Maroc. Serv. géol. Maroc Notes Mém. *181*, 135-152.

Ambroggi, R., 1963: Etude géologique du versant méridional du Haut Atlas occidental et de la plaine du Souss. Serv. géol. Maroc Notes Mém. *157*, 321 p.

Beauchamp, J., and Petit, J.-P., 1983: Sédimentation et taphrogènese triasique au Maroc: L'Exemple du Haut Atlas de Marrakech. Cent. Rech. Explor.-Prod. Elf-Aquitaine Bull. *7*, 389-397.

Bernasconi, R., 1983: Géologie du Haut Atlas de Rich (Maroc). Thèse, Univ. Neuchâtel.

Brechbühler, Y.-A., 1984: Etude structurale et géologique du Haut Atlas calcaire entre le Jebel Ayachi et Rich (Maroc). Thèse, Univ. Neuchâtel.

Brown, R. H., 1980: Triassic rocks of Argana Valley, Southern Morocco, and their regional structural implications. Am. Assoc. Petroleum Geologists Bull. *64*, 988-1003.

Busson, G., 1970: Le Mésozoïque saharien. 2ème partie: Essai de synthèse des données des sondages algéro-tunisiens. Centre Rech. Zones arides Pub. (C.N.R.S., Paris), Sér. géol., *11*, 2 vol., 340 p., 811 p.

Busson, G., 1982: Le Trias comme période salifère. Geol. Rundschau *71*, 857-880.

Choubert, G., and Faure-Muret, A., 1962: Evolution du domaine atlasique marocain depuis les temps paléozoiques. *In*: Livre mémoire P. Fallot. Soc. géol. France Mém. (hors sér.) *1*, 447-527.

Dercourt, J., et al. 1985: Présentation de 9 cartes paléogéographiques au 1:20,000,000 s'étendant de l'Atlantique au Pamir pour la période du Lias à l'Actuel. Soc. géol. France Bull. (8), *1*, 637-652, atlas 10 pl.

Dubar, G., 1938: Sur la formation de rides à l'Aalénien et au Bajocien dans le Haut Atlas de Midelt. Acad. Sci. Paris C. R. (D) *206*, 525-527.

Dubar, G., 1962: Note sur la paléogéographie du Lias marocain (domaine atlasique). *In*: Livre mémoire P. Fallot. Soc. géol. France Mém. (hors sér.) *1*, 529-544.

du Dresnay, R., 1971: Extension et développement des phénomènes récifaux jurassique dans le domaine atlasique marocain, particulièrement au Lias moyen. Soc. géol. France Bull. (7), *13*, 46-56.

du Dresnay, R., 1979: Sédiments jurassiques du domaine des chaînes atlasiques du Maroc. *In*: Symposium ''Sédimentation jurassique W-Européen.'' *A.S.F.* Publ. spéc. *1*, 345-365.

Froitzheim, N., 1984: Oberkretazische Vertikaltektonik im Hohen Atlas SW' von Marrakesch/Marokko—Rekonstruktion eines Bewegungsablaufes im Frühstadium der Atlas-Orogenese. N. Jb. Geol. Paläont. Mh. Jhrg. *1984*, 463-471.

Jenny, J., 1984: Dynamique de la phase tectonique synsédimentaire du Jurassique moyen dans le Haut Atlas central (Maroc). Eclogae geol. Helvetiae *77*, 143-152.

Jenny, J., and Le Marrec, A., 1980: Mise en évidence d'une nappe à la limite méridionale du domaine hercynien dans la boutonnière d'Ait-Tamlil (Haut Atlas central, Maroc). Eclogae geol. Helvetiae *73*, 681-696.

Jenny, J., Le Marrec, A., and Monbaron, M., 1981: Les couches rouges du Jurassique moyen du Haut Atlas central (Maroc): Corrélations lithostratigraphiques, éléments de datation et cadre tectono-sédimentaire. Soc. géol. France Bull. (7), *23*, 627-639.

Laville, E., 1981: Rôle des décrochements dans le mécanisme de formation des bassins d'effondrement du Haut Atlas marocain au cours des temps triasique et liasique. Soc. géol. France Bull. (7), *23*, 303-312.

Laville, E., and Harmand, C., 1982: Evolution magmatique et tectonique du bassin intracontinental mésozoïque du Haut-Atlas (Maroc): un modèle de mise en place synsédimentaire de massifs ''anorogéniques'' liés à des décrochements. Soc. géol. France Bull. (7), *24*, 213-227.

Le Marrec, A., and Jenny, J., 1980: L'accident de Demnat, comportement synsédimentaire et tectonique d'un décrochement transversal du Haut-Atlas central (Maroc). Soc. géol. France Bull. (7), *22*, 421-427.

Makris, J., Demnati, A., and Klussmann, J., 1985: Deep seismic soundings in Morocco and a crust and upper mantle model deduced from seismic and gravity data. Ann. Geophysicae *3*, 369-380.

Manspeizer, W., 1982: Triassic-Liassic basins and climate of the Atlantic passive margins. Geol. Rundschau *71*, 895-917.

Mattauer, M., Proust, F., and Tapponnier, P., 1972: Major strike-slip-fault of late Hercynian age in Morocco. Nature *237*, 160-162.

Mattauer, M., Tapponnier, P., and Proust, F., 1977: Sur les mécanismes de formation des chaînes intracontinentales. L'exemple des chaînes atlasiques du Maroc. Soc. géol. France Bull. (7), *19*, 521-526.

Michard, A., 1976: Eléments de géologie marocaine. Serv. géol. Maroc Notes Mém. *252*, 422 p.

Monbaron, M., 1980: Le magmatisme basique de la région de Tagalft, dans son contexte géologique régional (Haut-Atlas central, Maroc). Acad. Sci. Paris C. R. (D), *290*, 1337-1340.

Monbaron, M., 1981: Sédimentation, tectonique synsédimentaire et magmatisme basique: L'évolution paléogéographique et structurale de l'Atlas de Beni Mellal (Maroc) au cours du Mésozoïque; ses incidences sur la tectonique tertiaire. Eclogae geol. Helvetiae *74*, 625-638.

Monbaron, M., 1982: Un relief anté-Bathonien enfoui sur la ride de Jebel La'bbadine (Haut Atlas central, Maroc); conséquences pour la chronologie de l'orogenèse atlasique. Ver. schweiz. Petroleum-Geol. Ing. Bull. *48*, 9-25.

Ovtracht, A., 1980: Tectonique salifère de l'Atlas (Maroc) (abstr.). 26th Congr. Géol. Internat., Paris 1980, Résumés, *1*, 370.

Petit, J. P., 1976: La zone de décrochement du Tizi n'Test (Maroc) et son fonctionnement depuis le Carbonifère. Thèse 3e Cycle, Montpellier.

Proust, F., 1973: Etude stratigraphique, pétrographique et structurale du bloc oriental du Massif ancien du Haut Atlas (Maroc). Serv. géol. Maroc Notes Mém. *254*, 15-53.

Proust, F., Petit, J. P., and Tapponnier, P., 1977: L'accident du Tizi n'Test et le rôle des décrochements dans la tectonique du Haut Atlas occidental (Maroc). Soc. géol. France Bull. (7), *19*, 541-551.

Proust, F., and Tapponnier, P., 1973: L'accident du Tizi n'Test (Haut Atlas, Maroc): Décrochement tardi-hercynien repris par la tectonique alpine (abstr.). Réun. Ann. Sci. Terre, Paris, Progr. Rés., 350.

Salvan, H.-M., 1974: Les séries salifères du Trias marocain. Caractères généraux et possibilités d'interprétation. Soc. géol. France Bull. (7), *16*, 724-731.

Schaer, J.-P., 1967: Interférence entre les structures du socle et celles de la couverture dans le Haut-Atlas marocain. Etages tectoniques (Colloque de Neuchâtel, 1966), 297-306.

Schaer, J.-P., and Persoz, F., 1976: Aspects structuraux et pétrographiques du Haut Atlas calcaire de Midelt (Maroc). Soc. géol. France Bull. (7), *18*, 1239-1250.

Stets, J., and Wurster, P., 1981: Zur Strukturgeschichte des Hohen Atlas in Marokko. Geol. Rundschau *70*, 801-841.

Studer, M., 1980: Tectonique et pétrographie des roches sédimentaires, éruptives et métamorphiques de la région de Tounfite-Tirrhist (Haut Atlas central, Maroc). Thèse, Univ. Neuchâtel.

Studer, M., and du Dresnay, R., 1980: Déformations synsédimentaires en compression pendant le Lias supérieur et le Dogger, au Tizi n'Irhil (Haut-Atlas central de Midelt, Maroc). Soc. géol. France Bull. (7), *22*, 391-397.

Sutter, J. F., and Smith, T. E., 1979: $^{40}Ar/^{39}Ar$ ages of diabase intrusions from Newark trend basins in Connecticut and Maryland: Initiation of central Atlantic rifting. Am. Jour. Sci. *279*, 808-831.

Tisserant, D., Thuizat, R., and Agard, J., 1976: Données géochronologiques sur le complexe de roches alcalines du Tamazeght (Haut-Atlas de Midelt), Maroc. Bur. Rech. Géol. Min. (Sect. 2), Bull. *3*, 279-283.

Tixeront, M., 1973: Lithostratigraphie et minéralisations cuprifères et uranifères stratiformes syngénétiques et familières des formations détritiques permo-triasiques du Couloir d'Argana, Haut Atlas occidental (Maroc). Serv. géol. Maroc Notes Mém. *249*, 147-177.

Van Houten, F., and Brown, R. H., 1977: Latest Paleozoic-early Mesozoic paleography, northwestern Africa. Jour. Geology *85*, 143-156.

Wildi, W., 1983: La chaîne tello-rifaine (Algérie, Maroc, Tunisie): Structure, stratigraphie et évolution du Trias au Miocène. Rév. Géol. dyn. Géogr. phys. *24*, 201-297.

Ziegler, P. A., 1982a: Geological atlas of Western and Central Europe. Shell Internat. Petroleum (The Hague), 2 vol., 130 p., 40 pl.

Ziegler, P. A., 1982b: Triassic rifts and facies patterns in Western and Central Europe. Geol. Rundschau *71*, 747-772.

Chapter 8

THE PAN-AFRICAN BELT OF WEST AFRICA FROM THE SAHARA DESERT TO THE GULF OF BENIN

R. CABY

Centre Géologique et Géophysique, Montpellier

Considered only a few years ago as a simple thermal rejuvenation, the Pan-African event appears today as a major orogenic cycle responsible for the cratonization of large parts of the Afro-Brazilian supercontinent.

The Pan-African belt, trending north-south and measuring more than 1,000 km wide, runs from the Sahara to the Gulf of Benin (Figure 8-1). To the west it is bounded by the West African craton, a stable block since 2,000 Ma. The belt is particularly well exposed within the Hoggar or Touareg shield in the mountain masses of Hoggar in Algeria, Adrar des Iforas in Mali, and Aïr in Niger. South of the Sahara the belt covers almost the whole of Benin, Togo, and Nigeria.

Since 1965, following the work of Lelubre (1952), detailed studies of key sectors have been undertaken in Hoggar. These have been supported by mapping on 1/50,000 air photographs; this mapping is not integrated into the 1/1,000,000 Hoggar geological map (Bertrand, Caby, et al. 1977). The groundwork done by Gravelle (1969), Caby (1970, 1983a), Boissonnas (1973), Bertrand (1974), Latouche (1978), and Vitel (1979), together with the unpublished work of the geologists of SONAREM, Algeria, has provided a basis for various syntheses of the area, such as those of Bertrand and Caby (1978), Bertrand et al. (1978), and Caby et al. (1981). A synthesis of 5 years of research and more than 65 man-months of work by a multidisciplinary team in the Adrar des Iforas, Mali, is presented in a 1/500,000 map and accompanying explanatory text (Fabre 1982b, Fabre et al. 1982).

The synthesizing articles of Trompette (1980) and Affaton et al. (1980) summarize modern ideas about the part of the belt running through Togo and Benin, and there is unpublished work by Affaton, Simpara, and the mining services of the two countries.

Thanks to geochronology, that indispensable tool of geological work in crystalline terrains, the Pan-African age of the high-grade terrains of the Dahomeyan belt of Togo-Benin has been recognized since 1960 (Bonhomme 1962, Black 1967), and also, little by little, areas of pre–Pan-African basement older than 2,000 Ma have been delineated within the Hoggar gneisses (Ferrara and Gravelle 1966, Allègre and Caby 1972, Bertrand and Lasserre 1976).

In the light of my own experience, obtained principally in the Saharan countries—regions of exceptionally good exposure, where rapid observation is possible—I here present a

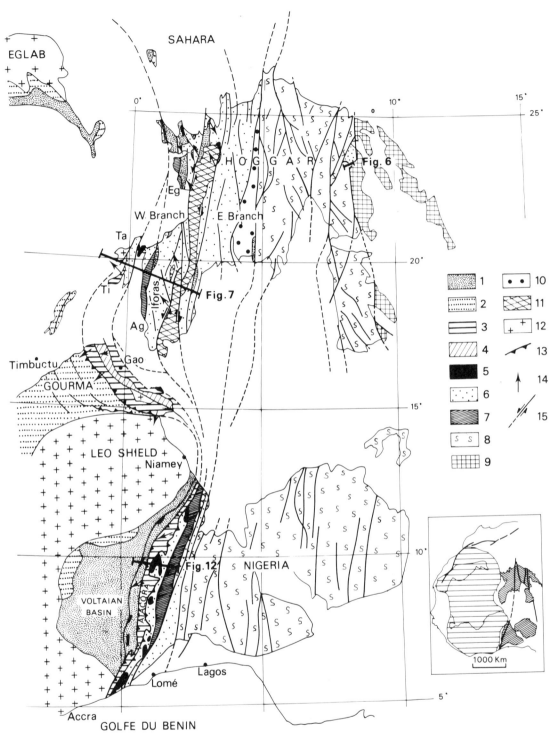

FIGURE 8-1. Summary map of the Pan-African belt of West Africa. Post-tectonic intrusive rocks have been omitted. 1: Molassic deposits. 2: Upper Proterozoic sediments deposited on the passive margin of the West African craton. 3: Atakora and Gourma metamorphic nappes (mainly quartzites and phyllites). 4: The same, with high pressure–low temperature metamorphic assemblages (in Gourma). 5: Main mafic-ultramafic massifs along the Pan-African suture. 6: Undifferentiated monocyclic

lithostratigraphic and structural summary of the Pan-African belt from the Sahara to the Gulf of Benin.

As is particularly clear in the Hoggar, and visible in LANDSAT photographs, the belt is dominated by major shear zones that trend north-south, dividing the belt into a number of major longitudinal units between which correlation is not always possible (Figure 8-1). A cryptic suture zone (Black et al. 1979, 1980) marks the abrupt boundary between the external parts of the belt, floored by the West African craton to the west, and the mobile zone of the Pan-African belt to the east. The former shows the characteristics of a passive continental margin; the latter, beginning in the upper Proterozoic, shows the style of tectonic evolution characteristic of an active continental margin.

I   MAJOR LITHOSTRATIGRAPHIC UNITS AND ASSOCIATED PRE-OROGENIC MAGMATISM

A   THE PRE–PAN-AFRICAN BASEMENT

An ancient basement giving ages greater than 2,000 Ma (Eburnean) has been identified in almost all areas.

A1 *Reactivated Basement*   Reactivated basement occurs as large orthogneiss masses enclosed within monocyclic gneisses (i.e., those that have been through only one orogenic cycle) of east-central Hoggar (Arechchoum Series, dated at 2,150 ± 100 Ma by Rb/Sr whole-rock isochron; see Bertrand and Lasserre 1976, Vialette and Vitel 1979). The early banding in these basement gneisses is cut by metamorphosed basic intrusives in deeper amphibolite facies (Bertrand 1974) to hornblende granulite facies (Sautter 1982).

The "polycyclic" gneisses of the "Kidalian assemblage" (as defined in Adrar des Iforas: see Boullier et al. 1978; Boullier 1982), in part granulites retrogressed to amphibolite-facies gneisses, are cut by numerous intrusive suites which themselves now show gneissic textures. The Ibadan gneisses in Nigeria had been interpreted as resulting from a Kibaran metamorphism at around 1,100 Ma (Grant et al. 1972). They have, however, given U/Pb zircon ages around 2,100 Ma (upper intercept on the concordia curve) and clearly also belong to the Eburnean basement, reworked during the Pan-African (Rahaman and Lancelot, ms.). Although critical U/Pb dating of zircons is lacking in east-central Hoggar, it is possible that similar conclusions apply to the metasediments of the Egéré and Aleksod Series, which,

---

metasediments and gneisses affected by Pan-African deformation. 7: The same, with high temperature–low pressure metamorphism. 8: Undifferentiated gneisses of central Hoggar and Nigeria (both monocyclic and polycyclic). 9: Undifferentiated rocks of the Djanet-Tafassasset domain (eastern Hoggar), stabilized before 725 Ma. 10: Undifferentiated rocks of the Silet-Abankor domain (central Hoggar), partly stabilized at about 840 Ma. 11: Pre-Pan-African granulite basement 2,000 Ma old. 12: West African craton 2,000 Ma old. 13: Main frontal overthrusting of the Gourma and Atakora nappes. 14: Main direction of movement. 15: Vertical shear zones and strike-slip faults. Eg: Egatalis. Ta: Taounant. Ti: Timetrine. Ag: Aguelhoc. Inset: Northwest Africa with location of the Hoggar (Touareg) and Nigerian shields (diagonal lines) and West African craton (horizontal lines).

on the strength of a whole-rock Rb/Sr isochron at 1,200 Ma and K/Ar dating on hornblendes in the range of 1,300 to 900 Ma, Bertrand and Lasserre (1976) claimed belonged to the Kibaran belt.

The Tamanrasset and Arefsa granulites (central Hoggar; see Ouzegane 1981) are retrogressively overprinted to varying degrees. They also had been assigned to the pre–Pan-African basement, and recent U/Pb dating on associated synkinematic granitoids has yielded an age of 2,075 ± 20 Ma, interpreted as that of the granulite-facies metamorphism (Bertrand, Michard, et al. 1984b). The Gour Oumelalen gneisses have been assigned to the pre–Pan-African basement (Latouche and Vidal 1974, Latouche 1978) and have given Rb/Sr whole-rock ages of 2,100 Ma.

In section II A we consider why the concept of a low-grade Pan-African overprint in many areas of east-central Hoggar (Bertrand 1974, Bertrand and Caby 1978) must be replaced by that of a high-grade main metamorphism of deeper amphibolite to granulite facies and Pan-African age. Eclogites formed in excess of 15 Kb seem to outline major low-angle shear zones (Sautter 1985).

In the Benin-Nigeria segment, the Pan-African age of the high-grade metamorphism is supported by the Rb/Sr method (Caen-Vachette 1979). The charnockites from southwestern Nigeria have also given U/Pb zircon ages between 620 and 580 Ma (Tubosun et al. 1984). A three-fold division of gneisses arises from the petrological and geochronological data. Potassium-poor gray gneisses of Ife-Ibadan type have yielded Rb/Sr whole-rock Archaen ages (Grant 1970), whereas granitic orthogneisses of Ibadan type have given Eburnean ages by the U/Pb method (Rahaman and Lancelot, ms.). On the other hand, the simpler structural evolution of most of the high-grade paragneisses with associated marbles and metaquartzite from Benin suggests that they underwent a single stage of metamorphic evolution during the Pan-African s.s. (Grant 1978, Fitches et al. 1985, Caby unpublished), during which high pressure–high temperature conditions were followed by isothermal decompression and anatexis (see section II B 5); they are thus thought to be lower to middle Proterozoic in age and possible time equivalents with the sequence of West Hoggar-Iforas (see section II B 3).

A2 *Partially Reactivated Basement*   Partially reactivated basement is seen in two units trending north-south: the In Ouzzal Granulite Unit (UGO) in western Hoggar and the Iforas Granulite Unit (UGI) in Adrar des Iforas. Both units are bounded to west and east by subvertical mylonite zones separating them from the neighboring Pan-African metamorphic rocks. Both are entirely composed of banded, mainly acid granulites with frequent metasedimentary intercalations (acid granulites, kinzigites, quartzites, marbles, and so on) and several suites of pre- or synmetamorphic intrusive rocks, including sub-alkaline leptynites, syenites, and abundant variegated charnockites and norites.

In western Hoggar Rb/Sr whole-rock ages of around 3,000 Ma have been obtained from the UGO (Ferrara and Gravelle 1966, Allègre and Caby 1972). This unit was almost totally unaffected by the Pan-African tectonism and metamorphism. Its uppermost Proterozoic andesitic cover in zeolite facies is preserved in a single half-graben structure (Caby 1970, 1983a; Chikhaoui et al. 1978). In Adrar des Iforas the UGI is almost everywhere cut by sinuous and generally subvertical mylonite zones and is affected by a penetrative Pan-African metamorphic imprint ranging from deeper greenschist to shallow amphibolite facies conditions (Boullier 1982) and cut by numerous Pan-African intrusives.

The age of the granulite-facies metamorphism of the UGO has been determined by U/Pb dating of zircons as 2,115 Ma (upper concordia intercept; see Lancelot et al. 1976). The structural relations of the UGO and UGI will be examined in section II B 1.

In the extreme northwest of Hoggar, the Tassendjanet is formed of an essentially granitic basement nappe capped by Proterozoic cover rocks (see section I C and Figure 8-5). This granite has, by a number of methods, given the age of 2,150 Ma.

B Middle Proterozoic Thick Basinal Sequences and Anorogenic Magmatism

In this section I will discuss the earliest sediments deposited after the Eburnean event. They form the base of the pile of metasediments that, together with the overlying upper Upper Proterozoic rocks, have been through only one tectonometamorphic event (Figure 8-2). They are probably of great lateral extent but have only been described in detail from western Hoggar and Adrar des Iforas (Caby and Andreopoulos-Renaud 1983).

They are only weakly metamorphosed in Ahnet (northwestern Hoggar), where they are composed of more than 3,000 m of deltaic and marine quartzites and pelitic schists conformably overlain by carbonates (see section I C) (Arène 1968). Elsewhere, they are more strongly metamorphosed and are now aluminous metaquartzites (containing kyanite and/or sillimanite) with associated, frequently peraluminous schists.

After the deposition of these thick basinal sequences, beneath which the basement is nowhere recognized, there was an important magmatic event during which alkaline rhyolites, numerous banded alkaline intrusives (now gneissic), and some hyperalkaline and syenitic rocks were emplaced. The petrological, mineralogical, and geochemical characteristics of these rocks (Dostal et al. 1979) are typical of areas of intracontinental anorogenic magmatism related to early rifting. Despite the kyanite-grade metamorphism, these intrusives have given U/Pb zircon ages (upper concordia intercept) of $1,755 \pm 10$ Ma from Hoggar and $1,827 \pm 5$ Ma from Adrar des Iforas (Caby and Andreopoulos-Renaud 1983).

In eastern Hoggar the Tazat quartzites (Blaise 1967, Bertrand et al. 1978), likewise cut by alkaline gneisses, are probably part of the same group of rocks, although their structural relations with those adjoining (Gour Oumelalen granulites, Latouche 1978; Issalane unit, Bertrand et al. 1978) are unknown. The quartzites of the internal nappes in Gourma (Caby 1979) and the Atakorian quartzites of Togo and Benin (Affaton et al. 1980) perhaps also belong to the same group (see section II B 5).

C Platform Sequences and the Stromatolite Series: Problems of the Middle Proterozoic–Upper Proterozoic Boundary

A platform sequence showing uniform sedimentary facies over large areas overlies the granulitic and granitic basement (section I A 2) in a number of regions (Figure 8-2).

The Stromatolite Series in northwest Hoggar (the Tassendjanet region; see Caby 1970,

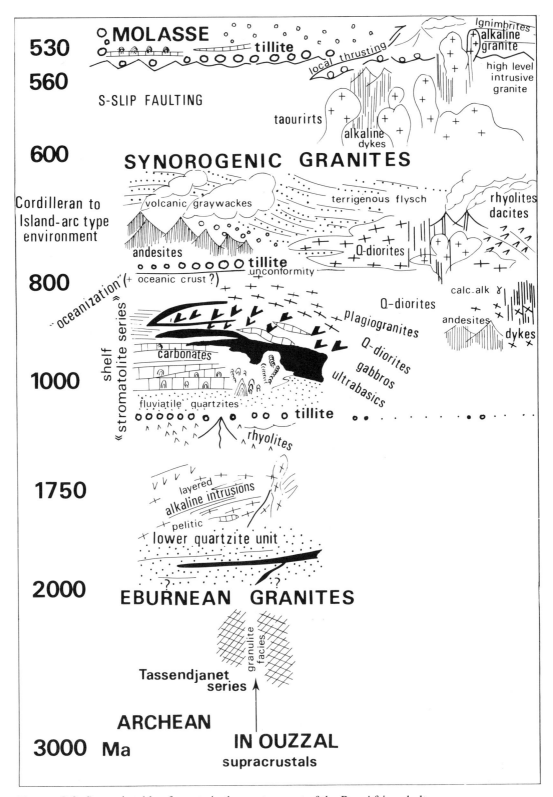

FIGURE 8-2. Synoptic table of events in the western part of the Pan-African belt.

1983a) is only very weakly metamorphosed. The sequence begins with a well-bedded basal sandstone unit 50 to 100 m thick that includes rounded quartz gravel ("grains de riz") near its base. This unit overlies both the Tassendjanet basement and, perhaps also discordantly, rare remnants of a red, molasse-like unit which itself directly overlies the reddish-weathered granitic basement.

North of Tassendjanet two isolated outliers show a progressive transition upward from the basal unit (interpreted as a beach sand) into uniform quartzites (Adrar Ougueda unit, 600 m) with intraformational conglomerate lenses, the pebbles being only quartzite. Aluminous gray pelitic schists similar to those in the Middle Proterozoic are intercalated, and potassic rhyolites are also present near the base.

There follows a calcareous-dolomitic formation that is more than 4,000 m thick at Tassendjanet and is characterized by beds of uniform and regular thickness over large distances. It begins with siliceous dolomites and dolomitic sandstones. A transitional boundary between these strata and the sandstones below suggests continuous sedimentation. Components include limestones (in part originally algal-laminated), stromatolitic limestones (with a basal horizon of polychromatic marbles with *Conophyton*), and alternating dolomites and chlorite-illite green and mauve shales with some quartzites toward the top.

The *Conophyton* marble band enclosed in black or purple pelites forms a good marker horizon, and the whole series with its basal "grains de riz" sandstone has been recognized in a number of areas of Hoggar and Adrar des Iforas, even within the more strongly metamorphosed zones.

Dating of the series is equivocal. Detailed studies in northwest Hoggar indicate a Vendian age for the stromatolitic association (Bertrand-Sarfati 1969, 1972). On this basis one can propose that the series is coeval with the Atar Group, which rests undisturbed and horizontal on the West African craton and which has given fine-fraction clay Rb/Sr ages ranging from 1,000 to 750 Ma. These ages have been interpreted as those of early diagenesis (Clauer 1976, Clauer et al. 1982). Microfossils that have been extracted from black shales near the base of the carbonates of the Stromatolite Series in northern Mali, however, include forms specific to the Middle Proterozoic (Amard 1983).

Quartz-diorite plutons that cut the equivalents of the Stromatolite Series in the region of Silet (central Hoggar) have given U/Pb zircon ages of 868 + 8/ − 5 Ma (Caby et al. 1982). At Tassendjanet pre-tectonic gabbro sills within the Stromatolite Series have given dates of 793 ± 32 Ma (Clauer 1976, Clauer et al. 1982). In Ahnet the carbonates of the Stromatolite Series are detached structurally and overlie stratigraphically the Ahnet deltaic quartzites (Arène 1968), which have been assigned to the Middle Proterozoic.

Only the Middle Proterozoic thick basinal quartzitic sequences are well defined; the nature of the basement beneath is not known. Alkaline magmatic rocks dated between 1,850 and 1,750 Ma cut these sequences, whose rocks therefore relate to the Eburnean (ca. 2,000 Ma) in the same way that the Lower Paleozoic sandstones of Tassili relate to the Pan-African. They are the first detrital deposits over the peneplaned surface formed subsequent to Eburnean orogeny.

As yet undated, the Stromatolite Series represents platform deposits directly overlying the Eburnean basement. Its basal quartzite unit is perhaps the reduced equivalent of the thick quartzitic sequences. The suggestion of an Upper Proterozoic age (1,000 Ma) and an equivalence with the Atar Group still requires confirmation. It is possible that the whole unit represents the lateral equivalents of the Middle Proterozoic quartzites, as Caby (1970, 1983a) proposed in the absence of any geochronological data.

D   Mafic and Ultramafic Magmatism: False Ophiolites and Thinning and Basification of the Continental Crust

There are a number of small metamafic and meta-ultramafic masses in west-central Hoggar, Adrar des Iforas, and Togo-Benin. In the high-grade metamorphic terrains they form layers and lenses, concordant with the banding of the surrounding gneisses and amphibolites, or sometimes isolated boudins of all sizes.

D1 *The Mafic Masses Marking the Suture*   The suture in North Mali is marked by a line of metamafic bodies (see section II A). To the south of Gao these are granulitic metagabbros; the "type" body of Amalaoulaou is described below.

This body was studied in detail by de la Boisse (1979). It is formed of ancient gabbroic cumulates of tholeiitic chemistry deformed and recrystallized under high-pressure granulite facies conditions (rutile-garnet-amphiclasites and pyriclasites). Its apparent age (800 ± 50 Ma, U/Pb on zircon; see de la Boisse 1979) has been interpreted as that of the original magmatic crystallization, but it is not clear whether the granulitic imprint took place during the original cooling or whether it records a lower crustal eo–Pan-African tectonometamorphic event. We consider the body as unrooted and detached, its lower contact dipping about 40° east above the metaquartzites of the internal nappes. Along the contact is a sole of complexly and intensely deformed metamafic rocks, including rocks with eclogitic assemblages above, amphibolites with blue-green hornblende or actinolite in the middle, and chlorite schists at the base. Within the last are banded hematitic jaspers of metasomatic origin, which contain magnetite, altered chromite, and a blue amphibole.

Within the chain in Togo and Benin, Menot (1980) distinguished a number of different types of metamafic bodies. The Kabré massif, which we have visited, has similarities with that of Amalaoulaou. It appears possible to delineate here, too, a granulite paragenesis (though the orthopyroxene could be primary magmatic as well) overprinted by kyanite-eclogite assemblages for which the pressures of formation would be around 15 to 18 Kb (by comparison with data in the literature; see Menot 1982). The base of the body dips gently 15 to 30° to the east and is marked by garnet-amphibolites separating it from the orthogneisses below. We believe these detached masses to be metacumulates emplaced in the base of the crust (perhaps of arc-type?).

D2 *Intrusions into a Continental Environment*   All the other mafic and ultramafic masses that we have studied, in both Hoggar and Adrar des Iforas, seem to us intrusions emplaced in areas that at the time had continental basement. This is particularly evident in those cases where the rocks are enclosed, for example, within limestones, and around which contact-metamorphic assemblages (with forsterite, grossular, diopside, and so forth) have not been totally destroyed during the Pan-African greenschist-facies metamorphism.

In the Tassendjanet zone (Caby 1970, 1983a) and the Silet zone (Gravelle 1969), in northwestern and central Hoggar, respectively, detailed mapping has demonstrated without ambiguity the intrusive nature of the serpentinites there. They generally appear as sills, but in areas of exceptionally good exposure one can see truncations of the country-rock strata. Where the Pan-African schistosity is only weakly developed, notably in the center of the thicker sills, one can pick out a magmatic layering. Where serpentinization is incomplete, one can recognize dunites, wehrlites, pyroxenites, and ampholites with a colorless high-temperature magnesian amphibole, all speckled with scattered, often chrome-poor spinel.

In northwestern Hoggar the Adrar Ougueda complex (ca. 700 km$_2$) includes a wide range of rock types from peridotites and tholeiitic gabbros to calc-alkaline rocks (Caby et al. 1980). The earliest generation includes granulitic metagabbros as veins and seams within the cumulate peridotites.

In the Silet zone, the surrounding rocks are quartzites, pelitic schists, marbles, and dolomites, within which are mixed basalt breccias and beds of basaltic pillow lavas. In Tassendjanet the surrounding rocks are, in contrast, siltstones, sandy quartzites, limestones, and stromatolitic dolomites of the typical platform facies.

There are also larger linear bodies of ultramafics along vertical fractures. Their tectonic contacts are the result of repeated diapirism, suggesting that they are unrooted bodies and as such do not mark suture lines (see Caby 1983b for an interpretation of the ultramafic rocks of the Arabian shield). One can furthermore recognize meta-harzburgites and chrome-rich meta-dunites in some bodies. The Timetrine mass described by Caby et al. (1981) (see Figure 8-3) belongs in this category.

Table 8-1 summarizes the characteristics of the diverse types of intrusions that are recognized. Whatever their geochemical affinities, which are clearly tholeiitic and on which geochemical work is still continuing, their field relations imply that they cannot be part of an allochthonous ophiolite suite s.s., that is, part of an obducted slab of Precambrian oceanic crust. In the whole Pan-African terrain of northwest Africa, only the Bou Azzer massif (Leblanc 1981) shows features similar to those of true ophiolites, though even here an emplacement by obduction has not yet been convincingly demonstrated.

One of the characteristics of these mantle-derived intrusions is the common occurrence, along the margins of the ultramafics, of ferruginous cherts (birbirites) and complex and brecciated iron-magnesium carbonates (listwänites) characteristically containing chrome-bearing minerals (mainly fuchsite) formed from the breakdown of chromite. We believe these rock types, well documented in the Soviet literature, to be metasomatic rocks resulting from complex hydrothermal processes associated with the emplacement and early serpentinization of the original ''mantle'' magma during its ascent into the higher levels of the hydrated continental crust.

E   VOLCANOCLASTIC SEQUENCES AND UPPER PROTEROZOIC CALC-ALKALINE MAGMATISM (800–600 MA)

In the mobile zone, the Upper Proterozoic (younger than 850 Ma) and the upper Upper Proterozoic (younger than 700 Ma) are of volcanoclastic facies; they result from early sedimentary recycling of penecontemporaneous calc-alkaline volcanics and intrusive rocks. This sequence and evolution are common to regions with newly formed eo–Pan-African crust, those

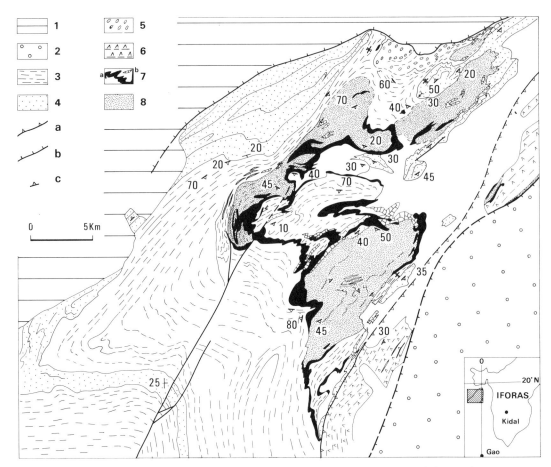

FIGURE 8-3. Simplified geological map of the Timetrine area (presumably allochthonous). 1: Cretaceous cover. 2: Post-Permian pre-Cretaceous red beds. 3: Sericite-chlorite schists. 4: Sericite quartzites. 5: Pillowed metabasalt and related diabase, apparently conformable on 3. 6: Diabase and metabasalt (with blue amphibole). 7a: Red hematite jaspers. 7b: Ca-Fe-Mg carbonates. 8: Ultramafic rocks. a: Thrust. b: High-angle fault. c: Strike and dip of cleavage.

with ancient crust, and, on the margin of the suture zone, a region with unique island-arc characteristics preserved to the northwest of Adrar des Iforas.

E1 *Eastern Hoggar*   The Tiririne Series (Bertrand et al. 1978) is composed of immature detrital deposits of arkose and graywacke type, often of greenish color. The petrographic nature of the rock components and the paleocurrent directions show that the sediments were derived almost entirely from the Djanet-Tafassasset domain to the east, which was uplifted and eroded just after its stabilization at around 730 Ma (Caby and Andreopoulos-Renaud, ms.) (see section II A 3, Figure 8-1, no. 9, and Figure 8-6).

This series, with a thickness up to 6 km along the 8°30′ shear zone, was intruded by a variety of hypabyssal and plutonic rocks (rhyolites, dacites, andesites, gabbros, diorites, granodiorites), generally in the form of sills. This pre-deformation magmatism has been dated at 660 Ma by the U/Pb method on zircons. The metamorphism localized along the 8°30′ shear

Table 8.1
Petrographic characteristics and mode of intrusion of mafic
and ultramafic complexes

Form of complex	Rock types	Mode of intrusion	Types of alteration or metasomatism
Asthenoliths	Metamorphic harzburgite Dunite and lherzolite Clinopyroxenite Foliated cumulates and dikes	Mantle diapir emplaced along hypabyssal strike-slip faults, with vertical foliation and vertical magmatic or late-magmatic lineation	Cold metasomatism: birbirite, listwänite Progressive cooling during diapirism Frequent veins and bodies of granulitized metagabbros, derived from the lower crust during initial cooling of gabbroic liquids
Lopoliths and laccoliths	Plagiogranite Granodiorite Dolerite Microgabbro Quartz-gabbros Layered gabbros Amphibolite	Layered intrusions with strong differentiation Multiple intrusions at any level of the continental crust Basification	Birbirite and listwänite around peridotites and amphibolites Cr-rich spinel
Sills	Troctolite Wehrlite Clinopyroxenite Dunite	100% cumulates	Contact metamorphism significant
Sills	Microgabbro Dolerite		Cryptic contact metamorphism ($<$ dm)

Basalts and dacites overlie a shelf-type cover
~ ~
Basement

zone and the associated synorogenic granites have, in contrast, been dated between 610 and 595 Ma by the same method (Bertrand et al. 1978).

E2 *Arefsa* The upper Upper Proterozoic forms a narrow belt of detrital volcanoclastic and volcanic material. The stratigraphic and structural relations between this and the high-grade gneisses of Aleksod are unknown, but the occurrence of staurolite strongly suggests a gradual transition from the schists into the higher-grade surrounding gneisses; hence, the synformal structure of the belt.

E3 *Eastern Branch of the Belt in Central Hoggar and Eastern Adrar des Iforas*   This branch is composed mostly, at least in its northern parts, of Upper Proterozoic rocks younger than 840 Ma, the date of the stabilization of the substrate in this area.

In the eastern part, where two small areas of retrogressed granulitic basement assigned to the Eburnean are also known, the upper Upper Proterozoic overlies deeply eroded and altered calc-alkaline and subalkaline batholiths dated, respectively, at 880 and 839 Ma. Detrital units roughly contemporaneous with lenticular dolomites that show features suggesting a marine origin interfinger laterally with volcanoclastics, rhyolites, and andesite-dacites. They are overlain by a terrigenous pelite formation, above which rhyo-dacitic volcanism becomes predominant.

In the eastern part and northern areas, the volcanoclastic facies (graywackes) is similar to that of the ''Série verte'' of northwestern Hoggar (see section I E 4). It is mixed with terrigenous sediments and intruded by a large number of diabase sills, themselves cut by vast, as yet undated, pre-tectonic calc-alkaline batholiths (quartz-diorites, microdiorites, granodiorites, plagiogranites, and so on). Possible unconformities exist in this unit, which is many thousand meters thick. As in the ''Série verte,'' one notes the presence of intraformational conglomerates with clasts of graywacke, epidotitic hornfels, and calc-alkaline plutonic rocks similar to those that cut the graywackes themselves.

The southwestern part of this branch is, for the most part, covered by a uniform terrigenous sequence, Gravelle's ''sandy-pelite series'' (1969), which that author also assigns to the base of the upper Upper Proterozoic.

In eastern Adrar des Iforas a number of series have been distinguished by Davison (1980) and Boullier (1982). We assign to the upper Upper Proterozoic a thick detrital and volcanoclastic series with associated pelites, at the base of which are conglomerates with volcanic and plutonic clasts resting directly on Middle Proterozoic quartzites and alkaline orthogneisses.

E4 *Western Branch of the Belt*   The ''Série verte'' of northwest Hoggar (Caby 1970, 1983a) is a thick and uniform flysch-like sequence of green volcanic graywackes. In the north it passes laterally into a sub-aerial volcanic complex composed of basic andesites, normal andesites, dacites, and, near its top, rhyolites, in all almost 6,000 m thick. The geochemical characteristics of these rocks are typical of calc-alkaline volcanics from ''orogenic zones,'' and in particular of active continental margins (Chikhaoui et al. 1978). Rare-earth patterns are consistent with this source and imply that the magmas in general underwent a period of low-pressure fractionation. The base of this sequence is formed of polygenic conglomerates with a basal tillite, within which are included clasts of the mafic and ultramafic rocks that cut the Stromatolite Series (Caby and Fabre 1981a).

In the In Zize synclinorium this unit rests directly, with probable angular unconformity, over the Middle Proterozoic quartzites, where the alkaline magmatism has been dated as 1,750 Ma. Here more than 6,000 m thick, it includes polygenic conglomerate units with essentially plutonic-volcanic calc-alkaline clasts, perhaps in part of fluvio-glacial origin (Caby and Fabre 1981a), but also intraformational conglomerates with volcanic graywacke clasts, suggesting an "auto-cannibalistic" evolution analogous to that seen in parts of the circum-Pacific belts.

E5 *Tafeliant Series*   This series is preserved within a synclinorium in central Adrar des Iforas and rests in places on marbles assigned to the Stromatolite Series, in places on a pre-tectonic quartz-diorite that intrudes the marbles and has been dated at $696 + 8/-5$ Ma (Caby and Andreopoulos-Renaud 1985, U/Pb on zircons, upper intercept) and elsewhere directly on the granulitic basement. The base of this series includes an unstratified continental tillite with large blocks of local origin, littoral calcareous sandstones, and a unit of black pelites with clasts interpreted as a marine tillite (Caby and Fabre 1981a). The rest of the series is formed of volcanic graywackes with abundant andesitic elements, in part mixed with semi-pelitic silt-stones and with arkoses derived from the sialic basement. A bimodal volcanic suite (basalts and rhyodacites) is found within the uppermost levels of this series. The andesite-basalt complex of Oumassene (Chikhaoui et al. 1978) is a probable time-equivalent to the Tafeliant Series.

Possibly at this time there was a progressive transition westward into an essentially calc-alkaline volcanic and plutonic domain that was later affected by an intense deformation associated with north-south, vertical, strike-slip shear zones. This calc-alkaline domain is now bordered on the west by a major north-south shear zone bounding the graben of the early molasse series of Tessalit-Anefis (see section I F 2). This shear zone sharply delimits the ensialic domain to the east from the Tessalit-Tilemsi unit with island-arc affinities to the west.

E6 *The Tessalit-Tilemsi Volcanic and Volcanoclastic Series of Island-Arc Affinities in North-western Adrar des Iforas*   This series delineates the suture zone (Black et al. 1979, 1980) and forms a wide belt around 100 km across with characteristics, both metamorphic and structural, distinct from all other parts of the region (Caby 1981a). It includes (Figure 8-4):

1. A bimodal volcanic series below (metabasaltic pillow lavas, dacites, and younger andesites).

2. A thick sequence of volcanic graywackes above (sedimentary structures testify to deep-water, in part turbiditic sedimentation).

3. A large volume (approximately 70%) of several generations of plutonic rocks, including a) metagabbros and ultramafic cumulates; b) microdiorite and diabase bodies, stocks, and dikes; and c) banded and differentiated gabbro-norite, tonalite, and granodiorite laccoliths and lopoliths.

The emplacement of these mantle-derived rocks was accompanied in the surrounding metasediments by early anatexis. During this event tonalitic gray gneisses and migmatites of low-pressure granulite facies were generated at 720 ± 15 Ma (Andreopoulos-Renaud, U/Pb on zircons, work in progress). The geochemistry of the plutonic rocks agrees with temporal evolution from oceanic tholeiites to calc-alkaline rocks. We interpret these massifs as deep-level magma chambers related to calc-alkaline volcanism. They are responsible for the positive gravity anomalies that enable us to delineate the suture zone (Bayer and Lesquer 1978, Caby et al. 1980). To the east, calc-alkaline batholiths were emplaced before the Pan-African

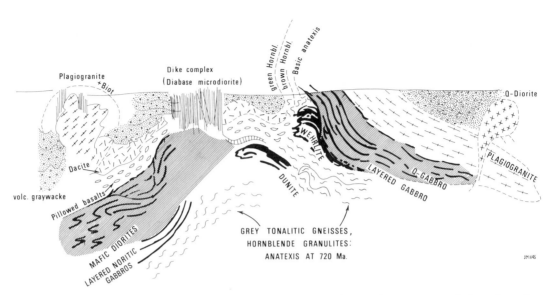

FIGURE 8-4. Diagrammatic section showing the relationships of layered sequences and plutonic rocks in the Tessalit-Tilemsi island arc (North Mali, northwestern Adrar des Iforas).

metamorphic and deformational events, and one batholith here has given an age of 633 Ma (Andreopoulos-Renaud, U/Pb on zircon, work in progress), which is interpreted as the age of magmatic crystallization.

The continual contribution of mantle-derived material in this domain is shown by the presence of syntectonic bodies of mafic diorite and quartz-diorite, in part synchronous with the emplacement of higher-level layered complexes (gabbro-troctolite-anorthosite). The distribution of all the plutonic complexes, particularly the pre-orogenic suite, and a progressive increase in K_2O and SiO_2 eastward suggest a magmatic polarity similar to those seen perpendicular to present-day subduction zones.

We believe therefore that the Tessalit-Tilemsi domain was an island-arc domain devoid of any continental crust, with a magmatic root of gabbroic lopoliths and basic gray gneisses. The arc was active from perhaps 800 to 630 Ma.

F   POST–PAN-AFRICAN MOLASSE SEQUENCES

As yet poorly dated, molasse sequences are, at least in part, Cambrian (''Série Pourprée''; see Caby 1967). They are discordantly overlain in Hoggar by the Lower Ordovician of Tassili. In some places they are preserved in grabens trending north-south and are often bordered by major north-south vertical mylonite zones; elsewhere they are preserved in ''residual'' basins situated immediately east of the suture zone. On the basis of their individual lithology and the nature of their substratum, it seems possible to separate these sequences into early and late molasse.

F1 *Early Molasse Sequences*   These sequences are volcanoclastic and do not contain any clasts of high-grade metamorphic rocks. Thus such rocks cannot yet have been exposed.

In northwestern Hoggar there is the lower formation of In Semmen (Caby and Moussu 1968), about 1000 m of arkoses and graywackes overlain discordantly by the Série Pourprée.

In Adrar des Iforas the Tessalit-Anefis molasse formation can be traced for almost 400 km along the 1° shear zone; it forms a narrow graben. To the north, no discontinuity has been identified at its base; it overlies concordantly more than 6,000 m of Upper Proterozoic terrigenous flysch-type graywackes. Its substratum is rarely exposed in the south and consists of low-grade greenschist-facies rocks of Upper Proterozoic age. Here the margins of the graben are formed by the high-grade gneisses of Aguelhoc, of low-pressure amphibolite to granulite facies (see II B 3), though these are never found as clasts.

The series is composed of fine sandstones and khaki siltstones at the base, then polygenic conglomerates with abundant volcanic clasts passing upward into a glacial facies (Caby and Fabre 1981a), and a volcanic sequence of basalts, andesites, and rhyolites toward the top. The sequence is cut by the later phases of the Iforas batholith and then cut by the hyperalkaline Tessalit granite, which is part of a province of ring complexes dated by Rb/Sr whole-rock isochrons between 560 and 540 Ma (Liegeois and Black 1984; see section II B 4[2]).

F2 *Molasse Sequence along the Adrar Fault*   This sequence forms the filling of a narrow graben more than 300 km long (Figure 8-7, eastern end). Younger than 600 Ma ± 2, it includes rhyolitic ignimbrites, arkosic sandstones, fine arkoses, purple or green shales, and some lenses of polygenic conglomerates (Fabre 1982a; Boullier 1982). It is intensely deformed into north-south trending, often steeply plunging folds with axial-plane flow schistosity (white mica, chlorite, epidote, actinolite). It shows its own unique pre-tectonic magmatic cycle, best displayed in southernmost Algeria, which includes diabases, some gabbros, gabbrodiorites, granodiorites, and some sub-volcanic banded granitoids. The eastern edge of the graben is a subvertical synmetamorphic shear zone, essentially a strike-slip fault (Caby et al. 1985).

There is effectively a metamorphic gradient between the epizonal rocks within the graben and the metapelitic series to the east (Boullier 1982). The domain to the east shows a low-pressure high-temperature metamorphism associated with synkinematic and synmetamorphic intrusives (olivine-bearing gabbro-norites and aluminous granitoids and associated aluminous anatexites), and hence coeval with the folding that affected the molasse graben.

In the west of the area showing the low-pressure metamorphism is the pre-tectonic sub-alkaline granite of Tamassahart. The margins of this granite are, like the surrounding metapelites, penetratively deformed and recrystallized. This granite has given an age of 581 + 7/ − 6 Ma (Caby et al. 1985, U/Pb on zircons, upper intercept), and the low-pressure metamorphism must be even younger. It should be recalled that this massif has given an Rb/Sr whole-rock age of 646 ± 37 Ma (Bertrand and Davison, 1981).

F3 *Pendjari Group of Benin and Togo*   Previously the Oti Series and in part the "Buem" (Grant 1969, 1978), this group is composed of arkosic sandstones and green micaceous shales, and a mixtite associated with banded ironstones, cherts, and phosphatites. This last formation has been correlated with the "triad" of the Taoudenni basin (Affaton et al. 1980).

The whole series is moderately deformed and contrasts markedly with the intensely deformed metamorphic schists with intercalated ultramafics that form the most external allochthonous units of the Dahomeyide chain (see section II B 5). We therefore suggest that this series is an early molasse, covered in part by the thrust units of the Dahomey chain. A Rb/Sr date for the series of 620 Ma (Clauer 1976, Clauer et al. 1982) is of ambiguous significance (Clauer et al. 1982), as the samples may have included undegraded detrital white micas of Pan-African age.

F4 *Late Molasse Sequences*   These formations are represented by the Série Pourprée of the Sahara. As redefined in northwest Hoggar (Caby and Moussu 1968, Caby and Fabre 1981b), this series includes arkosic sandstones and reddish continental pelites, green-khaki rocks rich in illite and locally in glauconite, and volumetrically significant glaciogenic units including tillites, some argillites, and varved argillites (Caby and Fabre 1981b).

The stratigraphic base of the Série Pourprée is formed by a "triad" of facies (tillites, argillites and cherts, limestones) which can be correlated very precisely with the "triad" of the Taoudenni basin (Caby and Fabre 1981b, Clauer et al. 1982). Two synsedimentary volcanic events have been identified: effusive rhyolites and ignimbrites at the base, alkaline basalts and syenites toward the top. The Ouallen and In Semmen grabens show the most complex evolution, with approximately 6,000 m of sediments preserved at many points. Their substratum includes the complete spectrum of structural levels of the chain, especially the Pan-African catazonal rocks. The age of 520 Ma, obtained from sediments at the diagenetic-anchizonal facies boundary (Clauer et al. 1982), clearly represents a younger limit for the age of sedimentation. It reflects perhaps the maximum burial of the rocks or the last tectono-thermal event that affected the graben in the Upper Cambrian, or both.

In Togo and Ghana, the Obosum Beds (Grant 1969), formed of arkosic sandstones and red continental conglomerates, are possible equivalents of a part of the Série Pourprée.

Although as yet incomplete and fragile, this lithostratigraphic synthesis of the molasse formations illustrates the complexities and problems inherent in the diachronous nature of the orogenesis in the Pan-African belt. To be sure, it may be only the age of the late phase of deformation that varies from one region to another, and not the age of the molasse strata affected by that deformation.

II   STRUCTURE AND METAMORPHISM

A   GENERAL POINTS

Although often affected by a polyphase structural and metamorphic history of Pan-African age, the various structural domains rarely show a complexity that defies full analysis, and the number of terrains younger than 2,000 Ma for which a true poly-orogenic history is recognized is in fact limited. It is convenient to distinguish (1) the early orogenic (and metamorphic) events, those younger than the stabilization of the Eburnean basement at ca. 2,000 Ma but not extending over the whole shield, and (2) the Pan-African orogenesis s.s., which affected with rare exceptions almost the whole of the Touareg and Nigerian shields.

A1 *The Problem of the Kibaran Event in Hoggar and Nigeria*   It was essentially on the basis of Rb/Sr whole-rock and K/Ar radiometric data that Grant et al. (1972) and Bertrand and Lasserre (1976) proposed the existence of an event between 1,200 and 900 Ma in east-central Hoggar and Nigeria. Nevertheless, in Hoggar the structural relations between the Aleksod area, where these older dates have been obtained, and the terrains considered to have suffered only a single orogenic phase (the graywackes and schists of Arefsa-Sersouf) do not indicate any marked tectonic break (Bertrand 1974). Biotite and staurolite can be found in these schists in the cores of synforms. The metamorphic grade reached in the Pan-African antiforms is not known to be less. Therefore, we believe all the extremely fresh metamorphic assemblages with kyanite and sillimanite in Arefsa, E'gere, and Aleksod to have been formed during the Pan-African. The Rb/Sr biotite ages of 600 to 570 Ma and the fission-track ages on sphene and zircon of around 550 Ma (Carpena 1982) are significant in this respect. The geological significance of the whole-rock isochrons of 1,150 Ma from the paragneisses (Bertrand and Lasserre 1976) will be understood only when U/Pb zircon ages are obtained from the same area.

In Nigeria the 1,150 Ma Rb/Sr whole-rock isochron obtained by Grant et al. (1972) from the polyphase Ibadan gneisses may not signify a pre–Pan-African metamorphism. Zircons extracted from the same Ibadan gneisses give points strung out between 2,150 and 600 Ma on the concordia diagram (Rahaman and Lancelot, ms.), clearly demonstrating a Pan-African overprint on Eburnean material.

A2 *The 870 Ma Event in Central Hoggar*   The tectono-metamorphic event at 870 Ma has been recognized only in the eastern part of the central branch (Gravelle 1969). It affects the ''lower Pharusian'' of Bertrand et al. (1966), which is composed partly of a platform-facies limestone-dolomite series with quartzites and metapelites—a series we correlate with the Stromatolite Series and which is probably younger than 1,000 Ma (see section I C)—and partly of metavolcanic rocks and intrusive peridotites and gabbros.

One older generation of isoclinal folds trending north-south has been recognized in this pile (Boullier and Bertrand 1981). The pile is cut by late-tectonic quartz-diorite and granodiorite batholiths dated by the U/Pb method at $868 + 8/-5$ Ma, then by sub-alkaline granitoids and some syenites in the same area giving dates by the same method of around 839 ± 4 Ma (Caby et al. 1982). This specific tectonic event, at least 870 Ma old, has not been recorded elsewhere in the shield. We tentatively interpret it as being linked to the collage of this domain alongside the central Hoggar paleocontinent.

A3 *The 730 Ma Event in Eastern Hoggar*   This event gave rise to the early ''cratonization'' of the Djanet-Tafassasset domain. Although as yet little studied, this domain contains deeper greenschist- and amphibolite-facies rocks that form a number of vertical belts trending north-south. The rocks are lithologically varied and of unknown age. They include a thick metapelitic and graywacke formation with marble lenses rich in organic matter intruded by strongly deformed north-south bands of, in part, pre-metamorphic granitoids. Another formation includes marbles and calc-silicate gneisses, some amphibolites, metagabbros, and serpentinites. The final components are numerous and syntectonic granitoids. Late kinematic quartz-diorite and granodiorite batholiths cut this group, as do post-tectonic granites of the Djanet type that predate the deposition of the Tiririne Series (see section I E 1).

The dating of 730 Ma was obtained by the U/Pb method on zircon of a late kinematic porphyritic granite (Caby and Andreopoulos-Renaud 1985). Although giving only one bracket, this dating fixes a limit for the age of the early cratonization of this domain at around 730 Ma. Its extension eastward is unknown. It is only along the extreme southern margins that it, together with the unconformable Tiririne Group (see section I E 1), has been affected by the Pan-African deformation along the Tiririne branch (Bertrand et al. 1978). The Tiririne Group remains subhorizontal away from the 8°30′ shear zone. It can thus be interpreted as the molasse related to the 730 Ma event.

A4 The 720 Ma Event in Western Adrar des Iforas   This event has been recognized in the western Tessalit-Tilemsi domain, considered to be an island arc (see section I E 6). It affects deep levels of the volcano-volcanoclastic assemblage and appears intimately associated with the emplacement of early basic plutonic rocks that form vast differentiated laccoliths (Figure 8-4). All intermediates exist between practically undeformed metagraywackes with hornstones and intensely deformed plagioclase gray gneisses with hornblende and biotite, folded by complex folds suggestive of flow folding and traversed by numerous generations of dioritic and granodioritic mobilizates and intrusives. The mineral assemblage brown hornblende + ilmenite + clinopyroxene ± orthopyroxene + calcic plagioclase ± quartz, typical of hornblende-granulite facies, has been observed in both the gray gneisses and the amphibolites. Lenses of foliated dioritic rock with flow folding may have formed during this episode of anatexis. They are found against dioritic anatexites and migmatites displaying textures characteristic of catazonal terrains. A few calc-silicate bands (grossular, calcic plagioclase and secondary epidote, quartz) represent original bands with carbonate cement, which are characteristic of the volcanosedimentary series.

The age 720 ± 15 Ma was obtained by U/Pb dating of zircons extracted from both a blue quartz-bearing plagiogranite mobilizate cutting gray banded gneisses and a foliated mafic tonalite (Caby and Andreopoulos-Renaud, in prep.). To the north, the Pan-African event s.s. is responsible in these rocks for at most only a partial retrogressive overprint to shallow greenschist facies associated with open folds with a subvertical flow schistosity, best developed in the low-grade metasediments. To the south, there is, in contrast, a gradual transition into the Pan-African Aguelhoc gneisses and, as described below, the complete absence of retrograde metamorphism enables us to distinguish them (see section II B 3).

We believe, therefore, that the 720 Ma event of the west Tessalit-Tilemsi domain happened at the base of a magmatic arc and was associated with the emplacement of differentiated rocks of mantle origin. Thus a first recycling of calc-alkaline volcano-sedimentary material younger than 800 Ma took place at 720 Ma within a subduction regime.

B   THE PAN-AFRICAN S.S. (CA. 600 MA)

The classical geodynamic interpretation of the Pan-African belt rests on the distinction of a number of deformation phases, recognized both in low-grade and amphibolite-facies or higher grade tectonometamorphic domains (Gravelle 1969; Caby 1970, 1979, 1983a; Bertrand 1974; Bertrand and Caby 1978; Boullier et al. 1978; Latouche 1978; Vitel 1979; Affaton et al. 1980; Caby et al. 1981; Boullier 1982). Most of these studies demonstrate that practically all domains show an early gently inclined schistosity interpreted as being the result of large-amplitude horizontal movements (Figure 8-5B). Similarly, the most usual metamorphic history described involves an initial intermediate- to high-pressure phase followed by lower pressure and/or higher temperature conditions (Caby 1970; Bertrand 1974; Boullier 1982).

B1 *Deep-Level Horizontal Structures and the Problem of the Early Pan-African Intracontinental Deformations in Hoggar and Adrar des Iforas*   In east-central Hoggar, we have seen the uncertainty regarding the age (Kibaran or Pan-African?) of the orogenic event responsible for the subhorizontal foliation, recumbent folds, and kyanite-sillimanite grade (Aleksod) or granulite facies (Arefsa, Tamanrasset) metamorphism. This uncertainty results from the paucity, except from late- or post-kinematic granites, of reliable radiometric dates from these complex terrains. U/Pb data from various granites has given a poorly defined alignment cutting the concordia (upper intercept) at 650 Ma (Picciotto et al. 1965). Within this

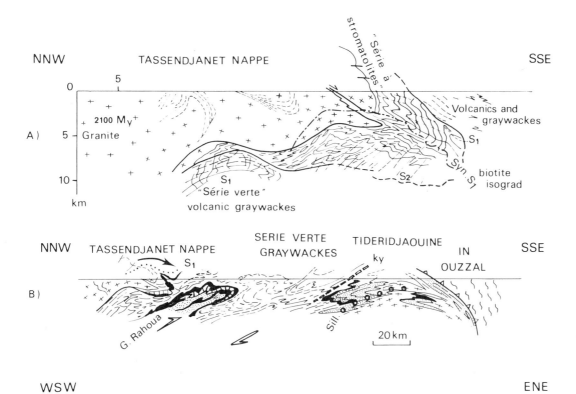

FIGURE 8-5. The early Pan-African collision in the western Hoggar at ca. 620 Ma. The Tassendjanet nappe is considered as part of a northern continent. S_1 is the upper limit of slaty cleavage. Ultramafic rocks (in black) associated with quartzites (dots) and gneisses (crosses) belong to a deeper structural unit with isoclinal folds facing north. Both units moved synchronously and are thought to be related to the closure of a sub-oceanic domain with a basified crust covered by many thousand meters of graywacke of the ''Série verte.''

group the polycyclic gneisses have been recognized by the presence of discordant mafic dikes that have subsequently been metamorphosed to garnet-amphibolite or granulite facies (Bertrand et al. 1978, Sautter 1982). Eclogites have been observed within garnet-amphibolite lenses enclosed both in orthogneisses and in metasediments, indicating very high-pressure conditions (15 kb at 700°C) prior to the lower-pressure anatexis that affected the surrounding rocks (Sautter 1982, 1985). It is clear that without further radiometric calibration we cannot classify these rocks with certainty as either poly-orogenic (Bertrand 1974) or "monocyclic" (Sautter 1982).

According to Boullier and Bertrand (1981), in the central Hoggar, southwest of Tamanrasset, only the great flat retrograde overthrusts affecting granulite facies rocks are of Pan-African age. The partial retrogression of the granulites in these zones has even been used to argue that these rocks were produced by a pre–Pan-African metamorphism (Latouche 1978, Vitel 1979, Ouzegane 1981). Recent U/Pb data on zircon from granulites define an upper intercept age of 2075 ± 30 Ma, which favors a pre–Pan-African granulite-facies metamorphism (Bertrand, Michard, et al. 1984b). Nearly concordant zircon and sphene at 615 ± 5 Ma from late kinematic granites, together with the retrogression of granulites along low-angle thrusts, relate to the Pan-African events.

In eastern Hoggar, high-grade gneisses of Issalane (Figure 8-6A) that overthrust the Tiririne branch of the Pan-African (Bertrand et al. 1978) may, in the total absence of radiometric dating, represent early Pan-African metamorphics later thrust over the Tiririne branch in a Himalayan type of structural arrangement (Figure 8-6B).

In Adrar des Iforas, the middle- to high-grade gneisses of central Iforas (the "Kidalian assemblage" of Boullier et al. 1978; Boullier 1982) correspond to the Pan-African middle and deep zones (including pre–Pan-African basement, cover, and pre-metamorphic intrusives) involved in large-scale horizontal movements of Pan-African age, all before 610 Ma. This zone with gently dipping foliation, 50 km wide, has been interpreted as originally lying beneath a crystalline nappe formed of the Eburnean granulites (UGI) (Boullier et al. 1978; Boullier 1982; Figure 8-8). It was principally from a detailed study of the northern contacts of the UGI that Boullier was led to suggest that the UGI was allochthonous. Indeed, the locally gently dipping foliation in only the frontal part of the UGI and the parallelism of northwest movement directions in the deformed granulites and the Kidalian gneisses is evidence for a northward movement of the granulites (Boullier 1982).

The eastern and western margins of the granulites, however, are vertical shear zones, and the constantly subhorizontal stretching lineations in the deformed rocks suggest that these zones are synmetamorphic strike-slip faults along which several granitoid intrusions have been identified (Boullier 1982; Lancelot et al. 1983). Shear zones within the UGI are also vertical and are marked by elongate masses of granitoids, diorites, and gabbros (Bertrand, Michard, et al. 1984a). The retrograde recrystallization of the granulites in amphibolite-facies conditions (mostly reaction rims) is conspicuous, whereas the Pan-African deformation is very heterogeneous, and large masses of charnockites and granulites behaved rigidly. The undetached Proterozoic cover is locally preserved and is affected by north-south open folding associated with subvertical axial-plane slaty cleavage with tiny chlorite and white mica crystals in shales, whereas biotite may appear close to granitoids.

In contrast to the purely allochthonist interpretation of Boullier (1982; see Figure 8-8),

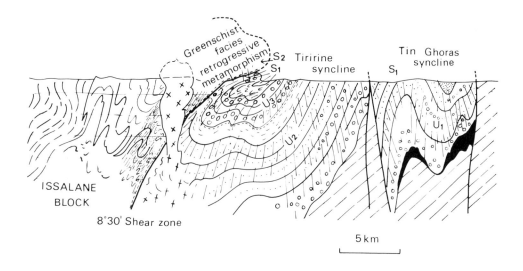

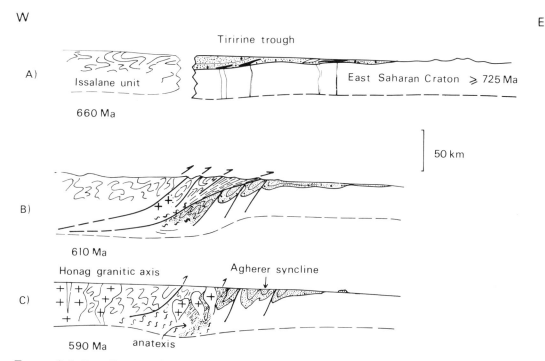

FIGURE 8-6. *Top*: Cross section of the late Pan-African Tiririne linear fold belt (eastern Hoggar). Note the superimposition of S_1 cleavage along the thrust in the Issalane basement and the cross-cutting high-level granite. A, B, and C: Three stages in the formation of the Tiririne fold belt.

we prefer to see the UGI as a gigantic flake of deep granulitic crust rooted and attached to its underlying continental mantle (Figure 8-7). It is involved in a system of north-south strike-slip faults of large displacement, as is its northern equivalent, the UGO (Caby 1970, 1983a). Only its northern part shows a limited northward thrusting over the Kidalian gneisses (Figure 8-8). Our interpretation is similar to that proposed for the Ivrea zone of the western

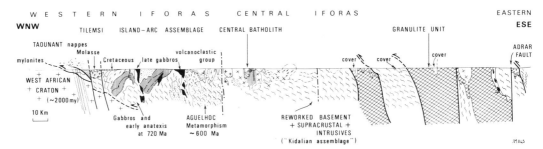

FIGURE 8-7. Simplified east-west section across central and western Adrar des Iforas. The frontal mylonites at the base of Taounant nappes are supposed to extend to depth (25 km), suggesting the allochthony of the island-arc assemblage in agreement with gravimetric data and the shallow dip of the suture zone to the west. The central batholith is delimited to the west by a huge vertical shear zone (possible back-arc cryptic suture). The pre–Pan-African granulites are here interpreted as rooted units that escaped strong deformation and metamorphism, their margins being essentially vertical lithospheric fractures of strike-slip type.

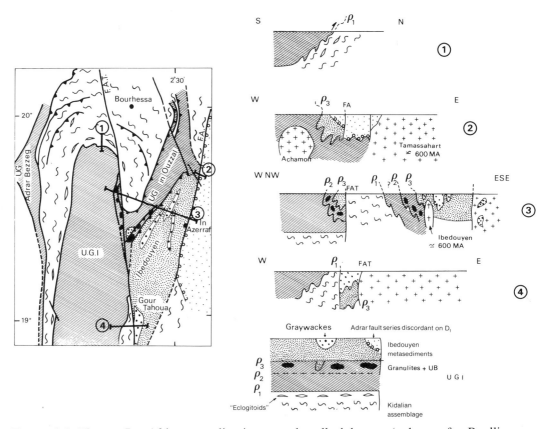

FIGURE 8-8. The pre–Pan-African granulites interpreted as allochthonous (redrawn after Boullier 1982). ρ_1, ρ_2, and ρ_3 are regarded as refolded low-angle tectonic contacts between metasediments above, granulites in between, and high-grade gneisses of the Kidalian assemblage below. Diagram at lower right of figure explains patterns.

Alps; a part of the southern Alps is still rooted and connected to its continental upper mantle, in contrast to the Dent Blanche nappe which has been detached from the upper mantle and alone is allochthonous, illustrating well the original partial overthrusting of the south Alpine domain.

B2 *Polyphase Development and Structural Level*   In northwestern Hoggar large horizontal displacements have been demonstrated both in high-level structures (Tassendjanet nappe; see Figure 8-5) and in deep-level domains with high-grade metamorphism (Caby 1970). The criteria of fold geometry have been used only for the classical distinction of different fold phases separated in time, and for the sense of the vergence of fold systems. It appears today, from the recent literature, that only features relating to non-coaxial (or "rotational") deformation may be used to give a true movement direction in crystalline terrains, and no study to this end has yet been undertaken in the Hoggar. The detailed study of the synclinorium of the "Série verte" of In Zize (northwestern Hoggar), however, has shown a progressive evolution from broad open folds (phase 3), in the core of the synclinorium, to a "polyphase" deformation (phases 1 and 2) in its deeper and marginal levels (Caby 1970, Figure 241B; 1983a).

In Adrar des Iforas the movement direction inferred from stretching lineations in the Kidalian assemblage and the frontal part of the UGI is toward the north-northeast (Boullier 1982). In the Tafeliant synclinorium, the youngest series affected by the Pan-African orogenesis (see section I E 5), one observes only a single fold generation trending north-northwest. The folds are fairly open and imply that east-west shortening did not exceed 30%. The folding is associated with a strong horizontal elongation lineation marked in particular by the intense stretching of conglomerate clasts, notably at the edges of the synclinorium. The study of the deformed rocks shows that this deformation is non-coaxial and consistent with a sinistral shear regime (Ball and Caby 1984). No data seem inconsistent with this deformation's being contemporaneous with the horizontal foliation of the Kidalian assemblage. Along the southeastern margin of the synclinorium one observes in the same series the transition from open folds with vertical axial planes to a regime with recumbent folds such as predominate in the Kidalian assemblage (Figure 8-9). This and the 610 to 620 Ma age obtained by U/Pb dating of a synkinematic adamellite (Ducrot et al. 1979, Andreopoulos-Renaud, unpublished data) enable us to bring forward this far the date for the formation of the folds in the neighboring Kidalian assemblage, at least for this sector of the belt.

In conclusion, we think that the idea of an "early Pan-African" event much before 620 Ma has not been adequately supported either by structural studies or by radiometric dating. At most, it must be acknowledged as a general rule that the deformation at depth is reflected by the gently inclined foliation and by recumbent and, in places, refolded folds, and that these structures were formed mostly under Barrovian metamorphic conditions with primary kyanite and the occasional presence of eclogitoids (Boullier 1982). The deformation implies a thickening of the crust as the tectogenesis began, probably only a little before the onset of uplift of the belt.

B3 *The Low-Pressure Pan-African Catazone*   Cropping out immediately to the east of the suture zone, the low-pressure Pan-African catazone has been recognized at Egatalis, Aguelhoc, and also in Benin (Figure 8-1). Detailed studies in the first two areas named have shown

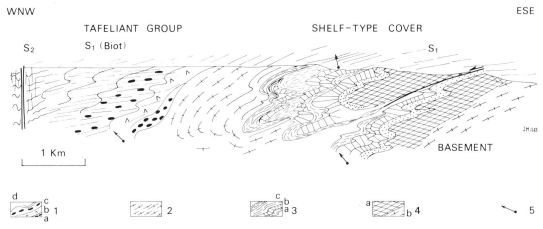

FIGURE 8-9. Geological section showing the similarity of tectonic style that affects both the upper Upper Proterozoic Tafeliant Group (younger than 693 Ma) and the pre–Pan-African basement capped by the shelf-type cover. 1: Tafeliant Group, with basal glaciogenic sediments (a), intrusive porphyries (b), marine tillite (c), and graywackes (d). 2: Foliated metadiorite, 793 Ma old, intruding 3 and 4. 3: Shelf-type cover of Upper Proterozoic age, with basal conglomeratic quartzites (a), limestones and dolomites (b), and phyllites (c). 4: Pre–Pan-African gneisses, unsheared (a) and sheared (b). 5: Direction of younging of strata.

that metamorphic isograds are vertical and extremely close together along thermal spines trending north-south. At Egatalis, in less than 1 km, chlorite-biotite pelitic schists grade into pelitic gneisses containing biotite + garnet + cordierite + spinel + sillimanite ± hypersthene. The gross structure is essentially controlled by tight folds with vertical axial planes and by synmetamorphic diapirs of plutonic rocks, including noritic gabbros. At Aguelhoc the situation appears the same, though here movements related to late uplift have destroyed the original isograd geometry. In both areas deep-seated synmetamorphic intrusions form boudinaged sills and apophyses with microstructures with triple points. The compositions of these intrusions vary from noritic to troctolitic, and hypersthene has been observed in the aluminous gneisses in their immediate vicinity.

It appears therefore that the metamorphism in this deep terrain is in part slightly later than the Barrovian mineral assemblages and was closely associated with mantle-derived hot synmetamorphic intrusions, which are absent everywhere else. Thus we interpret this metamorphism as resulting from a regional thermal anomaly parallel to the suture zone and entirely controlled by the rise of synkinematic gabbro-norite magmas.

The situation is similar in the Dahomeyan gneisses of Benin, where sillimanite pseudomorphs after kyanite are conspicuous. In northern Benin, late kinematic intrusions of norites and charnockites also intrude this assemblage (Lazzarotto et al. 1983). These intrusions are similar to those of southwestern Nigeria, which have U/Pb zircon ages between 620 and 580 Ma (Tubosun et al. 1984).

B4 *Summary of the Synorogenic and Post-Orogenic Pan-African Magmatism*   Pan-African granitoids form about 50% of the outcrop in many domains, notably in west-central Hoggar. In some catazonal domains, however, they are much less abundant, and they are totally absent west of the suture zone.

(1) SYNOROGENIC GRANITOIDS.   As a general rule parautochthonous granites with migmatitic margins predominate in amphibolite-facies terrains. Their chemistry frequently reflects that of the surrounding rocks, and as a consequence they are most often calc-alkaline in domains that have suffered the Pan-African tectogenesis only, or in those with abundant metagraywackes and metavolcanic rocks. Certain foliated masses possess in effect a whole spectrum of rock compositions, from quartz-diorites (sometimes mafic) to granodiorites. More potassium-rich varieties are abundant in east-central Hoggar. The early mafic parts of the spectrum are certainly of an ''infracrustal'' origin. Aluminous, true ''S''-type granitoids are found within pelitic series and are scarce.

The distinction between foliated, truly syntectonic masses with diffuse margins and unoriented masses with clear-cut contacts seems to result more from the level of emplacement than from the individual compositional affinities. This has been demonstrated for many batholiths in Hoggar, such as those of Immezzarene (Boissonnas and Gravelle 1961, Gravelle 1969) and Tin ed'Ehou (Caby 1970, 1983a). The central calc-alkaline batholith in west Iforas (Bertrand and Davison 1981, Fabre et al. 1982, Bertrand, Dupuy, et al. 1984, Liegeois and Black 1984) is unique in the chain. It is similar to the Andean batholith, runs parallel to the suture for more than 300 km, and is around 30 to 50 km wide. Tonalites form the earliest units, and an age of 613 ± 3 Ma (Ducrot et al. 1979, U/Pb on zircon, upper intercept) has been obtained from the southern part of the batholith. Porphyritic adamellites and fine-grained adamellites are easily the most abundant rocks and truncate the earlier units. Rb/Sr wholerock isochron ages are between 620 and 580 Ma (Liegeois and Black 1984).

(2) LATE AND POST-OROGENIC GRANITOIDS.   Examples of this type include the Taourirt granites in Hoggar (Boissonnas 1973), a homogeneous suite of restricted granites of calc-alkaline chemistry and concentric structure. An age of 560 ± 40 Ma (Rb/Sr whole-rock isochron) has been obtained on one massif of this province (Boissonnas et al. 1969). To the east of the 4°50' shear zone, there is also a family of high-level granites with tungsten-tin mineralization that cut gneisses of indeterminate age (see section II B 1).

Concentrically zoned alkaline to hyperalkaline granites also form an important suite in Adrar des Iforas, in part centered on the central batholith. Rb/Sr whole-rock isochron ages range from 560 to 540 Ma (Liegeois and Black 1984). The Kidal massif has been studied in detail (Ba et al. 1985). It shows a classical suite (600 to 580 Ma old) of petrographic facies little different from those of the anorogenic ''younger granites'' of Nigeria. The preliminary age of 590 ± 6 Ma (Ducrot et al. 1979, U/Pb on zircons, upper intercept) is inconsistent with 561 ± 7 Ma given by whole-rock Rb/Sr isochrons (Liegeois and Black 1984) and poses the question of the real geological significance of the two, since no thermal event appears to have affected the massif after its pre-Ordovician cooling, though there may have been brittle deformation connected with late north-south strike-slip faulting (partly of post-Paleozoic age).

The late magmatic suite (600 to 580 Ma old) that follows the Adrar fault (Figure 8-7; see section I F 2) partly predates the metamorphism in the molasse (Caby et al. 1985). It raises the possibility of an extremely young age (? 570–540 Ma) for the deformation in the whole of this domain.

The magmatism accompanying the various phases of infilling and graben formation in the late molasse sequences produced mainly acid lavas and hypabyssal granites, such as those of the In Zize massif in northwestern Hoggar, which have given an age of 520 Ma (Allègre and Caby 1972, Rb/Sr whole-rock). Undated, but probably still younger, syenites cut the uppermost units of the molasse (see section I F 4).

B5 *The Major Nappe Units Lying Upon the Passive Margin of the West African Craton*   Large-displacement, high-level nappes have been observed principally in Gourma and in Togo-Benin.

(1) GOURMA.   In Gourma the structures form a smooth convex arc through an angle of about 120° (Figure 8-10). The Gourma basin s.s. (Reichelt 1972) is a passive-margin basin

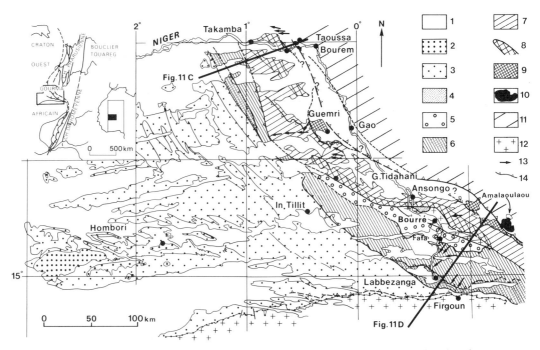

FIGURE 8-10. Simplified map of the Gourma region (modified from Reichelt 1972), showing the extension of allochthonous terranes at its eastern margin. 1: Alluvium and dunes. 2: Unconformable Bandiagara sandstones (Cambrian?). 3: Undifferentiated formations of the Gourma basin (assumed upper Upper Proterozoic age). 4: The same, parautochthonous. 5: Fafa parautochthonous unit. 6: External nappes with shallow greenschist-facies metamorphism. 7: Internal nappes with deep greenschist-facies metamorphism of high-pressure affinity. 8: The same, with high pressure–low temperature metamorphism including eclogitic micaschists. 9: Shales and quartzites of the Guemri window. 10: Amalaoulaou mafic-ultramafic massif that delineates the suture zone. 11: Cenozoic of the Gao trough. 12: Undifferentiated rocks, 2,000 Ma old, of the West African Craton and the Bourré inlier. 13: Fold axis. 14: Thrust fault.

that subsided rapidly and filled with mainly flysch-type terrigenous sediments; beneath it the basement is more than 12 km deep (Bayer and Lesquer 1978). An early evolution of aulacogen type has been proposed here on the basis of sedimentological (Moussine-Pouchkine and Bertrand-Sarfati 1978) and gravimetric (Lesquer and Moussine-Pouchkine 1980) data.

(a) The external nappes of eastern Gourma (Caby 1979) are composed of parautochthonous terranes supposedly derived from the basin itself, of structurally higher nappes composed entirely of chlorite-sericite phyllites and similar phyllonites, of a series with quartzites, marbles, and siliceous dolomites, and also of turbiditic terrigenous sediments of more internal origin than the Bourré window (Figure 8-11), which is formed of the Eburnean basement (de la Boisse and Lancelot 1977) capped by its autochthonous cover (Caby and Moussine-Pouchkine 1978, Caby 1979). In the northern part (Timetrine) the allochthonous sequence includes a series of quartzites overlain by chlorite-sericite schists and pillow lavas, in which are intruded the ultramafic rocks (see section I D 2) and near which blue amphibole and aegerine are found. These nappes rest in Niger directly on a thin, practically undeformed sandstone sedimentary cover on the Eburnean basement.

(b) The internal nappes include an aluminous quartzite series bearing white mica and sometimes kyanite (Reichelt 1972) and aluminous mica schists showing high-pressure–low-temperature mineral assemblages, for example, phengite + almandine + rutile $\pm$ bluish amphibole in Afarag (Caby 1979) or phengite + jadeitic pyroxene + pyrope-rich garnet + rutile in the eclogitic aluminous mica schists of Takamba, where de la Boisse (1981) has calculated the initial conditions of metamorphism at P equals 18 kb and T equals 800°C.

These internal nappes rest in tectonic contact over the metapelites, siltstones, and sandy-quartzites of Guemri (Caby 1979) and root beneath the metamafic masses of the Amalaoulaou type (de la Boisse 1979) that mark the suture. This last mass includes granulitic metagabbros dated at around 800 Ma by U/Pb on zircons (de la Boisse 1979; see section I D 1) and partially overprinted under eclogitic conditions and later retrogressive conditions at its base.

(2) TOGO-BENIN.   The work of Affaton (1975, summarized in Affaton et al. 1980) and of Simpara (1978) serves as a basis for the structural interpretation proposed by Trompette (1980). There is a complete parallelism between the development in Gourma and here, at least in north Togo, which I have been privileged to visit (Figure 8-12).

(a) The foreland shows a sequence with sandstones at the base ("Lower Voltaian") resting on the Eburnean basement. This subhorizontal, essentially sandy formation has given a date of 993 $\pm$ 65 Ma (Clauer 1976; Clauer et al. 1982, Rb/Sr isochron on clay minerals) from a horizon close to the base. It is unconformably overlain by the Oti series (Pendjari Group).

(b) The external nappes are composed both of presumed lateral equivalents to the Oti beds and of strongly deformed schists of the "Zone des collines" and Kandé in northern Togo. These schists, which include hematitites and a conglomerate level, contain serpentinite masses with typical cumulate structures (lherzolite, dunite) with which are associated ferruginous metasomatic schists with local blue amphibole, just as in the Timetrine area (see section I D 2).

(c) The gently inclined Atakorian quartzite nappe (Figure 8-12) includes micaceous quartzites with kyanite and rutile and biotite-albite pelitic schists sometimes with chloritoid (Menot and Seddoh, 1977). The quartzites are intensely deformed at the base of the nappe and

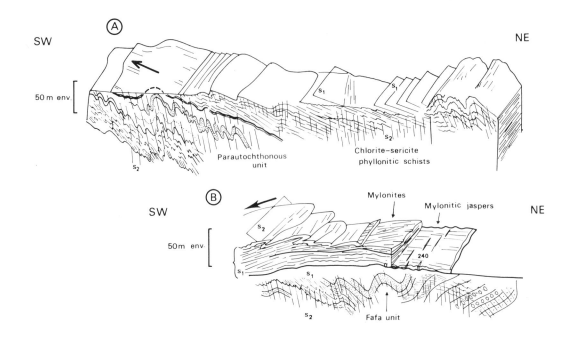

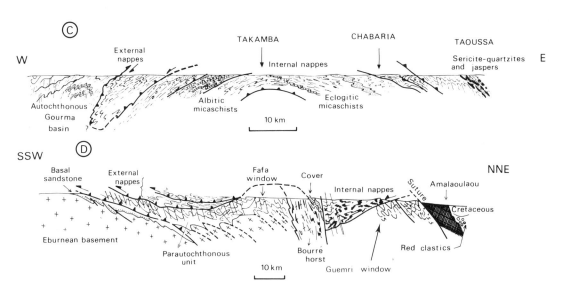

FIGURE 8-11. Block diagrams and cross sections in the Gourma region. A: Basal contact of the external nappe in the region of In Tillit, where it is folded by open folds that follow the Gourma arc. Note the S₁ schistosity refolded in both units, and the significant shortening in the autochthon. B: Basal contact of the external nappes south of the Fafa tectonic window. Note the basal mylonites and the importance of the upper truncation of the Fafa beds, which are affected by two phases of deformation. C: Section across the northern sector of the nappes at the level of the River Niger. The internal nappes trace out an anticlinorium (forming the axis of the Guemri tectonic window—

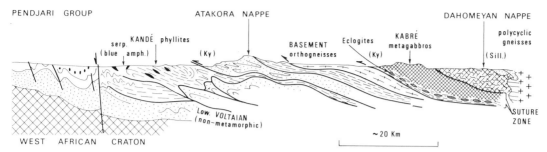

FIGURE 8-12. Simplified section across the frontal part of the Dahomeyan belt in north Togo. Ky: Kyanite isograd. Sill: Sillimanite isograd.

show strongly flattened sheath folds and a penetrative east-west lineation indicating west-ward-directed movement. They are overlain by muscovite quartzites and phengite-garnet–bearing pelitic gneisses with early mobilizates of unknown age. This nappe may be traced for more than 800 km from the Niger through to Ghana (Fitches 1970, Shackleton 1971).

(d) The Lamakara orthogneisses next above the Atakorian nappe are supposed to represent intensely deformed, ancient, two-mica granitoids belonging originally to the pre–Pan-African basement. They have given an age of 2,064 ± 90 Ma (Caen-Vachette 1979, Rb/Sr whole-rock isochron).

(e) The metamafic masses of the Kabré type are granulitic metagabbros analogous to those of Amalaoulaou (Mali), dated at around 800 Ma. They have, however, suffered an evolution through kyanite-eclogite facies conditions (P equals 18 kb) (Menot 1982, 1985), then garnet-amphibolite facies at their base. In Ghana the basic granulites a little distance away from the frontal thrust (described by Bondesen 1972) belong perhaps to the same suite.

(f) The Dahomeyan high-grade gneisses form the most internal nappe. They are cut by late-tectonic granitoids dated at 580 Ma (Bonhomme 1962, Rb/Sr whole-rock isochron) and may represent the Pan-African catazone with mainly high-temperature syn- and post-kinematic mineral associations. Strongly retrogressed kyanite eclogites and kinzigites with sillimanite pseudomorphs after kyanite testify to early high-pressure conditions, which we believe to be synchronous with those recorded in the Atakorian.

B6 *The Major North-South Trending Shear Zones*   The major shear zones are without doubt a special and unique feature of the Pan-African belt of West Africa. They show up magnificently in LANDSAT images, particularly in Hoggar and Adrar des Iforas. Lelubre (1952) believed them to be ancient faults that had accompanied the various stages of pre-orogenic

---

parautochthonous?). They are mantled to the west by the external nappes which dip beneath the autochthonous-parautochthonous terrains (an early drape structure?). D: Section across the eastern sector of the nappes, 250 km southeast of C. Important décollement from the basement is inferred from the strong shortening of the autochthonous-parautochthonous units. Prolongation of the Guemri tectonic window below the internal nappes in section C is inferred. Structure of the basement is speculative and deduced from that of the refoliated rocks of the Bourré inlier.

and orogenic evolution; Caby (1968) regarded them as essentially late Pan-African structures. The shear zones are marked by several hundred meters, or locally several kilometers, of mylonites and ultramylonites whose final evolution is in the greenschist facies; in places they are also marked by epidote cataclasites. Very often post-tectonic granites are themselves mylonitized.

Detailed studies in the mylonitic zones in central Hoggar have shown the pervasive presence of a subhorizontal elongation lineation associated with late folds with subvertical axial planes (Vitel 1979). Boullier and Bertrand (1981) have interpreted the zones as ancient thrust planes later made vertical. In fact, the penetrative and constant horizontal elongation lineation seems to us necessarily to imply lateral movements of considerable magnitude (a hundred kilometers or more), which would explain the difficulties of making correlations between the various domains. Possibly such large lateral displacements were synchronous with late horizontal movements, as indicated by some retrograde, gently inclined mylonites found away from the vertical ones. In northwestern Hoggar the distribution of metamorphic zones with truncated isograds implies in any case sinistral movements on two major fractures of 160 km and 65 km, respectively (Caby 1970, 1983a). In the same part of Hoggar there are also ultramylonites showing a vertical lineation, compatible with differential vertical uplift movements that offset earlier tangential structures very little and in part form the margins of the molasse grabens (Caby 1970, 1983a).

In eastern Hoggar, the 8°30′ shear zone is marked in the north by a narrow linear belt, the Tiririne belt (Figure 8-6). There is perfect continuity by very gradual transition from the vertical mylonites with a horizontal lineation (affecting the basement and the upper Upper Proterozoic Tiririne Group) on the south to a gently west-dipping thrust on the north, clearly demonstrating that pure strike-slip movement on vertical faults and horizontal tectonics were contemporaneous in this area between 604 ± 13 and 585 ± 14 Ma (Bertrand et al. 1978, U/Pb on zircons, upper intercepts from syn- and post-kinematic granites, respectively).

In Adrar des Iforas the vertical mylonites, delineating to the west the UGI (Boullier 1982), show everywhere a subhorizontal lineation, but the mylonitic schistosity is itself involved in polyphase folding with curved axes, complex geometries, and augen-shaped closures symptomatic of sheath folds. Granitoids as variably deformed and boudinaged bands and veins, and also as some elongate masses, cut the mylonites. One of these masses has given an age of 568 ± 8 Ma (Lancelot et al. 1983, U/Pb on zircon, lower intercept), though the cessation of the mylonitization took place here at 535 ± 5 Ma and 539 ± 5 Ma (Lancelot et al. 1985, concordant Ar_{39}-Ar_{40} ages on feldspars from the deformed granite). Note that these ages are, within their narrow error bands, little different from those obtained by a number of methods from the rocks of the late molasse sequences (section I F 4). The last movements along the major north-south fractures are therefore in part younger than the emplacement of the hyperalkaline granites of the Kidal type (section II B 4[2]) and are thus of Cambrian age.

It is worth noting here that the mylonite zones are never found to the west of the suture. They are therefore genetically related to the nature and rigid-plastic behavior of the largely reactivated continental crust of the eastern half of the belt during the indenting movement of the still rigid West-African craton (Bertrand and Caby 1978, Caby et al. 1981). Younger reactivation of these mylonite zones has been well documented in Sahara and Benin (see section IV).

III   GEODYNAMIC INTERPRETATION IN THE LIGHT OF THE HISTORIES OF OTHER OROGENIC BELTS

A   THE SUTURE ZONE: INITIAL GEOMETRY AND ACTUAL GEOMETRY

The suture zone may be defined geophysically by the string of positive gravity anomalies related to unrooted metamafic masses with an easterly dip (Bayer and Lesquer 1978, Ly et al. 1980) (Figure 8-13). Geologically the suture is, for the period of the upper Upper Proterozoic between 800 and 600 Ma, an abrupt boundary between, to the west, passive-margin metasediments (Caby 1978) and, to the east, rocks with an evolution of island-arc style (Tessalit-Tilemsi domain) and of active continental-margin style over a width, before the Pan-African shortening, of several hundreds of kilometers in Hoggar. The suture zone thus defined is

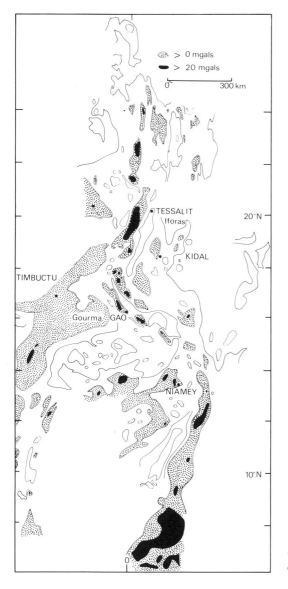

FIGURE 8-13. Simplified Bouguer anomaly map of the eastern margin of the West African Craton.

gently inclined in Togo-Benin, though subvertical in outcrop in Mali. In the latter area the geometry is a result of more recent, repeated post–Pan-African (Hercynian and Cretaceous-Eocene) isostatic readjustments (Caby 1981b) (see section IV).

B   EVIDENCE FOR THE EXISTENCE OF A WIDE PRE–PAN-AFRICAN OCEAN

We have seen that there is only indirect evidence for the opening of a wide pre–Pan-African ocean. The absence of ophiolites over nearly 3,000 km raises the question whether they really did not exist or whether they simply were not preserved (see section I D).

We believe the acceptance of indirect evidence for ocean spreading is itself more reliable than the hasty ascription of ultramafic masses to ocean crust, as has been done, notably on the Nubo-Arabian shield (see discussion in Caby 1983b). We consider the following indirect arguments for important ocean spreading to be the strongest:

1. Totally dissimilar continental margins with histories that cannot be correlated.

2. Wide belts formed by continental accretion in a manner identical to that seen along modern subduction zones.

3. Paired metamorphic belts, indicating differing thermal regimes on either side of the suture.

4. Evidence of extremely large (around 100 km) horizontal tectonic movements.

5. A gravimetric signature of the suture itself and an asymmetry of the crust across it.

C   EVIDENCE FOR A THICKENING OF THE CRUST

The evidence is of a tectonic and petrologic nature. First, the major eastward-dipping tectonic contacts along which the large-amplitude horizontal movements took place must necessarily extend at least to the base of the crust. Second, the pressure required for the formation of the mineral assemblages in certain units in which the metasediments themselves were recrystallized under eclogitic conditions (P equals 15-18 Kb; see de la Boisse 1981; see section II B 5) are those of conditions deeper than the normal base of the crust. The superposition of the low-pressure high-grade conditions in the Dahomeyan belt on the eclogites, themselves superposed on green schists (see section II B 5), is in many respects similar to the situation in the Himalayas (Le Fort 1975, Caby et al. 1983).

D   TIME SPAN OF THE PRE-OROGENIC EVOLUTION

We have seen that the platform sedimentary regimes and the passive-margin basinal sedimentation were superseded in the Late and latest Proterozoic (later than 800 Ma in North Mali and western Hoggar, but as early as 900 Ma in central Hoggar) by a ''cordilleran'' evolution. We can therefore estimate that the time span of the cordilleran period was between 200 and 300 Ma, and that that of the island-arc evolution was around 200 Ma, evidence of a subduction regime accompanying the life span of the Pan-African ocean up to its closure at around 600 Ma. This time span is comparable to those known for the Hercynian and Alpine chains.

E Duration of the Pan-African Orogenic Events s.s.

We have already discussed the hypothesis of an "early Pan-African" event well separated in time from that associated with the final continent-continent collision (see particularly Bertrand and Caby 1978; Caby et al. 1981; Boullier 1982; and section II B 1), an event for which an age between 700 and 650 Ma has been advanced. The problem is exactly that of rigorous radiometric dating of the event. Similar problems are found in many orogenic belts. Certainly the geochronological techniques (except the Ar_{39}-Ar_{40} method, which has not yet been applied to these rocks) permit indisputable dating in an orogenic belt not of the deformation itself, but rather only of the end of the thermal event. In many domains relics from conditions of extremely high pressure associated with the first horizontal tectonics occur, almost always in the midst of metamorphic assemblages formed at lower pressures (and/or higher temperatures).

Ultimately the question is whether intracontinental deformation only followed the closing of the oceanic domain or whether, on the contrary, a "subduction metamorphism" affecting continental material could or could not incite deformation and related metamorphism associated in time and space with the ending of subduction, or indeed with subduction itself. This latter possibility has been suggested for the Alpine chain, where the time gap between the eo-Alpine and the main Alpine events proper is of the order of 40 to 60 Ma.

F Magnitude of the Horizontal Movements: Are the Proterozoic Orogenic Belts Different from Those of the Phanerozoic?

The presence of domains with a foliation that formed and remained subhorizontal over large distances, and the constancy of the elongation lineation, which we read as giving the movement direction of the nappes, seems ample evidence to affirm the existence of large horizontal displacements in the belt. In the high-grade metamorphic piles (deeper amphibolite and higher-grade facies), the tectonic contacts are in part synmetamorphic and can be traced only by careful mapping to pick out the truncation of lithologies. The gently inclined retrograde phyllonites, however (i.e., those formed by the dynamic recrystallization of earlier, higher-grade assemblages at lower pressure and/or temperature), are certainly evidence of the importance of horizontal displacements of large amplitude. Recognizing and mapping them has permitted delineation of the Main Central Thrust in the Himalayas (Le Fort 1975) and also in the Massif Central of France (Burg and Matte 1978, Pin 1979). Unfortunately, the identification of such rocks has often been difficult for petrographers without a structural bias. The search for them must be pursued in all major metamorphic terrains; they are the key to the geometry of the deeper zones of mountain chains.

G The Illusory Problem of Ophiolites in This Orogenic Belt

Imagine the Alpine belt peneplained to a level close to that of the sea; all the characteristic ophiolites, which are seen only at high levels in the pile, would be eroded. In place of the allochthonous Piemontese zone would be outcrops of the underlying polyphase Hercynian basement, already exposed over wide areas in the actual internal crystalline massifs. The only remaining rocks of possible oceanic derivation would be serpentinite lenses, garnet-bearing amphibolites, or eclogites—difficult to interpret and found in narrow complex synforms more or less enclosed between gneiss masses representing the thoroughly reactivated Hercynian basement, as already seen in the Simplon-Tessin area and around Monte Rosa. The picture of the Alpine chain would then be that of an intracontinental orogenic belt within which intra-continental deformation would appear to become progressively stronger from the external zones through the deepest Alpine zone to the Insubric line and would then disappear completely on the other side of that line. This major line, which clearly represents the cryptic Alpine suture, reactivated many times, is itself not marked by ophiolites. The parallel with the Pan-African belt is striking.

IV   THE POST-CAMBRIAN EVOLUTION

It is known that the paleontologically dated Lower Ordovician (Beuf et al. 1971) ends the Precambrian and partly Cambrian molasse development. Relevant in this respect are fission-track ages from zircons, which cluster around 520 to 590 Ma (Carpena 1982).

The distribution of Paleozoic sedimentary basins and the paleocurrent data from Ordovician and Devonian strata show that the Hoggar area was completely buried under at least 3 km of Paleozoic sediments (Beuf et al. 1971). Along the north Saharan extension of the suture, the thickness of the Paleozoic, including Stephanian and Lower Permian, exceeds 6 to 7 km and even 10 km to the north of Sahara. This burial of the Pan-African belt (Fabre 1978) is now clearly demonstrated by fission-track ages on apatite; for western Hoggar, these ages all fall between 150 and 235 Ma (Carpena 1982). We should point out also the presence of clastic dikes of now brecciated Paleozoic sandstones, which were cored in the Precambrian along some north-south fractures of west-central Hoggar. In central Hoggar, by contrast, two apatite fission-track ages, 151 and 128 Ma, suggest a longer burial followed by a post-Jurassic uplift.

The uppermost Jurassic is continental and includes clasts of earlier Jurassic dolerites; it is deformed along the northern extension of the suture near 27°N (Conrad 1981). In the Tilemsi basin the continental infilling of Cretaceous-Eocene age could be around 800 m thick; it exceeds 2,000 m in the Gao trough. The reactivation in the Permian of the suture in northern Mali is demonstrated by the presence of an undersaturated alkaline province there (nephelinitic syenites, carbonatites with rare-earth mineralization, and so forth). An age of 262 ± 7 Ma has been obtained from the Tadhak massif from a Rb/Sr whole-rock isochron (Liegeois et al. 1983).

Later tectonic evolution also includes the post-Paleozoic reactivation of certain mylonite zones. The Ordovician sandstones of Tassili are cut by reverse faults, sometimes major (Beuf et al. 1971). The tectonism of the north Saharan basins is in part due to Jurassic movements as well (Beuf, personal communication; Conrad 1981). The Cretaceous itself may be deformed along major fractures, as is the Eocene at Tilemsi, thereby showing the Alpine age of some of the mobility on the platform localized along those fractures. Post-Quaternary throws are also observed. The arching, which has uplifted the Precambrian basement in Hoggar itself to an altitude of almost 2,000 m, is younger still and is in part associated with recent volcanism of post-Eocene and Quaternary age.

V   GENERAL SUMMARY

From central Sahara to the Gulf of Benin, the Pan-African belt (ca. 600 Ma old) is exposed in the Hoggar or Touareg shield and in the Nigerian shield. Recent data obtained in southern Algeria and in Mali make it possible to present a synthetic account of the structure and evolution of the Precambrian crust in this part of West Africa.

The pre–Pan-African basement (ca. 2,000 Ma) is exposed in several structural provinces both as uplifted blocks, upthrusted slices, and nappes and as polycyclic gneisses reactivated during the Pan-African metamorphism. During Middle Proterozoic times, ensialic subsiding areas were filled by thick orthoquartzite-pelitic sediments; this was followed by a large-scale anorogenic magmatic event of alkaline type, which has been dated at 1850 to 1750 Ma. Shelf-type Upper Proterozoic deposits are well characterized in northwestern Hoggar, where they include the limestones and dolomites of the Stromatolite Series, which unconformably overlie granites 2,150 Ma old. These sediments were deposited prior to mafic-ultramafic intrusions in the form of sills, laccoliths, and protrusions emplaced during extensive thinning of the continental crust. This magmatic event, dated at 800 Ma, was related to the opening of a wide Pan-African ocean located between the mobile zone and the West African Craton, buried at that time below a thin upper Proterozoic cover. The late Upper Proterozoic (800-600 Ma) is characterized essentially by volcanoclastic trough and marginal sea deposits indicative of a subduction-type environment. The volcanic graywackes derive from the destruction of penecontemporaneous andesitic volcanoes with calc-alkaline affinity, typical of continental-margin orogenic zones. Island-arc assemblages are also preserved along the suture zone; unrooted metamafic massifs that mark the suture are believed to be differentiated intrusions emplaced initially at the root of the volcanic arc. These intrusions, 800 to 700 Ma old, generated high thermal gradients responsible for a first anatexis within the volcanic-volcanoclastic pile, whereas autocannibalistic sedimentation of peri-Pacific type was going on together with calc-alkaline volcanism on land.

The Pan-African orogeny (ca. 600 Ma) affects the entire Hoggar and Nigerian shields. Large-scale horizontal movements and crustal thickening during Barrovian metamorphism partly predates north-south strike-slip movements along megashear zones outlined by vertical mylonite zones with overall horizontal stretching lineations. This major synmetamorphic and synmagmatic strike-slip system was still active throughout the final stages of the orogeny at shallower and shallower structural levels. West of the suture, thin overthrust sheets were transported more than 80 km westward onto less deformed parautochthonous units deposited at the edge of the West African craton. High pressure–low temperature metamorphic assemblages with pressures in excess of 15 Kb are found in the internal nappes in both pelitic and metamafic rocks south of Sahara.

Syntectonic and late tectonic molasse deposits are preserved in graben and residual basins in the western part of the belt. Isotopic dates obtained on late intrusive granites, on metamorphic minerals, and on molasse deposits bracket orogeny between 620 and 560 Ma. They testify to the Cambrian age of the late Pan-African molasse, which shows glacial influence and is well preserved in northwestern Hoggar.

Geodynamic interpretations are discussed with reference to several Phanerozoic belts with which the author is familiar. It is suggested that the lack of ophiolites s.s. is a false problem in old orogenic belts and is only to be ascribed to the deep erosion level. The post-Cambrian platform evolution is briefly described with reference to the stratigraphy of north Saharan basins and in the light of recent fission-track data on Precambrian apatites from Hoggar.

REFERENCES

Affaton, P., 1975: Etude géologique et structurale du nord-ouest Dahomey, du nord Togo, et du sud-est de la Haute-Volta. Univ. Aix-Marseille III, Lab. Sci. Terre Trav., Sér. B, *10*, 203 p.

Affaton, P., Sougy, J., and Trompette, R., 1980: The tectonostratigraphic relationships between the Upper Precambrian and Lower Paleozoic Volta basin and the Pan-African Dahomeyide orogenic belt (West Africa). Am. Jour. Sci. *280*, 224-248.

Allègre, C. J., and Caby, R., 1972: Chronologie absolue du Précambrien de l'Ahaggar occidental. Acad. Sci. Paris C. R. (D), *275*, 2095-2098.

Amard, B., 1983: Découverte de microfossiles dans le Protérozoïque métamorphique de l'Adrar des Iforas (Mali): Nouveaux éléments de datation dans la "série à stromatolites" en Afrique de l'Ouest. Acad. Sci. Paris C. R., II, *296*, 85-90.

Arène, J., 1968: Stratigraphie et évolution structurale du Précambrien dans la région de l'Adrar Ahnet (Sahara central). Acad. Sci. Paris C. R. (D), *266*, 868-870.

Ba, H., Black, R., Benziane, B., Diombana, D., Hascoet-Fender, J., Bonin, B., Fabre, J., and Liegeois, J. P., 1985: La province des complexes annulaires alcalins sursaturés de l'Adrar des Iforas, Mali. Jour. African Earth Sci. *3*, 123-142.

Ball, E., and Caby, R., 1984: Open folding and wrench movements, their relationships with horizontal tectonics in the Pan-African belt of northern Mali. *In*: Klerkx, J., and Michot, J. (eds), Volume en hommage à L. Cahen, Tervuren, Belgium, 75-89.

Bayer, R., and Lesquer, A., 1978: Les anomalies gravimétriques de la bordure orientale du craton ouest-africain: Géométrie d'une suture pan-africaine. Soc. géol. France Bull., Sér. 7, *20*, 863-876.

Bertrand, J.M.L., 1974: Evolution polycyclique des gneiss précambriens de l'Aleksod (Hoggar central, Sahara algérien): Aspects structuraux, pétrologiques, géochimiques et géochronologiques. Centre Rech. Zones Arides (Paris, CNRS), Sér. géol., *19*, 307 p.

Bertrand, J.M.L., Boissonnas, J., Caby, R., Gravelle, M., and Lelubre, M., 1966: Existence d'une discordance dans l'antécambrien du "fossé pharusien" de l'Ahaggar occidental (Sahara central). Acad. Sci. Paris C. R. (D), *262*, 2197-2200.

Bertrand, J.M.L., and Caby, R., 1978: Geodynamic evolution of the Pan-African orogenic belt: A new interpretation of the Hoggar shield (Algerian Sahara). Geol. Rundschau *67*, 357-388.

Bertrand, J.M.L., Caby, R., Ducrot, J., Lancelot, J. R., Moussine-Pouchkine, A., and Saadallah, A., 1978: The late Pan-African intracontinental linear fold belt of the eastern Hoggar (central Sahara, Algeria): Geology, structural development, U-Pb geochronology, tectonic implications for the Hoggar shield. Precambrian Res. *7*, 349-376.

Bertrand, J.M.L., Caby, R., et al., 1977: Carte géologique et gitologique du Hoggar au 1/1.000.000. SONAREM, Alger.

Bertrand, J.M.L., and Davison, I., 1981: Pan-African granitoid emplacement in the Adrar des Iforas mobile belt (Mali): A Rb/Sr isotope study. Precambrian Res. *14*, 333-362.

Bertrand, J.M.L., Dupuy, C., Dostal, J., and Davison, I., 1984: Geochemistry and geotectonic interpretation of granitoids from Central Iforas (Mali, W. Africa). Precambrian Res. *26*, 265-283.

Bertrand, J.M.L., and Lasserre, M., 1976: Pan-African and pre–Pan-African history of the Hoggar (Algerian Sahara) in the light of new geochronological data from the Aleksod area. Precambrian Res. *3*, 343-362.

Bertrand, J.M.L., Michard, A., Carpena, J., Boullier, A. M., Dautel, D., and Ploquin, A., 1984a: Pan-african granitic and related rocks in the Iforas granulites (Mali). Structure, geochemistry and geochronology. *In*: Klerkx, J., and Michot, J. (eds), Volume en hommage à L. Cahen, Tervuren, Belgium, 147-166.

Bertrand, J.M.L., Michard, A., Dautel, D., and Pillot, M., 1984b: Ages U/Pb éburnéens et pan-africains au Hoggar central (Algérie); Conséquences géodynamiques. Acad. Sci. Paris C. R., II, *298*, 643-646.

Bertrand-Sarfati, J., 1969: Etude comparative des édifices stromatolitiques de plusieurs horizons calcaires du Précambrien supérieur de l'Ahaggar occidental (Tanezrouft et Ahnet). Soc. hist. nat. Afrique Nord Bull. *60*, 21-37.

Bertrand-Sarfati, J., 1972: Stromatolites columnaires du Précambrien supérieur du Sahara nord-occidental. Centre Rech. Zones Arides (Paris, CNRS), Sér. géol., *14*, 82 p.

Beuf, S., Biju-Duval, B., Charpal, O. de, Rognon, P., Gariel, O., and Bennacef, A., 1971: Les grés du Paléozoïque inférieur au Sahara. Technip (Paris), 464 p.

Black, R., 1967: Sur l'ordonnance des chaînes métamorphiques en Afrique occidentale. Chron. Mines Rech. Min. *35*, 225-238.

Black, R., Bayer, R., and Lesquer, A., 1980: Reply to comment by Thomas et al: Evidence for late Precambrian plate tectonics in West Africa. Nature *284*, 192.

Black, R., Caby, R., Moussine-Pouchkine, A., Bayer, R., Bertrand, J.M.L., Boullier, A. M., Fabre, J., and Lesquer, A., 1979: Evidence for late Precambrian plate tectonics in West Africa. Nature *278*, 223-227.

Blaise, J., 1967: Le Précambrien du Tazat; sa place dans les structures du Hoggar oriental. Centre Rech. Zones Arides (Paris, CNRS), Sér. géol., *7*, 197 p.

Boissonnas, J., 1973: Les granites à structures concentriques et quelques autres granites tardifs de la chaîne pan-Africaine en Ahaggar (Sahara central, Algérie). Centre Rech. Zones Arides (Paris, CNRS), Sér. géol., *16*, 662 p.

Boissonnas, J., Borsi, S., Ferrara, G., Fabre, J., Fabriés, J., and Gravelle, M., 1969: On the early Cambrian age of two late orogenic granites from west-central Ahaggar (Algerian Sahara). Canadian Jour. Earth Sci. *6*, 25-37.

Boissonnas, J., and Gravelle, M., 1961: Un massif granitique de l'Ahaggar à mode de gisement hybride: Le Tihoiiarène. Soc. géol. France Bull., Sér. 7, *3*, 152-155.

Bondesen, E., 1972: On the structure of the Akwapim range and the Birrimian-Dahomeyan boundary in Ghana. Univ. Abidjan Ann., C, *8*, 85-98.

Bonhomme, M., 1962: Contribution à l'étude géochronologique de la plate-forme de l'Ouest africain. Univ. Clermont-Ferrand Fac. Sci. Ann. *5*, Géol-Minéral. fasc. 5, 62 p.

Boullier, A. M., 1982: Etude structurale du centre de l'Adrar des Iforas (Mali): Mylonites et tectogénèse. Thèse, Univ. Nancy Fac. Sci.

Boullier, A. M., and Bertrand, J.M.L., 1981: Tectonique tangentielle profonde et couloirs mylonitiques dans le Hoggar central polycyclique (Algérie). Soc. géol. France Bull., Sér. 7, *38*, 17-22.

Boullier, A. M., Davison, I., Bertrand, J.M.L., and Coward, M., 1978: L'unité granulitique des

Iforas: Une nappe de socle d'âge pan-Africain précoce. Soc. géol. France Bull., Sér. 7, *20*, 877-882.

Burg, J. P., and Matte, P. J., 1978: A cross section through the French Massif Central and the scope of its Variscan geodynamic evolution. Ztschr. deutsch. geol. Ges. *129*, 429-460.

Caby, R., 1967: Existence du cambrien à faciès continental (''série pourprée,'' ''nigritien'') et importance du volcanisme et du magmatisme de cet âge au Sahara central (Algérie). Acad. Sci. Paris C. R. (D), *264*, 1386-1389.

Caby, R., 1968: Une zone de décrochements à l'échelle de l'Afrique dans le précambrien de l'Ahaggar occidental. Soc. géol. France Bull., Sér. 7, *10*, 577-587.

Caby, R., ms., 1970: La chaîne pharusienne dans le Nord-Ouest de l'Ahaggar (Sahara central, Algérie); sa place dans l'orogénèse du Précambrien supérieur en Afrique. Thèse d'Etat, Univ. Montpellier, 335 p.

Caby, R., 1978: Palégéodynamique d'une marge passive et d'une marge active au Précambrien supérieur: Leur collision dans la chaîne pan-africaine du Mali. Soc. géol. France Bull., Sér. 7, *20*, 857-861.

Caby, R., 1979: Les nappes précambriennes du Gourma dans la chaîne pan-africaine du Mali. Rév. Géol. dyn. Géogr. phys. *21*, 365-376.

Caby, R., 1981a: Associations volcaniques et plutoniques pré-tectoniques de la bordure de la chaîne pan-africaine en Adrar des Iforas (Mali): Un site de type arc-cordillère au Protérozoïque supérieur (abstr.). Coll. African Geology, 11th, Milton Keynes 1981, Abstr., 30.

Caby, R., 1981b: La suture pan-africaine entre Sahara et la Golfe de Guinée: Arguments géologiques relatifs à sa géométrie initiale et à sa géométrie finale (abstr.). 11th Coll. African Geology, Milton Keynes 1981, Abstr., p. 28.

Caby, R., 1983a: La chaîne pharusienne dans le Nord-Ouest de l'Ahaggar. Algérie, Dir. géol., Publ. *47*, 290 p.

Caby, R., 1983b: Pan-African crustal evolution of the Tuareg shield (central Sahara) and the Arabian shield: A comparison. King Abdulaziz Univ. Fac. Earth Sci. Bull. *6*, 89-99.

Caby, R., and Andreopoulos-Renaud, U., 1983: Age à 1800 Ma du magmatisme sub-alcalin associé aux métasédiments monocycliques dans la chaîne Pan-Africaine du Sahara central. Jour. African Earth Sci. *1*, 193-198.

Caby, R., and Andreopoulos-Renaud, U., 1985: Etude pétrostructurale et géochronologie U/Pb sur zircon d'une métadiorite quartzique de la chaîne pan-africaine de l'Adrar des Iforas (Mali). Soc. géol. France Bull., Sér. 8, *1*, 899-903.

Caby, R., and Andreopoulos-Renaud, U., ms.: Le Hoggar oriental, block exotique cratonisé à 730 Ma dans la chaîne pan-africaine du Nord du continent africain.

Caby, R., and Andreopoulos-Renaud, U., in prep: Un arc magmatique d'âge Protérozoïque supérieur: Cadre géologique, géochronologie U/Pb sur zircon et conséquences paléogéodynamiques pour la chaîne pan-africaine en Afrique de l'Ouest.

Caby, R., Andreopoulos-Renaud, U., and Gravelle, M., 1982: Cadre géologique et géochronologique U/Pb sur zircon des batholites précoces dans le segment pan-africain du Hoggar central (Algérie). Soc. géol. France Bull., Sér. 7, *24*, 677-684.

Caby, R., Andreopoulos-Renaud, U., and Lancelot, J. R., 1985: Les phases tardives de l'orogenèse pan-africaine dans l'Adrar des Iforas oriental (Mali): Lithostratigraphie des formations molassiques et géochronologie U/Pb sur zircon de deux massifs intrusifs. Precambrian Res. *28*, 187-199.

Caby, R., Bertrand, J.M.L., and Black, R., 1981: Pan-African closure and continental collision in the

Hoggar-Iforas segment, central Sahara. *In*: Kröner, A., ed, Precambrian Plate Tectonics. Elsevier, 407-434.

Caby, R., Chikhaoui, M., Dupuy, C., Dostal, J., and Mevel, C., 1980: Magmatic evolution from oceanic tholeiites to calc-alkaline rocks: Geochemical and petrological evidences from upper Proterozoic rocks of Northwest Africa (abstr.). 26th Internat. Geol. Congr., Paris 1980, Abstr. *1*, 26.

Caby, R., Dostal, J., and Dupuy, C., 1977: Upper Proterozoic volcanic graywackes from northwestern Hoggar (Algeria)—geology and geochemistry. Precambrian Res. *5*, 283-297.

Caby, R., and Fabre, J., 1981a: Late Proterozoic to Early Paleozoic diamictites, tillites and associated glaciogenic sediments of the Série pourprée of western Hoggar, Algeria. *In*: Hambrey, M. J., and Harland, W. B., eds, Earth's pre-Pleistocene glacial record (IGCP Project 38: Pre-Pleistocene Tillites). Cambridge Univ. Press, 140-145.

Caby, R., and Fabre, J., 1981b: Tillites in the latest Precambrian strata of the Touareg shield (central Sahara). *In*: Hambrey, M. J., and Harland, W. B., eds, Earth's pre-Pleistocene glacial record (IGCP Project 38: Pre-Pleistocene Tillites). Cambridge Univ. Press, 146-149.

Caby, R., and Moussine-Pouchkine, A., 1978: Le horst birrimien de Bourré (Gourma oriental, République du Mali): Nature et comportement au cours de l'orogénèse pan-africaine. Acad. Sci. Paris C. R. (D), *287*, 5-8.

Caby, R., and Moussu, H., 1968: Une grande Série detritique du Sahara: Stratigraphie, paléogéographie et évolution structurale de la ''Série pourprée'' dans l'Aseg'rad et le Tenezrouft oriental (Sahara algérien). Soc. géol. France Bull., Sér. 7, *9*, 876-882.

Caby, R., Pécher, A., and Le Fort, P., 1983: Le grand chevauchement central himalayen: Nouvelles données sur le métamorphisme inverse à la base de la dalle du Tibet. Rév. Géol. dyn. Géogr. phys. *24*, 89-100.

Caen-Vachette, M., 1979: Orthogneiss birrimiens et migmatisation Pan-Africaine au Togo-Bénin (abstr.). Coll. Géologie africaine, 10th, Montpellier 1979, Rés., 22.

Carpena, J., 1982: Late thermal history of the Hoggar shield (western Africa) (abstr.). 5th Intern. Conf. Geochron. Cosmochron. Isotope Geology, Japan 1982, Abstr., 6-7.

Chikhaoui, M., Dupuy, C., and Dostal, J., 1978: Geochemistry of late Proterozoic volcanic rocks from Tassendjanet area (N. W. Hoggar, Algeria). Contr. Min. Pet. *66*, 157-164.

Clauer, N., 1976: Géochimie isotopique du strontium des milieux sédimentaires: Application à la géochronologie de la couverture du craton ouest-africain. Univ. Strasbourg Sci. géol. Mém. *45*, 256 p.

Clauer, N., Caby, R., Jeanette, D., and Trompette, R., 1982: Geochronology of sedimentary and meta-sedimentary Precambrian rocks of the West African craton. Precambrian Res. *18*, 53-71.

Conrad, J., 1981: La part des déformations post-hercyniennes et de la néotectonique dans la structuration du Sahara central algérien, un domaine relativement mobile de la plate-forme africaine. Acad. Sci. Paris C. R. (D), *292*, 1053-1056.

Davison, I., ms., 1980: Structural geology and Rb/Sr geochronology of Precambrian rocks from Eastern Iforas (Mali). Ph.D. thesis, Univ. Leeds.

de la Boisse, H., ms., 1979: Pétrologie et géochronologie de roches cristallophylliennes du bassin du Gourma (Mali). Conséquences géodynamiques. Thèse 3e cycle, Univ. Montpellier, 54 p.

de la Boisse, H., 1981: Sur le métamorphisme du micaschiste éclogitique du Takamba (Mali) et ses conséquences paléogéodynamiques au Précambrien supérieur. Soc. géol. France C. R. Somm. *1981*, 97-100.

de la Boisse, H., and Lancelot, J. R., 1977: A propos de l'évènement à 1,000 Ma en Afrique occidentale: l. Le ''granite'' de Bourré. Soc. géol. France Bull., Sér. 7, *19*, 223-226.

Dostal, J., Caby, R., and Dupuy, C., 1979: Metamorphosed alkaline intrusions and dyke complexes within the Pan-African belt of western Hoggar (Algeria): Geology and geochemistry. Precambrian Res. *10*, 1-20.

Ducrot, J., de la Boisse, H., Andreopoulos-Renaud, U., and Lancelot, J. R., 1979: Synthèse géochronologique sur la succession des évènements magmatiques pan-africains au Maroc, dans l'Adrar des Iforas et dans l'Est Hoggar (abstr.). 10th Coll. Géol. africaine, Montpellier 1979, Rés., 40-41.

Fabre, J., 1978: Introduction à la géologie du Sahara algérien. I. Couverture phanérozoïque. Société nationale d'Edition et de Diffusion, Alger.

Fabre, J., 1982a: Pan-African volcano-sedimentary formations in the Adrar des Iforas (Mali). Precambrian Res. *19*, 201-214.

Fabre, J., 1982b: Carte géologique (et gravimétrique) de l'Adrar des Iforas (1/500.000). Mali Dir. natl. Géologie Mines, Bamako.

Fabre, J., Ba, H., Black, R., Caby, R., Leblanc, M., and Lesquer, A., 1982: La chaîne Pan-Africaine, son avant-pays et la zone du suture au Mali. Notice explicative de la carte géologique et gravimétrique de l'Adrar des Iforas au 1/500.000. Mali Dir. natl. Géologie Mines, Bamako.

Ferrara, G., and Gravelle, M., 1966: Radiometric ages from western Ahaggar (Sahara) suggesting an eastern limit for the west African craton. Earth Planet. Sci. Lett. *1*, 319-324.

Fitches, W. R., 1970: "Pan-African orogeny" in the coastal region of Ghana. Nature *226*, 744-746.

Fitches, W. R., Ajibade, A. C., Egbuniwe, I. G., Holt, R. W., and Wright, J. B., 1985: Late Proterozoic schist belts and plutonism in NW Nigeria. Geol. Soc. London Jour. *142*, 319-337.

Grant, N. K., 1969: The late Precambrian to early Paleozoic Pan-African orogeny in Ghana, Togo, Dahomey, and Nigeria. Geol. Soc. America Bull. *80*, 45-55.

Grant, N. K., 1970: Geochemistry of Precambrian basement rocks from Ibadan, southwestern Nigeria. Earth Planet. Sci. Lett. *10*, 29-38.

Grant, N. K., 1978: Structural distinction between a metasedimentary cover and an underlying basement in the 600-m.y.-old Pan-African domain of northwestern Nigeria, West Africa. Geol. Soc. America Bull. *89*, 50-58.

Grant, N. K., Hickman, M. H., Burkholder, F. R., and Powell, J. L., 1972: Kibaran metamorphic belt in Pan-African domain of West Africa? Nature, Phys. Sci., *238*, 90-91.

Gravelle, M., ms., 1969: Recherches sur la géologie du socle précambrien de l'Ahaggar centro-occidental dans la région de Silet-Tibehaouine. Thèse d'Etat, Univ. Paris, 781 p.

Lancelot, J. R., Boullier, A. M., Maluski, H., and Ducrot, J., 1983: Deformation and related rediochronology in a late Pan-African mylonitic shear zone, Adrar des Iforas (Mali). Contr. Min. Pet. *82*, 312-326.

Lancelot, J. R., Vitrac, A., and Allègre, C. J., 1976: Uranium and lead isotopic dating with grain-by-grain zircon analysis: A study of a complex geological history with a single rock. Earth Planet. Sci. Lett. *29*, 357-366.

Latouche, L., ms., 1978: Le Précambrien de la région des Gour Oumelalen. Thèse d'Etat, Univ. Paris VII, 255 p.

Latouche, L., and Vidal, P., 1974: Géochronologie du Précambrien de la région des Gour Oumelalen (Nord-Est de l'Ahaggar – Algérie). Un exemple de mobilisation du strontium radiogénique. Soc. géol. France Bull., Sér. 7, *16*, 195-203.

Lazzarotto, V., Maraga-Piovano, F., Martinotti, G., et al., 1983: Carte géologique de la République du Benin Nord (Projet FED) au 1/200,000. Ist. Ric. BREDA, Borgo San Dalmazzo, Italy.

Leblanc, M., 1981: The late Proterozoic ophiolites of Bou Azzer (Morocco): Evidence for Pan-African plate tectonics. *In*: Kröner, A., ed, Precambrian Plate Tectonics. Elsevier, 435-451.

Le Fort, P., 1975: Himalayas: The collided range. Present knowledge of the continental arc. Am. Jour. Sci. *275A*, 1-44.

Lelubre, M., 1952: Recherches sur la géologie de l'Ahaggar central et occidental (Sahara central). Algérie Serv. Carte géol. Bull., Sér. 2, *22*, vol. 1: 354 p., vol. 2: 385 p.

Lesquer, A., and Moussine-Pouchkine, A., 1980: Les anomalies gravimétriques de la boucle du Niger. Leur signification dans le cadre de l'orogenèse panafricaine. Canadian Jour. Earth Sci. *17*, 1538-1545.

Liegeois, J. P., Bertrand, H., Black, R., Caby, R., and Fabre, J., 1983: Permian alkaline undersaturated and carbonatite province, and rifting along the West African craton. Nature *305*, 42-43.

Liegeois, J. P., and Black, R., 1984: Pétrographie et géochronologie Rb-Sr de la transition calco-alcaline-alcaline fini-panafricaine dans l'Afrar des Iforas (Mali): Accrétion crustal au Précambrien supérieur. *In:* Klerkx, J., and Michot, J. (eds), Volume en hommage à L. Cahen, Tervuren, Belgium, 115-146.

Ly, S., Albouy, Y., Chauvin, M., Foy, R., Lachaud, J. C., and Lesquer, A., 1980: Apport de la gravimétrie à la compréhension de la chaîne panafricaine dans l'Adrar des Iforas. ORSTOM Cahiers, Sér. géophys., *17*, 37-57.

Menot, R. P., 1980: Les massifs basiques et ultrabasiques de la zone mobile pan-africaine au Ghana, Togo et Bénin: État de la question. Soc. géol. France Bull., Sér. 7, *22*, 297-304.

Menot, R. P., 1982: Les eclogites des Mts. Lato: Un témoin de l'évolution tectonométamorphique de la chaîne pan-africaine du Sud-Togo (Afrique de l'Ouest) (abstr.). Terra Cognita *2*, 320.

Menot, R. P., 1985: The eclogites of the Lato Hills (South Togo, West Africa); relics from early tectonometamorphic evolution of the Pan-African orogeny. Chemical Geology *50*, 313-330.

Menot, R. P., and Seddoh, K. F., 1977: Données nouvelles sur la pétrographie de l'Atacorien au Togo (région des plateaux, Atakpamé). Soc. géol. France Bull., Sér. 7, *19*, 331-334.

Moussine-Pouchkine, A., and Bertrand-Sarfati, J., 1978: Le Gourma: Un aulacogène du Précambrien supérieur? Soc. géol. France Bull., Sér. 7, *20*, 851-857.

Ouzegane, K., ms., 1981: Le métamorphisme polyphasé granulitique de la région de Tamanrasset (Hoggar central). Thèse 3e cycle, Univ. Paris VII, 171 p.

Picciotto, E., Ledent, D., and Lay, C., 1965: Etude géochronologique de quelques roches du socle cristallophyllien du Hoggar (Sahara central). C.N.R.S. Coll. Internat. *151*, 275-288; Sci. Terre, *10*, 481-495 (1966).

Pin, C., ms., 1979: Géochronologie U-Pb et microtectonique des Séries métamorphiques anté-stéphaniennes de l'Aubrac et de la région de Marvejols (Massif central). Thèse 3e cycle, Univ. Montpellier.

Rahaman, O. M., and Lancelot, J. R., ms.: Continental crust evolution in South-Western Nigeria. Constraints from U/Pb dating of pre–Pan-African gneisses.

Reichelt, R., 1972: Géologie du Gourma (Afrique occidentale); un "seuil" et un bassin du précambrien supérieur: Stratigraphie, tectonique, métamorphisme. Bur. Rech. Géol. Min. Mém. *53*, 213 p.

Sautter, V., 1982: Première étude de paragénèse de hautes-pressions dans le socle du Hoggar central polycyclique (sud-Algérien) (abstr.). Terra Cognita *2*, 321.

Sautter, V., 1985: The eclogite paragenesis from the Aleksod basement, central Hoggar, South Algeria. Chemical Geology *50*, 331-347.

Shackleton, R. M., 1971: On the south-eastern increase in deformation and metamorphism at the margin of the Pan-African domain in Ghana. Leeds Univ., Res. Inst. African Geology Ann. Rept. *15*, 2-7.

Simpara, N., 1978: Etude géologique et structurale des unités externes de la chaîne Pan-Africaine (600

Ma) des Dahomeyides dans la région de Bassar (Togo). Univ. Aix-Marseille III, Lab. Sci. Terre Trav., Sér. B, *13*, 164 p.

Trompette, R., 1980: La chaîne panafricaine des Dahomeyides et le bassin des Volta (Bordure sud-est du craton Ouest-africain). Bur. Rech. Géol. Min. Mém. *92*, 9-62.

Tubosun, I. A., Lancelot, J. R., Rahaman, M. A., and Ocan, O., 1984: U-Pb Pan African ages of two charnockite-granite associations from Southwestern Nigeria. Contr. Min. Pet. *88*, 188-195.

Vialette, Y., and Vitel, G., 1979: Geochronological data on the Amsinassene-Tefedest block (central Hoggar, Algerian Sahara) and evidence for its polycyclic evolution. Precambrian Res. 9, 241-254.

Vitel, G., ms., 1979: La région Téfedest-Atakor du Hoggar central (Sahara). Evolution d'un complexe granulitique précambrien. Thèse, Univ. Paris VII, 323 p.

Part IV

SOUTH AMERICA

Chapter 9

TECTONIC EVOLUTION OF THE SOUTHERN ANDES, TIERRA DEL FUEGO: A SUMMARY

A. G. MILNES

Swiss Federal Institute of Technology, Zürich

The southernmost east-west trending segment of the Southern Andes (Figure 9-1) differs from the main Andean chain in being inactive and aseismic (Zeil 1979) and in showing many features generally associated with collision-type orogenic zones (Nelson et al. 1980). On Tierra del Fuego and neighboring islands, the chain can be subdivided into several morphotectonic zones in a north-south cross section (Figure 9-2).

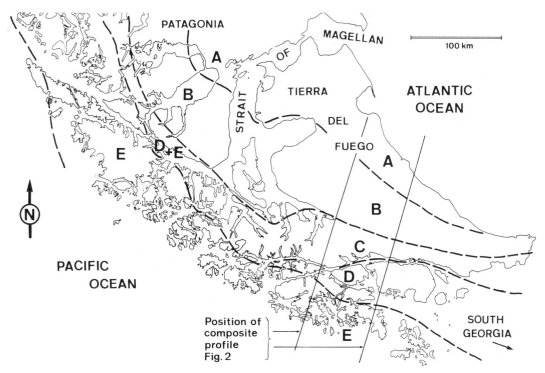

FIGURE 9-1. Sketch map of Tierra del Fuego (southern tip of South America), showing the boundaries of the main morphotectonic zones.

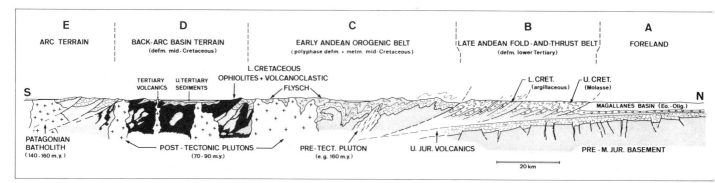

FIGURE 9-2. Composite cross section through Tierra del Fuego (see Figure 9-1 for position).

Zone A. Northern Plains (Foreland)

The Northern Plains is an undisturbed platform of pre-Mesozoic continental basement, overlain by a cover sequence consisting of thick Upper Jurassic volcanics, thin Cretaceous clastics, and thick Eocene-Oligocene molasse (Magallanes Basin, Natland et al. 1974).

Zone B. Subandean Foothills (Late Andean Fold-and-Thrust Belt)

This thin-skinned fold-and-thrust belt involves mainly thick Cretaceous sequences (argillaceous Lower Cretaceous, coarse clastic Upper Cretaceous). The present deformed wedge represents a telescoping of the Upper Cretaceous basin, which originally was 90 km wide, to about half its original width, mainly during early Tertiary times (Mingramm 1981, Winslow 1981).

Zone C. Cordillera Darwin (Early Andean Orogenic Belt)

In this zone mid-Cretaceous polyphase deformation and regional metamorphism affected the southern margin of the pre-Mesozoic continental basement, together with its cover of Upper Jurassic–Lower Cretaceous volcanics and related intrusions and dike swarms (Dalziel and Palmer 1979, Nelson et al. 1980). The whole edifice is cut by post-orogenic granite plutons, intruded 70-90 Ma ago (Figure 9-3).

Zone D. Inner Islands and South Georgia (Back-Arc Basin Terrain)

This fold-and-thrust belt involves thick volcanoclastic Lower Cretaceous flysch and associated ophiolites (Katz and Watters 1966, Winn 1978, Bruhn 1979, Saunders et al. 1979, Andrews-Speed 1980, Tanner 1982). The age of deformation and granite intrusion is the same as in zone C. In contrast, however, some much younger events have been identified here, including Tertiary volcanism and the development of a small upper Tertiary sedimentary basin.

Zone E. Outer Islands (Volcanic Arc Terrain)

This zone consists mainly of calc-alkaline plutonic rocks (Patagonian Batholith)—as well as associated agglomerates and volcanoclastics—of Upper Jurassic age (Bruhn et al. 1978, Suarez 1979).

FIGURE 9-3. View of the Mount Shipton massif (highest point 2,469 m), Cordillera Darwin, looking east along the Beagle Channel (*right*).

The pre-Jurassic history of southernmost South America has recently been described in some detail (Dalziel 1982). The tectonic evolution of the Southern Andes proper can be thought of as beginning in the Upper Jurassic, when crustal distention occurred over wide areas of the Gondwanan continental platform. Extensive normal faulting and silicic-intermediate volcanism took place behind a main volcanic arc (sited above the present Patagonian Batholith, zone E), above a subduction zone dipping toward the continent. During Early Cretaceous times, extension became concentrated in a narrow zone immediately behind the main arc, resulting in a rapidly subsiding back-arc basin partially floored with newly formed oceanic crust (zone D, see also Katz 1973, Dalziel 1981). A compressional regime was established in the Middle Cretaceous, destroying the back-arc basin (Bruhn and Dalziel 1977), partially thrusting basin sediments and ocean floor over the continental margin, and involving the margin itself in a deep ductile shear zone with a complicated movement history (early Andean orogeny, zones C and D). These orogenic movements and the subsequent Late Cretaceous granite intrusions are presumed to be associated with continuing subduction of Pacific oceanic crust under the continent from the south.

The early Andean orogen was consolidated during the Upper Cretaceous and suffered rapid uplift and erosion, shedding debris toward the continent into the developing Magallanes foredeep. Throughout early Tertiary times, the foredeep clastic successions were successively involved in a foreland fold-and-thrust belt (zone B). The main décollement lay within the

Lower Cretaceous argillites and descended southward into an intra-basement shear zone, at depth below the early Andean orogen. The plate-tectonic situation is now more complicated and is possibly dominated by major transform motion (Dalziel et al. 1975, de Wit 1977, Winslow 1982) that leads over to the late Tertiary development of the North Scotia Ridge.

 Because the plate-tectonic framework is potentially determinable in considerable detail, and because the geological record is unusually complete, the Andean mountain chain of Tierra del Fuego is likely to become a unique test case for relating tectonic features on many different scales. Many of its features (polyphase ductile deformation of both cover and basement rocks, regional metamorphism, strong post-orogenic uplift, and so forth; cf. Milnes 1978) suggest a comparison to collisional orogens such as the Alps. However, one collisional phenomenon is strikingly absent; there seems to have been no intense dismembering (e.g., nappe, thrust sheet, imbricate zone, or mélange formation) at any stage in the orogeny.

REFERENCES

Andrews-Speed, C. P., 1980: The geology of central Isla Hoste, southern Chile; sedimentation, magmatism and tectonics in part of a Mesozoic back-arc basin deformation in the Andes of Tierra del Fuego. Geol. Mag. *117*, 339-349.

Bruhn, R. L., 1979: Rock structures formed during back-arc basin deformation in the Andes of Tierra del Fuego. Geol. Soc. America Bull. *90*, I-998–I-1012.

Bruhn, R. L., and Dalziel, I.W.D., 1977: Destruction of the Early Cretaceous marginal basin in the Andes of Tierra del Fuego. *In*: Talwani, M., and Pitman, W. C. (eds), Island Arcs, Deep Sea Trenches and Back-Arc Basins. Am. Geophys. Union, Maurice Ewing Series, *1*, 395-405.

Bruhn, R. L., Stern, C. R., and de Wit, M. J., 1978: Field and geochemical data bearing on the development of a Mesozoic volcano-tectonic rift zone and back-arc basin in southernmost South America. Earth Planet. Sci. Lett. *41*, 32-46.

Dalziel, I.W.D., 1981: Back-arc extension in the southern Andes: A review and critical reappraisal. Royal Soc. London Phil. Trans., Ser. A, *300*, 319-335.

Dalziel, I.W.D., 1982: The early (pre-Middle Jurassic) history of the Scotia Arc region: A review and progress report. *In*: Craddock, C. (ed), Antarctic Geoscience. Univ. Wisconsin Press, Madison, 111-126.

Dalziel, I.W.D., Dott, R. H., Winn, R. D., Jr., and Bruhn, R. L., 1975: Tectonic relations of South Georgia Island to the southernmost Andes. Geol. Soc. America Bull. *86*, 1034-1040.

Dalziel, I.W.D., and Palmer, K. F., 1979: Progressive deformation and orogenic uplift at the southern extremity of the Andes. Geol. Soc. America Bull. *90*, I-259–I-280.

Katz, H. R., 1973: Contrasts in tectonic evolution of orogenic belts in the Southeast Pacific. Royal Soc. New Zealand Jour. *3*, 333-362.

Katz, H. R., and Watters, W. A., 1966: Geological investigations of the Yahgan formation (upper Mesozoic) and associated igneous rocks of Navarino Island, southern Chile. New Zealand Jour. Geol. Geophys. *9*, 323-359.

Milnes, A. G., 1978: Structural zones and continental collision, central Alps. Tectonophysics *47*, 369-392.

Mingramm, A.R.G., ms., 1981: Evolution scheme of Fuegian Andes along Los Cerros—P. Hardy section. Shell Research.

Natland, M. L., Gonzales, E., Canon, A., and Ernst, M., 1974: A system of stages for correlation of Magallanes Basin sediments. Geol. Soc. America Mem. *139*, 126 p.

Nelson, E. P., Dalziel, I.W.D., and Milnes, A. G., 1980: Structural geology of the Cordillera Darwin: Collisional-style orogenesis in the southernmost Chilean Andes. Eclogae Geol. Helvetiae *73*, 727-751.

Saunders, A. D., Tarney, J., Stern, C. R., and Dalziel, I.W.D., 1979: Geochemistry of Mesozoic marginal basin floor igneous rocks from southern Chile. Geol. Soc. America Bull. *90*, I-237–I-258.

Suarez, M., 1979: A late Mesozoic island arc in the southern Andes, Chile. Geol. Mag. *116*, 181-190.

Tanner, P.W.G., 1982: Geologic evolution of South Georgia. *In*: Craddock, E. (ed), Antarctic Geoscience. Univ. Wisconsin Press, Madison, 167-176.

Winn, R. D., Jr., 1978: Upper Mesozoic flysch of Tierra del Fuego and South Georgia Island: A sedimentological approach to lithosphere plate restoration. Geol. Soc. America Bull. *89*, 533-547.

Winslow, M. A., 1981: Mechanisms for basement shortening in the Andean foreland fold belt of southern South America. *In*: McClay, K. R., and Price, N. J. (eds), Thrust and Nappe Tectonics. Geol. Soc. London Spec. Publ. *9*, 513-528.

Winslow, M. A., 1982: The structural evolution of the Magallanes Basin and neotectonics in the southernmost Andes. *In*: Craddock, C. (ed), Antarctic Geoscience. Univ. Wisconsin Press, Madison, 143-154.

de Wit, M. J., 1977: The evolution of the Scotia Arc as a key to the reconstruction of southwestern Gondwanaland. Tectonophysics *37*, 53-81.

Zeil, W., 1979: The Andes; a Geological Review. Beitr. Reg. Geol. Erde, Borntraeger, Berlin, Stuttgart, *13*, 260 p.

Chapter 10

STRUCTURE AND EVOLUTION OF THE PERUVIAN ANDES

François Mégard

Centre Géologique et Géophysique, Montpellier

The Peruvian Andes record the phenomena that have constructed and deformed the western margin of the South American continent during the last two billion years. In the central Andes of Peru and Bolivia and the southern Andes of Chile and Argentina, excluding the Magellan area, the Mesozoic and Tertiary Andean orogenic belt lies entirely on a Precambrian sialic basement partly covered by Paleozoic sediments that were deformed during the Hercynian. No ophiolitic suture of post-Paleozoic age has been identified; it is thus by definition a "liminal" orogen, formed on the edge of the continent in an active margin regime, but without accretion of exotic terranes. Such is not the case in the Ecuadorian and Colombian Andes where, as in Canada and the United States, island arcs and subduction complexes were accreted.

This paper aims at providing a synthesis of the present state of knowledge of the Peruvian Andes, highlighting tectonic aspects and emphasizing data obtained since the publications of Audebaud et al. (1973), Mégard (1973, 1978), Dalmayrac et al. (1980), and Cobbing et al. (1981).

STRUCTURAL ZONING

The present features of the Andes are basically related to their Cenozoic evolution. The range was in existence in the Miocene, and the Subandean Zone was incorporated into the Andes during the Pliocene. Intense Quaternary tectonic deformation indicates that the Andes are presently active. The main longitudinal morphostructural zones of the Peruvian Andes— Coastal Zone, Western Cordillera, Altiplano, Eastern Cordillera, and Subandean Zone—are the product of Cenozoic evolution, although their boundaries are commonly old, reactivated structural features (Figure 10-1).

The *Coastal Zone* exists on land only north of 6° S and south of 13°30′ S, where it is from 50 to 100 km wide. It includes a positive coastal ridge, where the Precambrian basement crops out in southern Peru and a Hercynian and perhaps Caledonian substratum is exposed in the north. This old axis has acted as a geanticline since the Paleozoic, with the result that its sedimentary cover is thin, possesses major stratigraphic gaps, and has not been much de-

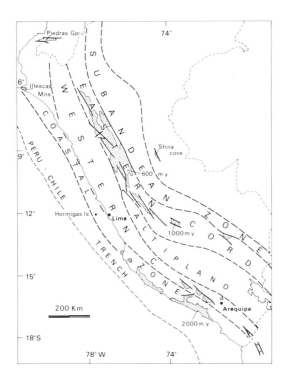

FIGURE 10-1. Structural zoning of the Peruvian Andes and location of the main Precambrian outcrops (stippled areas).

formed. During the Mesozoic, it constituted the Paracas geanticline, which formed the western boundary of the West Peruvian trough. After the folding and subsequent uplift of the Western Cordillera, Tertiary shelf basins developed between the Paracas geanticline and the western edge of the Western Cordillera; between 5° S and 15° S the basins are marine and most are submerged; farther south they are continental.

The *Western Cordillera* is very asymmetrical. Its western slope is from 100 to 150 km wide and extends from 0 to over 5,000 m in elevation between 6° S and 13°30′ S, whereas its eastern slope is only a few kilometers wide and overhangs the Altiplano by 500 to 1,000 m. This morphological dissymmetry reflects a structural dissymmetry: the eastern edge is a fault zone that was active until the late Tertiary; south of 13°30′ S, the Western Cordillera is merely the western limit of the Altiplano, on which young volcanoes are superimposed. Only Mesozoic and Tertiary rocks are exposed in the Western Cordillera. The Mesozoic sequence thins from west to east; rich in volcanic intercalations in the west, it becomes almost entirely sedimentary to the east. North of 13°30′ S, strain becomes more intense from west to east and shortening is concentrated on the northeastern edge of the Cordillera.

The *Altiplano* corresponds to high plateaus extending from 10°30′ S to northern Argentina; in central Peru these plateaus have been deeply dissected by erosion. Their altitudes range from 3,800 to 4,500 m. The Paleozoic substratum crops out locally and is covered by relatively thin shelf strata of Jurassic and Cretaceous age. During the Tertiary, longitudinal intermontane basins, by and large rapidly subsiding, developed on the Altiplano; they received as much as 8,000 m of volcanic and clastic sediments. North of 10°30′ S, the Altiplano disappears and thrust sheets of the northeastern edge of the Western Cordillera are in direct contact with the Eastern Cordillera. On the whole, folding is moderate and structures are simple on the Altiplano.

The *Eastern Cordillera*, locally rising to 6,000 m, is a vast anticlinorium cut into blocks by numerous subvertical longitudinal and transverse faults. The Precambrian basement and Paleozoic substratum crop out extensively. The Mesozoic cover has been preserved only in a few down-faulted blocks; it is not very thick, and sedimentation was frequently interrupted by hiatuses, disconformities, or even angular unconformities. Andean deformation is expressed in the Eastern Cordillera by open folds and steep thrusts commonly dipping to the southwest.

The *Subandean Zone* is characterized by hills with elevations reaching 2,000 m near the Eastern Cordillera and decreasing to the east. Farther east lies a very flat basin within which the great Amazonian rivers meander. The basement crops out in the Shira uplift (see Figure 10-8, section B; this figure is discussed in more detail in the subsection entitled "The Structure of the Andes," below). The cover is thick near the foot of the Andes and conformable to disconformable from the Paleozoic to the Pliocene, forming an eastward-tapering wedge. A recently recognized fold-and-thrust belt of Pliocene age occupies most of the Subandean Zone.

THE PRECAMBRIAN AND HERCYNIAN SUBSTRATUM

THE PRECAMBRIAN OF PERU

South of the equator, the Andean range is located on the periphery of the Brazilian shield. The western part of the shield is covered by a sedimentary wedge of Paleozoic to Quaternary age that thickens progressively westward to 5,000 to 10,000 m in the Subandean Zone. Within the Andes, Precambrian outcrops are numerous in the entire Eastern Cordillera and on the coast, but in the Western Cordillera they are known only south of 15° S in southern Peru (Figure 10-1). In the Brazilian shield and in the Andean Precambrian massifs, a distinction can be made between old cores—generally with high-grade metamorphism—with ages between 3,000 and 1,000 Ma and Upper Precambrian rocks that suffered deformation and middle- or low-grade metamorphism around 600 Ma.

The main old core in the Peruvian Andes is the Arequipa massif (Mégard et al. 1971, Shackleton et al. 1979, Dalmayrac et al. 1980). The older rocks are staurolite and andalusite schists and granulite-facies migmatitic gneisses containing hypersthene and sillimanite; they are related to a metamorphic event M1 that has been dated at about 1,900 Ma by the Rb/Sr whole-rock method (Cobbing et al. 1977, Shackleton et al. 1979) and at about 1,950 Ma by the U/Pb method on zircons (Dalmayrac et al. 1977). According to Shackleton et al. (1979), the episode of intrusion and metamorphism that follows M1 is about 450 Ma old. It must be noted, however, that no deformation or metamorphism of this age is known in the lower Paleozoic of the Eastern Cordillera, and that other data (Stewart et al. 1974, Dalmayrac et al. 1977) point to a late Precambrian event at about 600 Ma in the Arequipa massif. In the Eastern Cordillera near 12°30′ S and 73°30′ W, an isolated outcrop of granulite-facies gneisses yielded zircons that give an age of about 1,000 Ma by the U/Pb method (Dalmayrac et al. 1980). Other granulitic cores of small dimensions are scattered in the late Precambrian schistose belt; they may also represent remnants of the old sialic basement that were the source rocks of the late Precambrian clastic sediments.

The Upper Precambrian is well exposed in the Eastern Cordillera. Along the coast, the metamorphic rocks of the Hormigas Islands, located 70 km west of Lima (Kulm et al. 1981), the Illescas Mountains at 6° S latitude (Caldas et al. 1980), and the Piedras Group in southernmost Ecuador (Feininger 1982) may also be of upper Precambrian age. In the Eastern Cordillera the series consists mainly of schists, quartzites, and metagraywackes with scarce marble layers; intercalations of greenschist are common.

Higher-grade metamorphic rocks are also present, and granulite-facies conditions have been reached locally; this high-grade metamorphism has been dated at about 600 Ma by the U/Pb method on zircons (Dalmayrac et al. 1980). Four episodes of ductile deformation are recorded in the late Precambrian rocks: the first is contemporaneous with the main metamorphic pulse, which is Barrovian; the second is contemporaneous with the second and last metamorphic pulse, which is of the Abukuma type (Dalmayrac et al. 1980). Some granitic bodies were intruded before the third pulse of deformation, which transformed them into cataclastic orthogneisses. Highly serpentinized ultramafic complexes that form outcrops from several hundred meters to more than a kilometer in size have been identified between 9°15′ S and 11°15′ S latitude. According to Carlier et al. (1982), they were emplaced as thrust slices during the second phase of deformation. The Late Precambrian metamorphic rocks belong to a belt that follows approximately the present Andean trend; folding and deformation in this belt are of about the same age (about 600 Ma) as folding and deformation in the Brasilide orogens of Brazil and in the Pan-African or Baikalian orogens (Dalmayrac et al. 1980).

THE HERCYNIAN BELT IN PERU

Marine strata of Paleozoic age overlie peneplained rocks of the Late Precambrian orogenic belt. In southern Peru and Bolivia these sedimentary rocks occupy a subsiding trough—the Peru-Bolivia trough—300 to 500 km wide, trending approximately northwest-southeast and bounded by the Brazilian shield to the east and the Arequipa massif to the west. North of 14° S, the western limit of the trough is unknown; its eastern extension is not known precisely either, since the Paleozoic deposits have been eroded in a large part of the Peruvian Amazon basin. At times the Peru-Bolivia trough was directly connected to the east-west trending Amazonian trough of Brazil. A few rocks of pre–Middle Ordovician age in the Eastern Cordillera have been described by Marocco and García-Zabaleta (1974); these rocks consist of volcanic and sandstone layers. The first well-dated Paleozoic strata are graptolite-bearing shales, of Arenigian to Llanvirnian age, overlying basal conglomerates and sandstones that overlap all the older metamorphic and non-metamorphic units. Sedimentation was continuous from the Middle Ordovician to the Upper Devonian except for a short hiatus during the uppermost Ordovician and/or the lowermost Silurian. At many places in the Eastern Cordillera of southern Peru, the post-hiatus transgression is of Llandoverian age and begins with a marine tillite (Laubacher 1978). All the other Ordovician, Silurian, and Devonian sedimentary rocks are shale, siltstone, and sandstone. Both deltaic and flysch-type sediments are common. During Ordovician and Silurian times, both the Brazilian shield and the Arequipa massif were source areas of clastics (Dalmayrac et al. 1980). In the Devonian most of the Arequipa Massif was covered by a marine transgression. The resulting aggregate thickness of Ordovician, Silurian, and Devonian strata is 10,000 to 15,000 m along the trough axis, which coincides approximately with the Eastern Cordillera in southern Peru. A substantial amount of crustal thinning is obviously required to account for such a great thickness, but no genuine oceanic basin ever opened, as demonstrated by the almost complete lack of synsedimentary magmatism.

Strata of Ordovician to Devonian age were strongly deformed during the eo-Hercynian phase, which occurred during the latest Devonian or the Early Mississippian (Mégard 1967, Mégard et al. 1971). The eo-Hercynian orogen (Figure 10-2) has been recognized from 7°30′ S to 33° S latitude (Mégard et al. 1971, Dalmayrac et al. 1980, Martinez, 1980). In Peru and northern Bolivia, at least as far south as 19°30′ S, it is clearly an intracontinental belt, built entirely upon sialic crust. Deformation is related to closure of a northwest-trending trough floored by a thinned crust. The best exposed and most complete cross section of this orogen is the northeast-southwest section between Lake Titicaca (near the southeastern corner of Figure 10-2) and the Subandean Zone in southern Peru (see Figure 10-8, section C; see also Figure 10-2), studied by Laubacher (1978). In this section eo-Hercynian folding begins approximately at the boundary between the Subandean Zone and the Eastern Cordillera. Axial surfaces of the folds form an asymmetrical fan. To the northeast, they dip 80 to 70° SW, farther west they become vertical, and even farther west they are inclined northeastward. Folds are tight and of chevron-type and are associated with an axial-plane fracture cleavage. Near the crest of the Eastern Cordillera, axial surfaces dip 45 to 30° NE, and similar folds develop with overturned limbs 4 to 7 km long; axial-plane slaty cleavage is widespread and chlorite, sericite, and even some biotite appear. Toward Lake Titicaca, strain decreases markedly, cleavage disappears progressively, and folds are parallel and upright. Elsewhere in the eo-Hercynian belt upright or slightly asymmetrical chevron folds are the commonest structures.

Eo-Hercynian metamorphism is at most places of greenschist facies, but mineral assemblages containing andalusite + cordierite or andalusite + staurolite + garnet and in some cases fibrous sillimanite formed around syntectonic granites in southern Peru and northern Bolivia (Bard et al. 1974; Laubacher 1978). However, eo-Hercynian granitoids as a whole have little extent. One of these granites, the Amparaes orthogneiss, located 53 km north of Cuzco, has been dated at 330 Ma by the U/Pb method (Marocco 1978). Post-tectonic grani-

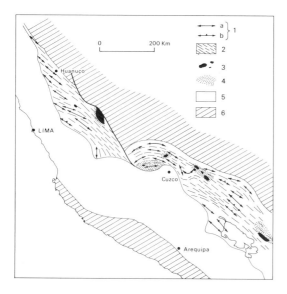

FIGURE 10-2. The eo-Hercynian belt in Peru and adjacent Bolivia (modified from Dalmayrac et al. 1980). 1: Trend of the main folds, including upright or slightly asymmetric folds (a) and recumbent folds (b) (barb denotes dip of axial plane). 2: Trends of cleavage. 3: Synkinematic granitoids. 4: Metamorphic aureole of 3. 5: Areas with no Paleozoic outcrops. 6: Areas where upper Paleozoic and Mesozoic strata are conformable or disconformable on the lower Paleozoic deposits.

toids are scarce; an undeformed adamellite from central Peru has been dated at 346 Ma (K/Ar on biotite; Maluski, in Mégard 1978).

Continental post-tectonic conglomerates and sandstones of Mississippian age rest unconformably upon the eroded folds of the eo-Hercynian belt. In central Peru silicic welded tuffs and some andesites are intercalated in their upper part. Marine sedimentation resumed during the Pennsylvanian and continued until the end of the lower Leonardian. As much as 3,000 m of a sequence of shale and sandstone capped by about 1,000 m of carbonates accumulated within a narrow trough centered on the Eastern Cordillera between 12° S and 14° S, but aggregate thicknesses of 1,000 to 2,000 m prevail elsewhere. Late Permian to Early Triassic continental red clastics interspersed with volcanic rocks mainly of andesitic and rhyolitic composition overlap the marine Carboniferous and Early Permian strata. This overlap follows a late Hercynian orogeny that consisted largely of vertical movements along normal faults and probably also horizontal movements along wrench faults; Carboniferous and Early Permian strata folded during this orogeny have been recognized north of Lake Titicaca (Audebaud et al. 1976) and in central Peru around Lircay (13° S, 74°45′ W; see Mégard et al. 1983). Late Permian magmatism, which was much wider in extent than eo-Hercynian magmatism, took place in an extensional tectonic regime, as shown both by the occurrence of peralkaline rhyolites and sub-alkaline basalts and by faulting involving large vertical components. Emplacement of Permo-Triassic batholiths and stocks followed the initial period of volcanism in the Eastern Cordillera. They form a suite of crustally derived sub-alkaline granitoids with initial intrusion ages ranging between 260 and 230 Ma (Lancelot et al. 1978) and uplift ages ranging between 220 and 200 Ma (Kontak et al. 1985).

The Late Permian to Early Triassic was a period of transition between an entirely intracontinental Hercynian evolution and the Andean cycle, most of which corresponds to an active margin evolution. The tensile strain of this epoch may be related either to back-arc spreading or to continental rifting (Noble et al. 1978; Dalmayrac et al. 1980). The recent data of Kontak et al. (1985) support the hypothesis of a typical ensialic rift zone in the Eastern Cordillera at this time.

THE ANDES (MESOZOIC TO PRESENT)

Andean evolution began in the Triassic and is still going on. Present knowledge strongly suggests that it has been related to subduction since the Liassic, and in this respect it is a new phenomenon for this part of America, but it is also true that some Andean structures were inherited from earlier Hercynian or even Precambrian structures. The main northwest strike of the Upper Cretaceous and Cenozoic Andean belt is not very different from the Hercynian or late Precambrian structural trends, and Hercynian faults were frequently reactivated during Andean orogenesis.

Andean evolution may be divided into a long period of relative tectonic quiescence, extending from the Triassic to the Late Cretaceous, and a shorter orogenic period, corresponding to the main phases of compressive deformation, which alternated or were associated with sedimentation, plutonism, and volcanic activity.

The Period of Sedimentation

The last movements that may be related to the Hercynian orogeny occurred a short time before the Norian transgression; Norian carbonate rocks locally rest unconformably upon faulted red beds of Upper Permian to Lower Triassic age. Sedimentary rocks of well-established Late Triassic age crop out north of 13° S latitude in the Altiplano, the Eastern Cordillera, and the Subandean Zone, and some isolated outcrops are also known in the Coastal Zone near Chiclayo (7° S). They consist of carbonate rocks that were deposited in a shallow shelf environment to the east and deep shelf conditions to the west (Loughman and Hallam 1982). Loughman and Hallam (1982) assert that the Norian neritic basin was widely open to the west, but data collected south of 10° S show that it was at least partly isolated from the open sea there (Mégard 1978).

Paleogeography of the Liassic sea is better known, especially during the Sinemurian (Mégard 1978; Vicente 1981; Loughman and Hallam 1982). In central and northern Peru (Figure 10-3A) deep shelf conditions with restricted bottom-water circulation led to deposition of bituminous shales. In central Peru some tuff and ash beds indicate the proximity of active volcanic centers probably located to the west. In southern Peru, a carbonate shelf basin lying east of Arequipa was bounded to the west by a volcanic belt containing carbonate reefs. Geochemical investigations of James et al. (1975) show that this Sinemurian andesitic volcanism is of volcanic-arc type. Carbonate shallow-shelf conditions prevailed during the upper Liassic over most of the arc earlier covered by the Sinemurian sea in central and southern Peru. In southern Peru the volcanic arc seems to have been active until at least the beginning of the Toarcian (Vicente 1981).

Marine Middle and Upper Jurassic series are well developed and well dated in southern Peru (Jenks 1948, Caldas 1978, Vicente 1981). After a period of quiescence during which shallow-shelf carbonates locally containing bioherms were deposited over most of southern Peru, a northwest-trending subsiding trough developed during the Malm (Beaudoin et al. 1982, Vicente et al. 1982), bounded to the southwest by a volcanic arc located along the present coast. During the Callovian (Figure 10-3C), a submarine turbiditic fan was pouring sand-size clastics southeastward along the volcanic arc while shales were being laid down farther to the northeast. The filling of this basin was gradual and ended, in the Tithonian, with the deposition of tidal and lagoon-type sequences.

In southern Peru, the Upper Jurassic is overlain disconformably by Lower Cretaceous deltaic and red-bed strata. The Albian transgression gave rise to shelf carbonates, which were deposited from the Albian to the Coniacian. These rocks underlie regressive evaporitic red beds of Santonian age. The Cretaceous shelf basin extended widely northeastward over the Cuzco–Lake Titicaca Altiplano, where continental and brackish-water strata were laid down (Figure 10-3D).

Evolution of central and northern Peru during the Middle and Late Jurassic may have been similar to that of southern Peru, but no rocks older than the Tithonian crop out between 7° S and 14° S latitude in the Coastal Zone or the Western Cordillera. Nevertheless, we know that a paleotectonic pattern completely different from that of the Liassic developed during the Middle Jurassic and later conditioned sedimentation during the Late Jurassic and the Cretaceous up to the Santonian. This pattern (Figure 10-4) consists of (1) an East Peruvian trough

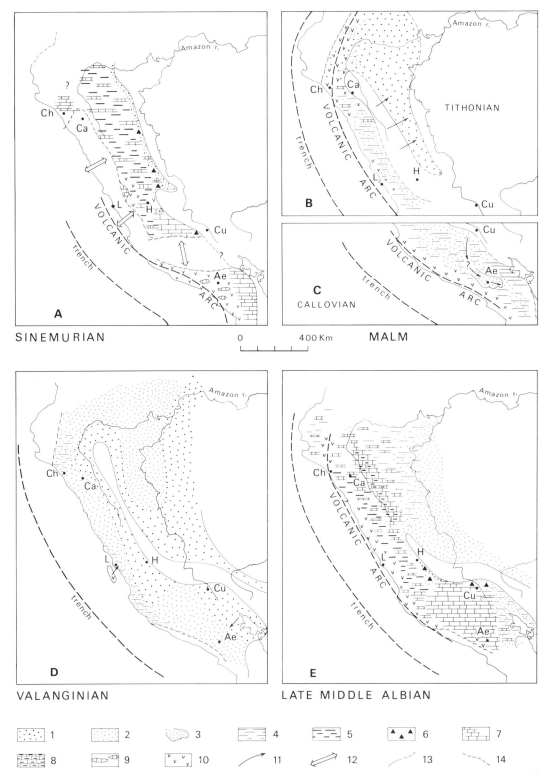

FIGURE 10-3. Lithofacies maps of Peru during the Mesozoic. 1: Continental terrigenous clastics. 2: Marine terrigenous clastics. 3: Turbiditic fans, where identified. 4: Shales and mudstones.

located over the southwestern edge of the Brazilian shield within which largely continental sediments accumulated, (2) the Marañon geanticline, corresponding to the present Eastern Cordillera, (3) a marine West Peruvian trough coinciding with the present Western Cordillera and part of the Coastal Zone, and (4) a western threshold termed the Paracas geanticline (Benavides 1956, Wilson 1963, Mégard 1973, Myers 1974, Cobbing 1978). The boundaries of most of these main structural units and of some second-order units they enclose are longitudinal growth-fault zones, some of which probably acted as volcanic conduits (Figure 10-4). The western basin can be divided into (1) a southwestern eugeosynclinal part, in which from 5 to 12 km of volcanic and volcanoclastic rocks with a minor proportion of terrigenous clastic strata accumulated, and (2) a northeastern miogeosynclinal part, which was filled by about 1.5 km of terrigenous and carbonate sediments, and which was flanked to the northeast by a shallower shelf platform. Atherton et al. (1983) and Atherton, Warden, and Sanderson (1985) interpret the ''western geosyncline'' as a marginal basin on the basis of geochemical and geophysical evidence. Lavas in the western part of the western basin are mainly basalts and basaltic andesites, tholeiitic to calc-alkaline in composition (Atherton et al. 1983). In some areas and at some discrete epochs, positive volcanic ridges formed between the miogeosyncline and the eugeosyncline (Webb 1976, Cobbing 1978).

Figures 10-3B and 10-3E illustrate the paleogeography of central and northern Peru, respectively, for the Tithonian and the late middle Albian; volcanism was extremely active during this latter epoch, and as much as 10,000 m of pillow lavas, tuffs, and hyaloclastites may have accumulated locally (Bussell 1975, Child 1976). Late Albian folding of the eugeosynclinal prism is documented north of Lima (Myers 1974, Webb 1976), where it is bracketed by radiometric ages on both pre- and post-tectonic intrusive bodies. Afterward, pyroclastic rocks were deposited unconformably in the eugeosyncline, whereas shelf carbonates of Upper Cretaceous age were laid down conformably in the miogeosyncline. The early contention that there is a genetic connection between the plutons of the Coastal batholith and their volcanic cover (Pitcher 1978, Cobbing et al. 1981) is challenged by Atherton, Sanderson, et al. (1985), mostly on the ground of their distinct geochemical features.

In summary, data concerning the successive Mesozoic volcanic arcs indicate that sedimentation during the first part of the Andean cycle took place along an active continental margin under which oceanic lithosphere was being subducted. The Mesozoic trench was almost parallel to the present Peru trench, at least between 7° S and 18° S latitude, from the Late Jurassic and perhaps the Liassic. Sedimentary troughs also paralleled the trench, and the western trough included part of the volcanic arc. Subsidence in the troughs occurred along longitudinal faults that were most probably listric normal faults (Figure 10-4). To comply with the laws of isostasy, the accumulation of as much as 12 km of relatively low-density sediments and volcanic materials in the western basin must have been compensated for by thinning of the crust, presumably involving plastic extension at depth (Mégard 1973, Cobbing 1976). Ac-

5: Bituminous shales and mudstones. 6: Evaporites. 7: Limestones. 8: Marls and marly limestones. 9: Reefal limestones. 10: Volcanic deposits. 11: Current directions. 12: Possible connections between apparently different basins. 13: Limits of emerged land (dashed where only probable). 14: Limits of areas with no outcrops. Ch: Chiclayo. Ca: Cajamarca. L: Lima. H: Huancayo. Cu: Cuzco. Ae: Arequipa.

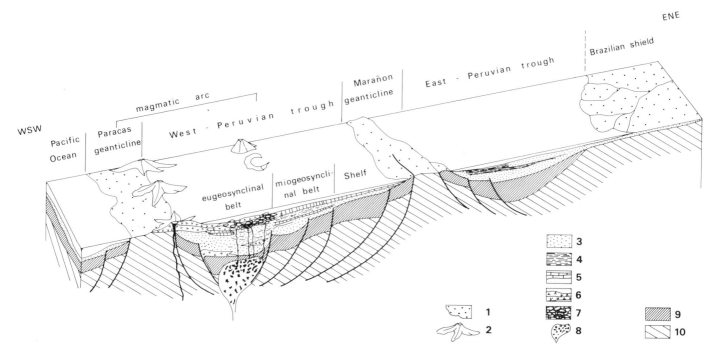

FIGURE 10-4. Diagrammatic representation of the paleogeographic setting in central and northern Peru during the Lower Cretaceous. *Physiographic features in Albian times*: 1: Emerged land. 2: Volcanoes. *Rocks of Albian and Neocomian age*: 3: Sandstones. 4: Shales. 5: Limestones. 6: Volcanic flows and sills. 7: Pillow lavas. 8: Gabbros and meladiorites. *Older rocks*: 9: Deposits of Liassic and Triassic age. 10: Older rocks, including the Precambrian basement.

cording to Atherton et al. (1983) and Atherton, Warden, and Sanderson (1985), this basin was a genuine marginal basin in central Peru in middle Albian times.

THE ANDEAN TECTOGENIC PERIOD

What may be termed the Andean tectogenic period began in the Albian (see, e.g., Cobbing et al. 1981) in a relatively restricted area presently located along the coast of central Peru. The Peruvian phase, which is the first widespread episode of deformation recorded in the Cordillera of Peru, is Santonian in age. It caused the emersion of both the West and the East Peruvian troughs. Related to this phase was the deposition of red continental siltstones, sandstones, and conglomerates, which overlie, either conformably or disconformably, various units of Albian to Santonian age. Westward, these red beds rest unconformably upon older strata (Benavides 1956). The conglomerates commonly include cobbles of Early Cretaceous quartzites derived from the west and indicating a deeper level of erosion, one that has been related by many authors to the Peruvian pulse of folding and subsequent uplift and erosion. In southern Peru the Cincha-Lluta overthrust in the vicinity of Arequipa (Vicente et al. 1979; see western part of section C on Figure 10-8) was emplaced during the Santonian. In parts of the Eastern Cordillera and Altiplano of central Peru the red beds rest unconformably upon eroded folds affecting strata of late Paleozoic to Late Cretaceous age (Mégard 1973). From the Santonian onward, the Cordillera became the major source area for clastic sediments in the East Peruvian trough (Figure 10-5A), where sedimentation was exclusively continental except for a short-lived Oligocene marine transgression.

In the Western Cordillera, red-bed sedimentation ended in the middle or late Eocene, when these beds were folded and thrust during the Incaic phase and subsequently overlain unconformably by terrestrial volcanic rocks and clastic strata dated at about 40 Ma near their base (Noble, McKee, and Mégard 1979). In the Altiplano, deposition of red beds continued during the Oligocene and at least part of the Miocene, and the supply of volcanic clasts from the west increased markedly. After the first Quechua pulse of deformation in the early to middle Miocene, lacustrine and fluviatile sediments of reduced thickness were laid down in the Andes as minor intercalations in dominantly volcanic units, with the exception that in some longitudinal subsiding intermontane basins in central Peru as much as 5,000 m of sedimentary and volcanic rock accumulated. It has been generally considered that subsidence of these basins was caused by extension, but in the Ayacucho basin of central Peru, compression was active during at least part of the deposition in the late Miocene (Mégard et al. 1984). Quechua pulses 2 and 3, dated at about 9 Ma and 6 Ma, respectively, resulted in folding in the intermontane basins and brittle tectonics elsewhere.

Like pre-orogenic Mesozoic volcanism, the Tertiary volcanic activity of the orogenic periods, which extended from the latest Cretaceous to the Recent, was arc volcanism. But the Tertiary magmatic arcs were centered on the Western Cordillera and Altiplano (Noble, McKee, et al. 1984) (Figure 10-5B), and volcanism was exclusively terrestrial. The rocks are

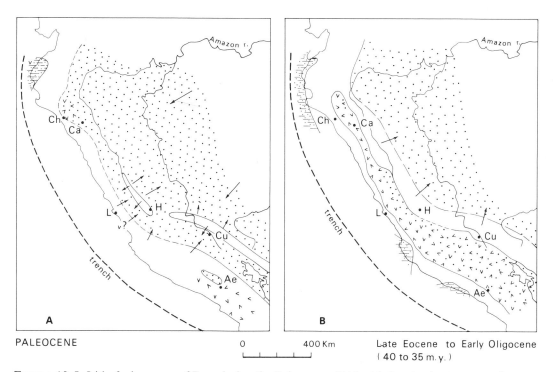

PALEOCENE   0   400 Km   Late Eocene to Early Oligocene
                                                           (40 to 35 m.y.)

FIGURE 10-5. Lithofacies maps of Peru during the Paleogene. (Abbreviations and patterns are the same as those in Figure 10-3.)

of intermediate and silicic composition, ranging from basic andesites with olivine to rhyolites, with andesites and dacites being most widespread. Northeast of the outcrops of the calc-alkaline rocks, lava flows of shoshonitic composition are found (Noble et al. 1975, Lefèvre 1979). Farther to the northeast is found a belt of highly peraluminous S-type volcanic and plutonic rocks of middle to late Cenozoic age (Clark et al. 1983; Noble, Vogel, et al. 1984). Volcanism remained active during the whole Tertiary, but with a widely variable intensity. It may have started as early as the Paleocene (Cobbing et al. 1981); it was very intense during the late Eocene and early Oligocene (40 to 35 Ma) and again during the Miocene and Pliocene, with peaks around 22 Ma and 10 Ma (Noble, McKee, et al. 1974, 1979). Quaternary volcanism is known only south of 13° S (Noble and McKee 1982 and unpublished data). At least part of the Tertiary volcanics in the Western Cordillera are probably linked genetically to the youngest batholithic plutonics (Noble et al. 1975).

Plutons of Andean age (see Figure 10-6) underlie large areas of the Peruvian Andes. The largest is the coastal batholith that occupies a strip 1,600 km long and generally 50 to 70 km wide; its axis is located between 180 and 220 km inland from the present Peru trench.

In the Huancabamba deflection, north of 7° S, the batholith is emplaced into the pre-Andean substratum and its relatively thin Mesozoic cover. It systematically cuts all structures in the entire region, and especially the Cajamarca east-west regional trend. Between 7° S and 13° S, it is emplaced along the axis of the eugeosyncline of the western part of the West Peruvian trough into the thick sequence of Mesozoic volcanic and volcanoclastic strata (Pitcher and Bussell 1977). Around 13° S, it cuts the Pisco-Abancay deflection and, farther south, is intruded into the Precambrian basement or its Mesozoic cover. In summary, the batholith as a whole was emplaced independent of the main structures of the Cordillera, but it utilized these structures where they were parallel to the batholithic trend.

The batholith and the enclosing series have been studied in detail during the last decade by Pitcher, Cobbing, and their numerous collaborators (e.g., Cobbing and Pitcher 1972, Myers 1975b, Pitcher 1978, Cobbing et al. 1981, Pitcher et al. 1985). According to these authors, the intrusions forming the batholith were emplaced over a period of 70 Ma between 100 and 32 Ma ago, with a peak of magmatic activity between 100 and 80 Ma. At many places the envelope is only slightly older than the individual plutons. The emplacement occurred at very shallow levels in the crust by cauldron subsidence and associated piecemeal stoping, involving almost no deformation of the host rocks. The older plutons are composed of hornblende gabbros that subsequently were cut by great lenticular bodies of tonalites and granodiorites; granites are generally younger and commonly occur in centered complexes, where they are associated with ring dikes.

From 35 to 7 Ma, magmatic activity migrated northeastward and gave rise to plutons of gabbroic to granodioritic composition. All are of stock size except the Cordillera Blanca batholith, which is 150 km long and 15 km wide.

Although there is no systematic genetic link between the Cretaceous and Tertiary volcanics and the plutons, their common localization and their emplacement over nearly the same period suggest at least a partly common origin at depth. In particular, a strong case can be made for contemporaneous arc volcanism and batholithic emplacement in the middle and late Miocene (Noble et al. 1975).

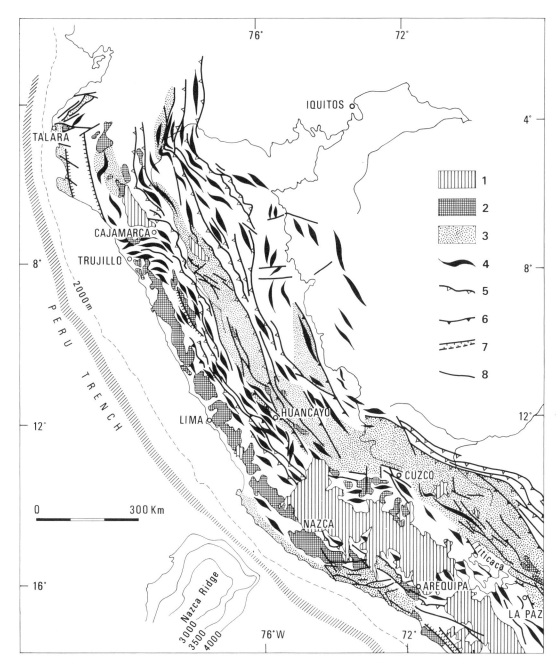

FIGURE 10-6. Structural sketch map of Peru and adjacent Bolivia. 1: Undeformed or slightly deformed late Cenozoic volcanic deposits. 2: Andean intrusives. 3: Outcrops of pre-Mesozoic rocks. 4: Main folds. 5: Main overthrusts. 6: Main steep thrusts. 7: Main normal faults. 8: Main wrench faults.

THE STRUCTURE OF THE ANDES

Age of the Tectonic Events   Figure 10-7 is a diagram showing the correlation between tectonic, volcanic, and sedimentary events in central Peru from 120 Ma to the present. It shows that deformation has been migrating systematically toward the foreland through time, but that the zone affected by one phase is, to a great extent, superposed on zones affected by previous phases. It should be noted in particular that the large longitudinal fault zones were readily reactivated, particularly the zone limiting the Western Cordillera on the east, where the effects of the Incaic and Quechua 1, 2, and 3 phases have been recorded with decreasing intensities. It can also be seen that the red beds, playing the same part as Alpine molasse, began to form everywhere during the Santonian, but that their tops become younger and younger as one moves toward the foreland. The fold-and-thrust belts also have migrated. The Marañon belt is essentially of Incaic age and was afterward partially reactivated by the Quechua 1, 2, and 3 phases, whereas deformation in the Subandean belt is entirely due to the Quechua 3 phase of late Miocene to Pliocene age.

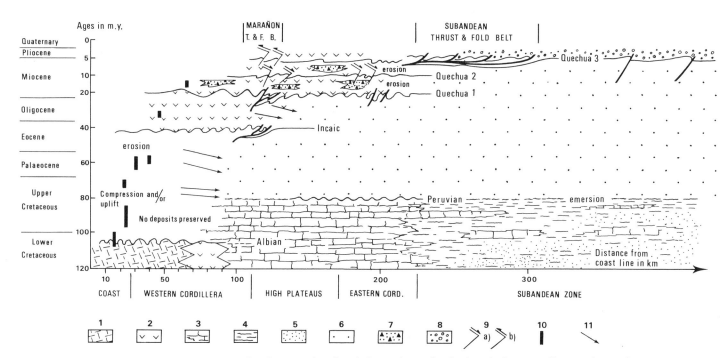

FIGURE 10-7. Composite diagram showing deformation episodes in relation to sedimentation and magmatism in central Peru (10°30′ S to 13° S latitude). Data pertaining to the ages of batholithic intrusions north of Lima are from Cobbing et al. (1981), and part of the data concerning volcanicity in the coastal area are after the same authors (modified). Age scale for the period 0 to 10 Ma is doubled. *Volcanic rocks*: 1: Submarine. 2: Terrestrial. *Marine sediments*: 3: Limestones. 4: Shales. 5: Sandstones. *Continental sediments*: 6: Red beds. 7: Infilling of inter-montane basins. 8: Alluvial fills. *Other symbols*: 9: Longitudinal wrench faults, specifically right-lateral (a) and left-lateral (b). 10: Location and time of emplacement of the main batholithic intrusions. 11: Source areas of the clastics, with arrows directed toward the sedimentary basins.

Andean Structures in Northern and Central Peru   The average trend of the Andean structures in northern and central Peru is N30° W to N45° W. Around 7° S, the Cajamarca-Huancabamba deflection begins, and the structures trend east-west before turning again and finally assuming a north-northeast trend in Ecuador. This complex zone will not be analyzed here. The present section is essentially a comment on cross sections A and B in Figure 10-8 and on the map in Figure 10-6.

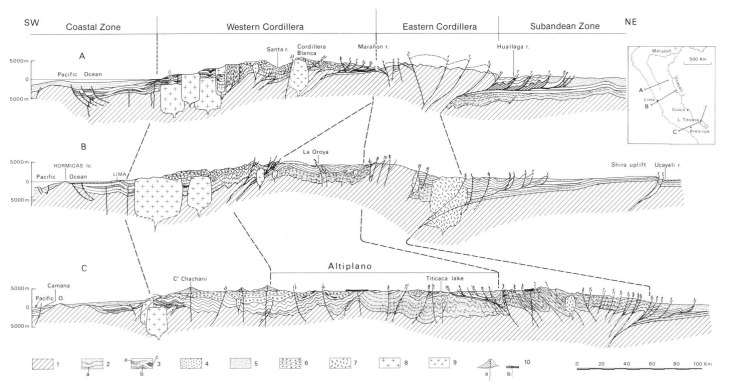

FIGURE 10-8. Geological cross sections and location map. Section A is after Mégard and Paredes (in Audebaud et al. 1973), based partly on data of Wilson et al. (1967), and modified by including data of Myers (1974, 1975a) and Kulm et al. (1981). Section B is modified from Mégard (1973, 1978). Section C is after Audebaud et al. (1973, 1976) and modified by including data of Vicente et al. (1979) in the area west of Cerro Chachani in the western part of the section. 1: Precambrian rocks, which may include Paleozoic rocks where covered by thick younger deposits. 2: Lower and middle Paleozoic deposits, including eo-Hercynian slaty cleavage (a). 3: Upper Paleozoic and Mesozoic deposits, including volcanic sills and flows (a), slaty cleavage (b), and evaporitic beds (c). 4: Red beds of Upper Cretaceous to Eocene age in the Western Cordillera and the Altiplano. 5: Red beds of Upper Cretaceous to Pliocene age in the Subandean Zone and coeval marine sedimentary rocks in the Coastal basins. 6: Volcanic and continental deposits of Oligocene to Miocene age in the Western Cordillera and the Altiplano; the black elongated spots in the central part of section C denote olistostromes of Cretaceous rocks. 7: Plutonic rocks of Hercynian age. 8: Plutonic rocks of the coastal batholith of Cretaceous to Oligocene age. 9: Other Andean plutonic rocks west of the batholith, at most places Miocene in age. 10: Plio-Quaternary volcanic rocks occurring in andesitic stratovolcanoes (a) or welded tuff plateaus (b).

Typically, the *Western Cordillera* begins in the west with concentric open upright folds, commonly with a wavelength of more than 5 km, affecting the essentially volcanic eugeosynclinal series of the western part of the West Peruvian trough. Dips are generally low. Subvertical fracture cleavage is commonly associated with the folds; locally there is flow cleavage and metamorphism of the greenschist or even garnet-amphibolitic facies (Myers 1974). In this western zone deformation is partly of Albian and probably partly of Santonian age. South of Lima, the marine upper Eocene lies unconformably upon folds affecting the Cretaceous. The transition between eugeosynclinal and miogeosynclinal series has not been clearly observed in the field because of discontinuities brought about by the batholith and Tertiary cover. Locally, a near-vertical fault (Webb 1976, Romani 1982) brings volcanic rocks west of the fault into direct contact with sedimentary strata of about the same age on the east. The zone where there should be an intertonguing of the two facies is missing, which suggests either tangential tectonics, implying a minimum shortening of about 10 km, or large lateral movement. However, there is presently no direct evidence for such tectonics.

Farther east, in the western part of the miogeosyncline with exclusively sedimentary filling, folds are generally tighter, in all cases upright or slightly asymmetrical, and commonly of chevron-type. Their axes are subhorizontal and their limbs have dips ranging from 30 to 70°. Minor folds are common; axial-plane cleavage—generally near-vertical—is commonly associated with these folds. Locally, as in the valley of the Río Santa (Figure 10-8, section A), the axial planes of the folds display a fan-like pattern.

Farther east still is the eastern limit of the Western Cordillera, which corresponds to the watershed between 10° S and 12° S (the Cordilleras of Huayhuash, Raura, La Viuda, and Tuyujuto). It is a fold-and-thrust belt 25 to 30 km wide, called the *Marañon fold-and-thrust belt* (Mégard 1984), within which the thick miogeosynclinal sequence overrides the much thinner shelf series to the east. This belt is very well exposed to the east of the Cordillera Blanca (Figure 10-8, section A), where it has been studied by Wilson et al. (1967). It comprises an inner southwestern zone and an outer northeastern zone, which is a typical imbricate fan. In the inner zone, the oldest known members of the miogeosynclinal sequence—namely Tithonian slates and sandstones—are exposed; folded into chevrons, with steep axial-plane cleavage that dips either northeast or southwest; and overlain by Cretaceous strata preserved in several more open synclines. There are a few reverse faults that dip 45 to 60° southwest. The boundary between the inner and outer zones is a major thrust fault dipping 60° southwest at the surface; nearby, secondary northeast-dipping reverse faults could be its conjugates. In the outer zone the only formations known to crop out are Tertiary red beds, Late Cretaceous carbonates, and Early Cretaceous sandstones, as well as carbonates, intercalated with evaporites exposed beneath the Cretaceous in the cores of some anticlines or along the thrusts. This zone is characterized by thrusts delimiting thrust sheets 2 to 3 km thick. The westernmost thrusts dip 45° southwest, whereas the easternmost dip 20°. Within the westernmost thrust sheets, the formations have been folded. The thrusts are most commonly parallel to the stratification in the upper thrust sheet, and thrusts appear in many cases to have formed by cutting through the crests of anticlines. The easternmost thrust sheets are composed exclusively of virtually unfolded Cretaceous strata that dip regularly 20° southwest and thus parallel the thin Cretaceous and Jurassic cover that has remained attached to the Hercynian and/or Precambrian substratum. Locally, the bedding-plane thrusts are seen to ascend along ramps from one décollement level to the next, for example, from Jurassic calcareous and evaporitic rocks to Albian marls. The general geometry of the Marañon fold-and-thrust belt presents obvious analogies with that, for instance, of the Canadian Rockies (Bally et al. 1966, Dahlstrom 1970, Price 1981), although the scale differs and an aggregate shortening in the region of 30 km, as proposed by Wilson (oral communication), seems likely. The lack of subsurface data—drilling or seismic—prevents our constructing precise balanced sections and specifying the amount of shortening. The surface relations show that the outer zone is detached from the autochthonous series. In the inner zone thrusts are rare and a great part of the deformation in Tithonian strata is penetrative. The gravity-gliding interpretation of Wilson et al. (1967) and Myers (1975a) for this fold-and-thrust belt implies either an area in extension in the inner parts of the range or an area of tectonic denudation, yet nothing of the kind is known, since well-dated middle Albian strata crop out continuously from the coast to the Marañon Valley. As in most ranges, the basement must therefore have been shortened by an amount equal to that observed at the surface; this has been shown in section A of Figure 10-8 by placing the root of the basal décollement surface of the intermediary and outer zones a little behind the boundary of the inner and intermediary zones. It must also be noted that this boundary coincides with that of the outcrops and probably that of the subsurface extension of Tithonian strata; similarly, the limit between the steeper southwestern thrust sheets and the flatter northeastern ones in the outer zone commonly corresponds to the transition from miogeosynclinal to shelf facies in the Lower Cretaceous. These faults could thus be listric faults related to subsidence (Figure 10-4) and contemporaneous with the sedimentation, which were reactivated as overthrusts during subsequent compression (Mégard 1978, Cobbing et al. 1981).

The Marañon fold-and-thrust belt extends at least as far north as Cajamarca (Reyes 1980, Janjou et al. 1981) and can be followed southward in the Huayhuash (Coney 1971), Raura (Romani 1982), and La Viuda Cordilleras and into the Canete high valley at 12°15′ S (Mégard 1978). In the last area greenschist-facies conditions were reached locally in the fold-and-thrust belt, and the thrust sheets were later refolded by concentric coaxial folds with steeply plunging axes (Mégard 1973, 1978). The Upper Cretaceous-Eocene red beds are everywhere involved in the thrusts and folds. Between 11°30′ S and 12°30′ S, volcanics 40 Ma old unconformably overlie these structures. Thus the age of folding and thrusting is probably middle to late Eocene and corresponds to the Incaic phase of compression (Noble, McKee, and Mégard 1979). But at least some of the thrusts were reactivated during the Miocene Quechua phases. At 10°40′ S latitude both coarse alluvial deposits and later slope breccias that are younger than a first pulse of folding were folded and overthrust successively by two discrete, later compressive pulses (Romani 1982). At least one of these pulses is Miocene in age. The volcanics that are 40 to 20 Ma old, where present, are thrown into broad open folds in most of the Western Cordillera and are cut by reactivated overthrusts in the fold-and-thrust belt (Mégard 1973, 1978).

Drag folds with steeply plunging axes and late, nearly horizontal striae on slickensides along the major overthrusts also indicate strike-slip movements after the Incaic phase (Soulas and Mégard, in prep.). It has been shown recently (Noble, McKee, and Mégard 1979) that the intensity of the Incaic folding decreased dramatically east of the fold and thrust belt. For these reasons, the presently folded Altiplano and also probably the shelf area that bordered the Marañon geanticline to the west (Figure 10-4) were still stable basement areas with a flat-lying sedimentary cover when the bedding-plane thrusts of the eastern part of the fold-and-thrust belt were emplaced. From central to northern Peru, one notes a northeastward migration of the fold-and-thrust belt; at 12° S it lies 280 km to the northeast of the Peru trench axis, and at 9° S the distance is 350 km. The position of the present fold-and-thrust belt reflects the left-lateral en échelon pattern of the growth faults that were the common boundary of the miogeosyncline and the shelf in the eastern part of the West Peruvian trough, particularly during the Late Jurassic.

In central Peru the *Altiplano* is present only south of 10°30′ S and is about 70 km wide. Jurassic and Cretaceous shelf strata crop out extensively in the Altiplano; they are thrown into open parallel folds (Figure 10-8, section B). In the southwestern part of the Altiplano these folds also affect the red beds of Late Cretaceous to Eocene age and the conformably or disconformably overlying lacustrine, volcanoclastic, and volcanic strata dated as Oligocene and early Miocene (Noble, McKee, and Mégard 1979; McKee and Noble 1982). Folding occurred between about 19 and 17 Ma during the Quechua 1 phase (McKee and Noble 1982). In the northeastern part of the Altiplano, folds in Mesozoic strata are tighter but still parallel, and they have been coaxially refolded after deposition of unconformable thin red beds of late Eocene to Miocene age; thus the first phase of folding is possibly the Late Cretaceous Peruvian folding.

Some Mesozoic and/or Tertiary strata have been preserved in graben in the *Eastern Cordillera* (see, e.g., Wilson et al. 1967). The Mesozoic sedimentary rocks rest with slight disconformity over late Paleozoic sedimentary rocks. They are well exposed south of section B (Figure 10-8) in central Peru. Folds in these rocks are open, but slaty cleavage is well developed, particularly in the abundant shales. Folding is mostly Late Cretaceous in age in the Eastern Cordillera of central Peru, as red beds dated in part as Late Cretaceous overlie the folds unconformably. However, the most common structures are high-angle thrusts with throws that generally are larger than 1 km. In northern Peru these thrusts display a fan-like pattern, but in central Peru they are regularly vergent to the northeast (Figure 10-8, sections A and B, respectively).

Natural seismicity data indicate that the present tectonics at depth under the eastern part of the Eastern Cordillera consists of thrusting along faults trending northwest and dipping southwest (Suárez et al. 1983). If one assumes that the presently active fault planes correspond to old reactivated thrusts, the steep southwest-dipping thrusts may also flatten to about 45° at depth, as shown on section A in Figure 10-8. Along some of the steep thrusts near Huancayo, syntectonic metamorphism of the shallow greenschist facies has been recognized. The age of the thrusting is not precisely known. In northern Peru compressive deformation in the Eastern Cordillera occurred after the formation of the late Eocene fold-and-thrust belt in the Western Cordillera. It is thus related to the Quechua phases. The easternmost thrusts of the Eastern Cordillera may even be of Pliocene age, like those of the Subandean Zone.

In the *Subandean Zone* sedimentary rocks of lower Paleozoic to Pliocene age are exposed. The Hercynian tectonism did not affect rocks of this zone and the lower Paleozoic is therefore almost conformable with its Mesozoic cover. The sedimentary rocks together form an eastwardly tapering wedge. Until recently, few data were available for the Subandean Zone because of the dense rain forest. The best-known structures at surface are very broad flat-bottomed synclines separated by narrow parallel anticlines with steep limbs that commonly extend for tens of kilometers. These anticlines are commonly vergent to the east and thrust along their crests or in their forelimbs. These thrusts are very steep, and in previous interpretations (Mégard 1973, 1978), I believed them to continue with the same dip to depth. The subsurface interpretation presented in sections A and B is different, since it involves a generalized décollement of the Subandean Zone, now appearing as a foreland fold-and-thrust belt, no longer attached to a substratum sinking under the Eastern Cordillera. This new interpretation is based on data from Pardo (1982) on the Santiago-Nieva basin (6° S), where the Tertiary and Mesozoic and part of the Paleozoic are detached at a depth of about 9 km from a Paleozoic substratum covering the Precambrian basement. The steep thrusts known at the surface flatten progressively downward and reach a dip of 10 to 15° to the west at a depth of 9 to 10 km. According to Pardo, seismic reflection data from the Huallaga zone, which is crossed by section A of Figure 10-8, are similar to those of the Santiago-Nieva area. The existence of a fold-and-thrust belt in the Huallaga zone is also suggested by maps of Benavides (1968) and Rodríguez and Chalco (1975). Décollement planes are probably provided by the thick saline beds located at the base of the Mesozoic sequence or slightly deeper at the top of the Pennsylvanian. Most of the salt diapirs are located along the crests of the anticlines and along the steep thrusts. Subandean structures around 11° S (Figure 10-8, section B) are interpreted in the same way, since they are very similar at the surface, and since the base of the Mesozoic also includes gypsum. However, the extension of the flat thrusts to the west, where they would shear the San Ramon batholith of Late Permian age, is hypothetical. All these structures are of upper Miocene or Pliocene age. In northern Peru seismicity is very active in the Subandean Zone, but at depths between 8 and 38 km; it can be supposed that the active faults dipping 45° west in the basement are connected to the flat décollement planes known in the sedimentary cover (Suárez et al. 1983).

Comparison between the Tectonics in the Northern and Central Peruvian Andes and That in Southern Peru   Section C of Figure 10-8 appears at first glance to differ from sections A and B. The Andes proper, that is, reliefs higher than 1,500 m, are much wider (470 as opposed to 250 km), but this is partly compensated by a narrow Subandean Zone (30 as opposed to 60 to 130 km). In southern Peru the Western Cordillera and Altiplano are morphological rather than structural units, as the boundary between them is very poorly defined, there being no fold-and-thrust belt on the northeastern edge of the Western Cordillera.

The marine strata of the western Mesozoic basin have been folded into parallel synclines and anticlines with a wavelength from 1 to 10 km. Thrusts are very rare. The only major structure that implies shortening on the order of 10 to 20 km is the Cincha-Lluta overthrust described by Vicente et al. (1979). It is a shear surface that brings the Precambrian basement and its Jurassic and Cretaceous cover over dissimilar Jurassic and Cretaceous facies. The thrust is very flat and appears to cut previously folded strata. Thrusting took place between the Coniacian and Maastrichtian during the Peruvian phase, which was the major tectonic event in the western slope of the Andes of southern Peru. Younger tectonic events formed broad parallel folds with a wavelength of 30 km or more in Eocene, Oligocene, and Miocene continental and volcanic strata. It is possible to recognize discrete effects of two Paleogene pulses, one of which is the late Eocene Incaic phase (Noble et al. 1985), and the effects of at least two Miocene Quechua phases (Audebaud et al. 1976).

The Late Cretaceous phase did not affect rocks of the Altiplano. Andean tectonics in this area is related mostly to the late Eocene Incaic phase and to late Quechua coaxial folding and thrusting of lesser intensity. Around Lake Titicaca, structures are locally very complex, and various interpretations have been proposed (Newell 1949, Chanove et al. 1969, De Jong 1974, Audebaud et al. 1976). With the exception of De Jong, who favors a purely gravity-gliding mechanism, the other authors support interpretations based on the presence of fold-and-thrust belts. The main belt of this type, discussed by Newell (1949) and mapped by Audebaud (1973 and unpublished maps) and by geologists of the Empresa Petrolera Fiscal (presently PETROPERU), lies northeast of Lake Titicaca and trends northwest-southeast about 300 km between 14° S and 15°30° S. It is represented on section C of Figure 10-8. In the interpretation that I propose in section C the rather steep southwest-vergent thrusts flatten at depth, where they merge with a large décollement surface. This explains rather satisfactorily the abrupt changes of facies observed by Audebaud et al. (1976) in the Middle Cretaceous from the southwest to the northeast in two adjacent thrust sheets. Gypsum occurs at the base of the Mesozoic and at different higher levels and, indeed, provides a lithology very favorable to décollements. The folds that develop in the fold-and-thrust belt are flexural-slip folds with the southwestern limb shorter than the northeastern limb. Their axes are horizontal. In the Pirin area at the northwestern end of Lake Titicaca complex structures, involving flat thrusts affecting both the Devonian and its unconformable Mesozoic cover and folds with large flat-lying inverted limbs, have been described by Newell (1949) and Chanove et al. (1969). Although grossly contemporaneous with the nearby northwest-trending fold-and-thrust belt described above, these structures are northeast-trending and northwest-vergent; probably northwest-trending strike-slip faults form the boundary between the two belts. Because of their limited size and complex structures, the Pirin thrust sheets are not shown on section C. Gravity-gliding structures giving rise to olistostromic complexes within clastic and volcano-clastic continental strata of Oligocene and Miocene age, both near Lake Titicaca (De Jong 1974) and west of it (Audebaud et al. 1976), appear to be related to failure of cliffs created by Incaic folding and uplift.

In the Eastern Cordillera almost all the rocks exposed are lower Paleozoic strata deformed by eo-Hercynian folding. The thrusts dipping 40 to 60° northeast that are common in its southwestern slope are probably eo-Hercynian, but they seem to have been reactivated during the Incaic and/or Quechua phases of compression, for they affect upper Paleozoic, Mesozoic, and Tertiary inliers (Laubacher 1978). Geometry of the opposite-verging thrusts of the northeastern slope of the Eastern Cordillera is very similar to that of the steep thrusts that cut Tertiary strata in the Subandean belt, and all the thrusts in the Paleozoic rocks either are related to the younger Quechua 3 phase or have been reactivated by it. The overall structural pattern of the Subandean belt is similar to that in central Peru (this paper) or in central and southern Bolivia (Martinez 1980, Roeder 1982, Jordan et al. 1983), and we interpret it likewise as a Pliocene fold-and-thrust belt with downward-flattening thrusts and a basal décollement.

In summary, the belt of Andean deformation in southern Peru is wider than in central and northern Peru, but the intensity of deformation is less. The vergence of the range is not so well defined; it includes both northeast-vergent overthrusts, for example, near Cerro Chachani in the Western Cordillera or in the Subandean Zone, and a southwest-vergent fold-and-thrust belt, for example, near Lake Titicaca; in the remainder of the southern Peruvian Andes most folds are upright.

Conclusions on the Structure of the Andes   I have already emphasized, at the beginning of this section, the migration of deformation toward the foreland from the Albian, when tectogenesis began, to the Late Cenozoic. The migration is clearly illustrated in Figures 10-9A, 10-9B, 10-9C, and 10-9E, which show the new or reactivated structures belonging to the Albian and Peruvian (Santonian), Incaic (upper Eocene), Quechua 1 (lower Miocene), and Quechua 3 (uppermost Miocene) phases, respectively. The migration would be even more obvious if we considered only the new structures involved in each phase. This is especially valid for phases with a northeast-southwest or east-west direction of shortening, that is, nearly perpendicular to the Andes trend. The Quechua 2 phase (Figure 10-9D), around 9 Ma (Mégard et al. 1984), which had a north-south compressive stress direction (Soulas 1977), mainly reactivated longitudinal faults which then acted as right-lateral strike-slip faults. Few new structures were created, except where the northwest-southeast strike-slip faults changed to east-west thrust faults, accompanied by east-west trending folds (Soulas and Mégard, in prep.). Thus it took no clear part in the migration of phases toward the foreland. A problem is raised by the presence of folds produced during the Santonian Peruvian phase in an isolated Eastern Cordillera massif and the eastern part of the central Peru Altiplano. It can partly be explained by the 6,000 to 8,000 m of Carboniferous and Permian strata to which were added 1,000 to 2,000 m of Mesozoic strata. The Carboniferous and Permian sediments accumulated in a trough, the limits of which coincided roughly with those of the fold zone and which could have developed on a thinned crust. The increase in temperature associated with the rise of granites could also have played a part by ''ductilizing'' the crust in that part of the Eastern Cordillera (Mégard 1973, 1978).

The way the migration took place gives the impression that zones folded and faulted during one phase would thus become more rigid and transmit the stresses during the next phase to zones farther east, which would then deform. This interpretation must obviously be qualified when we take into account (1) the possibility for each structural zone or subzone to deform more intensely than the surrounding areas—well illustrated by the Tertiary intermontane basins, which commonly are markedly folded between rigid zones where only a few faults were weakly active (Mégard et al. 1984), and by the folding within the Eastern Cordillera between 10°30' and 14° S during the Peruvian phase—and (2) the partial reactivation of structures formed during previous phases, particularly the great overthrusts (Figure 10-9).

As for deep structural interpretations, objective data are scarce, largely because of the lack of seismic reflection information. It is now certain that the Subandean Zone is overthrust by the Eastern Cordillera (Pardo 1982), but it is not known to what extent. The new opportunities for petroleum exploration opened up by the likely presence of structural as well as sedimentary traps in this region should help to answer that question. Another factor influencing interpretation at depth is the possibility that the ''basement'' underwent ductile deformation during the Andean phases, obviously depending in part on basement lithology, which is widely variable. In general, we can say that (1) schistosity and Andean folds could have affected early and middle Paleozoic strata deformed during the Hercynian under greenschist-facies conditions, or even without metamorphism, as in much of the region; (2) Andean duc-

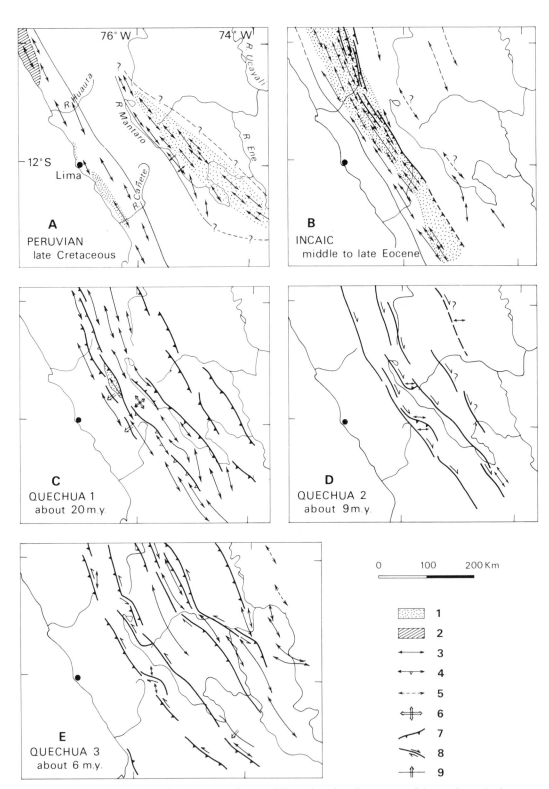

FIGURE 10-9. Paleotectonic sketch maps of central Peru showing the extent of the various Andean compressional phases and the structures related to them. 1: Areas within which cleavage is present. 2: Areas within which an Albian phase of deformation has been recognized. 3: Upright and slightly

tile deformation did not affect the Precambrian massifs made of high-grade metamorphic rocks, for example, the two-billion-year-old granulites of southern Peru, which were clearly a rigid basement. This is the reason why in section B of Figure 10-8, in which Hercynian terranes are shown extending under the Western Cordillera, Andean schistosity is depicted as passing downward into the substratum, which would therefore be ductile. Unless decoupling, which allows for a thorough change in deformation style, took place at depth everywhere in the Western Cordillera, a large part of the deformation very probably occurred in a ductile manner in the substratum of the chevron-folded, subvertical cleavage zone of the Western Cordillera in central and northern Peru.

Differences in basement ductility also account, to a great extent, for the differences between the Andes in southern Peru and those in central and northern Peru, especially in the Western Cordillera. In central Peru, at 12° S, the eastern part of the Hercynian belt is exposed in the Eastern Cordillera and on the Altiplano. Its most deformed central part is exposed a few kilometers east of the eastern edge of the Western Cordillera fold-and-thrust belt (Figure 10-8, section B), and it can thus be supposed that the western part of the Hercynian belt forms the substratum of the Western Cordillera. Since this range trends about N 60° W, the same is probably true farther north. This "basement" is evidently more ductile than the partly granulitic, gneissic, rigid basement beneath the Western Cordillera in southern Peru (Figure 10-8, section C), which was sheared at low temperature to produce the Cincha-Lluta overthrust.

CONCLUDING REMARKS

The knowledge we have acquired during the past few years about the West Peruvian trough (Myers 1974, Webb 1976, Cobbing et al. 1981, Atherton et al. 1983) and its equivalent in southern Peru (Vicente et al. 1982) demonstrates that some criteria do not distinguish adequately between "geosynclinal" and "liminal" ranges. The absence of flysch in the latter is not a valid criterion; the Jurassic includes characteristic turbiditic flysch strata in southern Peru, and turbiditic graywackes are known in the Cretaceous of the West Peruvian trough and the Lancones basin on the Peru-Ecuador border. Therefore, the presence of flysch does not imply later, Alpine-type tectonic evolution. Sutures, and particularly ophiolitic sutures, seem a much surer criterion, since their absence rules out the possibility of collision tectonics, be it accomplished through accretion or obduction.

As to deformation style and origin, the bulk of the observed deformation cannot be explained by superficial deformation of a sedimentary cover that detached and flowed in response to vertical movement of crustal blocks bounded by subvertical faults with concurrent normal and reverse movement (Myers 1975a). Vertical movements were certainly involved in the coastal region, especially during the emplacement of the batholith (Pitcher and Bussell 1977, Cobbing et al. 1981), but such movements could not have produced the Marañon and Subandean fold-and-thrust belts, in the absence of tectonic denudation, extension, or else a

---

asymmetrical folds. 4: Asymmetrical and recumbent folds (barb denotes dip of axial plane). 5: Folds that are supposed to belong only to a discrete phase. 6: Domes. 7: Thrusts. 8: Wrench faults. 9: Flexures.

buoyant metamorphic core giving rise to gravity spreading to compensate the shortening. We must, therefore, invoke a shortening of the basement equivalent to that of the cover. Despite the lack of precise data, the overall amplitude of the thrusts in central and northern Peru is clearly less than in other orogenic belts, for example, the Canadian Rockies (see Price 1981, Bally 1981), where it varies between 160 and 240 km. In central and northern Peru the aggregate width of the Marañon and Subandean fold-and-thrust belts is around 120 km, and shortening is a few tens of kilometers for each belt. If we take into account the fold-and-thrust belts, then, the figure of 100 km for the overall shortening across central Peru (Mégard 1973, 1978) must be increased by 30 to 50 km. This does not solve the problem of the sialic root of the Peruvian Andes, the volume of which is known approximately at the latitude of Lima (James 1971). In fact, shortening of 150 km, which seems a maximum, accounts for only 75% of the volume of the root, assuming that the crust was, on average, 30 km thick before shortening. It must be pointed out, however, that when we discuss shortening in the Andes, we lack points of reference on the Pacific side; the shortening we are discussing brought the Paracas geanticline nearer the Brazilian shield by 150 km, but what happened west of the Paracas geanticline is unknown. Because subduction has been active since the Malm and perhaps since the Liassic, we would expect to find a voluminous accretionary prism west of the Paracas geanticline; in fact, there is almost none. At around 6°30′ S the crystalline basement even appears to extend westward to about 30 km from the trench. It must be concluded, therefore, that most of the sediments that should have been accreted since the Liassic, and perhaps also the western part of the Paracas geanticline, have been drawn under the continent by subduction and have contributed toward enlarging the root of the Andes by tectonic underplating.

The evolution of the liminal Andes with time raises the problem of the relationship between (1) subduction and extension and (2) subduction and compression. During the long initial sedimentation period of Andean evolution and probably at least part of the periods of sedimentation and igneous activity between compression phases, subduction was associated with a tensile regime along the entire continental margin. During the compression phases, and during these only, the evolving Andean orogenic belt was included, geometrically speaking, between a western northeast-dipping oceanic or "B" subduction zone (in the terminology of Bally and Snelson 1980) and an eastern southwest-dipping continental or "A" subduction zone. In addition, it should be noted that two subduction zones of "A" type have been successively active: the first, during the middle and/or late Eocene, was located under the Marañon fold-and-thrust belt and was reactivated mainly during the Quechua 1 phase around 20 Ma ago; the second, during the Pliocene and Quaternary, was situated under the Subandean fold-and-thrust belt and the eastern edge of the Eastern Cordillera. At these times the belt undergoing deformation appeared as a "ductilized" zone squeezed between two more rigid jaws represented by the Nazca plate and either the Marañon geanticline (Incaic phase) or the Brazilian shield (Quechua 3 phase) (see also Douglas 1920). To explain how subduction of oceanic lithosphere under the western edge of the South American continent could induce extension in some cases in the continent but compression in others, I have suggested a hypothesis (Mégard 1973, 1978) whereby the dip of the subducted oceanic plate controls the mechanical coupling between this plate and the overriding continental plate. A low dip of the subducted plate would correspond to high friction over large areas of contact between both plates; high stress transmission to the overriding plate; and, hence, compressive deformation.

A steep dip would correspond to low friction; low stress transmission to the overriding plate; and intercalation of an asthenospheric wedge between the two plates, allowing for the existence of thermal diapir phenomena likely to create and/or maintain volcanism and extension. Dips of the subducted plate, in turn, are probably related to convergence rates; age of subducted lithosphere; and absolute motion of the overriding plate, the latter factor probably being the dominant one, according to Dewey (1980), Cross and Pilger (1982), and Uyeda (1982). Actually, both the low- and high-dip scenarios can be considered extreme; real situations doubtless have often been transitional and more complex, as is well shown by analysis of the present situation along the Peru-Chile trench, where the subduction zone of ''B'' type is longitudinally segmented. North of 15° S, for instance, it dips slightly; this is expressed by the absence of recent volcanism and the dominance of compression at the surface (Mégard and Philip 1976, Barazangi and Isacks 1979). To the south, it plunges markedly, producing essentially extensional tectonics and active Quaternary volcanism (see also Lavenu 1978, Sébrier et al. 1985). A similar segmentation occurs farther south, under the Bolivian, Chilean, and Argentinian Andes (Stauder 1973, Barazangi and Isacks 1976, Jordan et al. 1983). Yet, since these very different situations all exist at one time, along one plate boundary, with very similar convergence rates, secondary factors such as the curvature of plate margins or the subduction of aseismic ridges must introduce perturbations into higher-order scenarios; they will have to be taken into account when analyzing old situations.

In summary, the Peruvian Andes appear to be the very type of liminal fold belts which are exclusively related to subduction processes. Their most characteristic features are (1) the large volume of pre- and synorogenic volcanism and plutonism—mostly represented by calc-alkaline rocks—that was emplaced parallel to the then-existing subduction zone, which in the case of Peru was also parallel to the present active margin; (2) sedimentation in troughs that were either parallel (Peru) or slightly oblique (Chile) to the ancient subduction zone and probably had a strongly thinned crustal basement; (3) the existence of structures that indicate a relatively small amount of shortening, despite the fact that fold-and-thrust belts of moderate extension commonly developed; and (4) the absence of ophiolitic sutures that would record any type of collision. In most of these aspects the liminal or Andean-type belts are strikingly different from the Alpine-type belts formed by continent-continent collision, or from the Cordilleran-type belts formed at least in part by collision and subsequent accretion to a continent of blocks of smaller size and varied origins. Nevertheless, it must be stressed that, during part of their development, at least some structural zones of the Alpine and Cordilleran belts had an Andean-type evolution that we can better understand by using our knowledge of the central Andes.

SUMMARY

The Andes of Peru are an excellent example of a ''liminal'' mountain belt, that is, a belt built at the margin of a continent but entirely upon a sialic basement and by processes related almost exclusively to subduction of oceanic lithosphere beneath continental lithosphere.

The Precambrian basement, exposed in the Eastern Cordillera and along the coast, includes a belt about 600 Ma old that is composed mainly of low- to medium-grade metamorphic rocks, and also older granulitic cores 2,000 Ma and 1,000 Ma old. An early Hercynian fold belt developed over this basement about 350 Ma ago. Late Hercynian block movements are associated with widespread volcanism during the Upper Permian and the Lower Triassic. The tensile strain of this period might be related to continental rifting.

Andean evolution began with deposition of thick prisms of sediments in elongated basins parallel to the present trend of the Andes and to a subduction zone that probably has been continuously active since the Liassic. In central and northern Peru as much as 12 km of Mesozoic sediments and subduction-related volcanics piled up in the West Peruvian trough. About 10 km of conformable Paleozoic to Pliocene strata accumulated in the East Peruvian trough. Both of these troughs were asymmetrical, with steep slopes on the west and gentle slopes on the east.

The tectonic period began in the Late Cretaceous or, at least locally, as early as the Albian. During this period, arc-type volcanism and plutonism were widespread, and sedimentation in the Andes was restricted to intermontane continental basins. Five phases of compression proceeded from southwest to northeast, progressively incorporating into the Andes paleogeographic units located farther and farther east. The late Eocene phase created the Marañon fold-and-thrust belt within which the thick miogeosynclinal prism of the eastern part of the West Peruvian trough is thrust to the northeast over eastward-tapering shelf strata. The latest Miocene to Pliocene Quechua 3 phase gave rise to the Subandean fold-and-thrust belt. According to recent data, the steep thrusts known at the surface in the Subandean Zone dip only 10° westward at a depth of 10 km in northern Peru. It seems logical to extend the same interpretation to all the similar structures known in the Subandean belt farther south.

In southern Peru a narrow Subandean fold-and-thrust belt is also present, but distribution and style of deformation are strikingly different in the main Andes themselves, where the main structural features are (1) widespread open concentric folds of varying ages, (2) a Late Cretaceous northeast-vergent overthrust in the region of Arequipa, and (3) a late Eocene and/or Miocene southwest-vergent fold-and-thrust belt in the vicinity of Lake Titicaca.

The amount of shortening calculated near the surface, between 100 and 150 km, cannot be accounted for by vertical basement-block movements that would cause gravity-induced flow folds and thrusts in the sedimentary cover. Shortening of the same amount has to be accommodated in the sialic crust. The thick sialic root of the Andes has resulted from this crustal shortening and also probably from underplating with material derived from tectonic erosion of both the accretionary prism and the sialic western edge of the continent along the trench. Both the Andean sedimentation and tectogenic periods must be explained by the single process of subduction of the Nazca and proto-Nazca plates under the American plate. As tensile strain seems to predominate during the first period and compressive strain during the second, variations in the subduction process have to be considered. I suggest that (1) predominating tension within the continent is related to a steeply dipping subduction zone and (2) predominating compression is related to a gently dipping subduction zone.

ACKNOWLEDGMENTS

Fieldwork was supported by the Centre National de la Recherche Scientifique, the Office de la Recherche Scientifique et Technique Outre-Mer (ORSTOM), the Institut Français d'Etudes

Andines, and the Geological Survey of Peru (now INGEMMET). Part of the logistical assistance in Peru was provided by Cia. de Minas Buenaventura.

Many of the data reported here result from a long-term collaboration with R. Marocco, G. Laubacher, and B. Dalmayrac of ORSTOM; D. C. Noble of the University of Nevada-Reno; and E. Audebaud of the University of Grenoble.

I thank D. C. Noble and the editors of this volume for their helpful comments and for their assistance in revising an early version of the manuscript.

REFERENCES

Atherton, M. P., Pitcher, W. S., and Warden, V., 1983: The Mesozoic marginal basin of central Peru. Nature *305*, 303-306.

Atherton, M. P., Sanderson, L. M., Warden, V., and McCourt, W. J., 1985: The volcanic cover: Chemical composition and the origin of the Calipuy group. *In*: Pitcher, W. S. et al. (eds), Magmatism at a Plate Edge: The Peruvian Andes. Blackie, Glasgow, 273-284.

Atherton, M. P., Warden, V., and Sanderson, L. M., 1985: The Mesozoic marginal basin of Central Peru: A geochemical study of within-plate-edge volcanism. *In*: Pitcher, W. S. et al. (eds), Magmatism at a Plate Edge: The Peruvian Andes. Blackie, Glasgow, 47-58.

Audebaud, E., 1973: Geología de los cuadrángulos de Ocongate y Sicuani. Perú Serv. Geol. Min. Bol. *25*, 72 p.

Audebaud, E., Capdevila, R., Dalmayrac, B., Debelmas, J., Laubacher, G., Lefèvre, C., Marocco, R., Martinez, C., Mattauer, M., Mégard, F., Paredes, J., and Tomasi, P., 1973: Les traits géologiques essentiels des Andes centrales (Pérou-Bolivie). Rév. Géog. phys. Géol. dyn. *15*, 73-113.

Audebaud, E., Laubacher, G., and Marocco, R., 1976: Coupe géologique des Andes du Sud du Pérou de l'océan Pacifique au bouclier brésilien. Geol. Rundschau *65*, 223-264.

Bally, A. W., 1981: Thoughts on the tectonics of folded belts. *In*: McClay, K. R., and Price, N. J. (eds), Thrust and Nappe Tectonics. Geol. Soc. London Spec. Publ. *9*, 13-32.

Bally, A. W., Gordy, P. L., and Stewart, G. A., 1966: Structure, seismic data, and orogenic evolution of southern Canadian Rocky Mountains. Bull. Can. Petroleum Geology *14*, 337-381.

Bally, A. W., and Snelson, S. L., 1980: Realms of subsidence. *In*: Miall, A. D. (ed), Facts and Principles of World Petroleum Occurrence. Can. Soc. Petroleum Geologists Mem. *6*, 9-75.

Barazangi, M., and Isacks, B. L., 1976: Spatial distribution of earthquakes and subduction of the Nazca plate beneath South America. Geology *4*, 686-692.

Barazangi, M., and Isacks, B. L., 1979: Subduction of the Nazca plate beneath Peru: Evidence from spatial distribution of earthquakes. Roy. Astron. Soc. Geophys. Jour. *57*, 537-555.

Bard, J. P., Botello, R., Martinez, C., and Subieta, T., 1974: Relations entre tectonique, métamorphisme et mise en place d'un granite éohercynien à deux micas dans la Cordillère Real de Bolivie (massif de Zongo-Yani). ORSTOM Cah., Sér. géol., *6*, 3-18.

Beaudoin, B., Léon, I., Chavez, A., and Vicente, J. C., 1982: The Arequipa basin (Peru): Carbonates and submarine fan (abstr.). Internat. Assoc. Sedimentologists Congress Abstr., 41.

Benavides, V., 1956: Cretaceous system in northern Peru. Am. Mus. Nat. Hist. Bull. *108*, 355-493.

Benavides, V., 1962: Estratigrafía pre-Terciaria de la región de Arequipa. Soc. Geol. Perú Bol. *38*, 5-63.

Benavides, V., 1968: Saline deposits of South America. Geol. Soc. America Spec. Paper *88*, 249-290.

Bussell, M. A., 1975: The structural evolution of the Coastal Batholith in provinces of Ancash and Lima, central Peru. Ph.D. thesis (unpublished), Univ. Liverpool.

Caldas, J., 1978: Geología de los cuadrángulos de San Juan, Acarí y Yauca. Perú Serv. Geol. Min. Bol. *30*, 78 p.

Caldas, J., Palacios, O., Pecho, V., and Vela, C., 1980: Geología de los cuadrángulos de Bayovar, Sechura, Le Redonda, Punta La Negra, Lobos de Tierra, Las Salinas y Morrope. Perú Inst. Geol. Min. y Metal., Ser. A, Bol. *32*, 92 p.

Carlier, G., Grandin, G., Laubacher, G., Marocco, R., and Mégard, F., 1982: Present knowledge of the magmatic evolution of the eastern Cordillera of Peru. Earth-Science Rev. *18*, 253-283.

Chanove, G., Mattauer, M., and Mégard, F., 1969: Précisions sur la tectonique tangentielle des terrains secondaires du massif de Pirin (nord-ouest du Lac Titicaca, Pérou). Acad. Sci. Paris C. R. (D) *268*, 1698-1701.

Child, R., 1976: The Coastal Batholith and its envelope in the Casma region of Peru. Ph.D. thesis (unpublished), Univ. Liverpool.

Clark, A. H., Palma, V. V., Archibald, D. A., Farrar, E., Arenas F., M. J., and Robertson, R.C.R., 1983: Occurrence and age of tin mineralization in the Cordillera Oriental, southern Peru. Econ. Geology *78*, 514-520.

Cobbing, E. J., 1973: Geología de los cuadrángulos de Barranca, Ambar, Oyon, Huacho, Huaral y Canta. Perú Serv. Geol. Min. Bol. *26*, 172 p.

Cobbing, E. J., 1976: The geosynclinal pair at the continental margin of Peru. Tectonophysics *36*, 157-165.

Cobbing, E. J., 1978: The Andean geosyncline in Peru, and its distinction from Alpine geosynclines. Geol. Soc. London Jour. *135*, 207-218.

Cobbing, E. J., Ozard, J. M., and Snelling, N. J., 1977: Reconnaissance geochronology of the crystalline basement rocks of the Coastal Cordillera of southern Peru. Geol. Soc. America Bull. *88*, 241-246.

Cobbing, E. J., and Pitcher, W. S., 1972: The Coastal batholith of central Peru. Geol. Soc. London Jour. *128*, 421-460.

Cobbing, E. J., Pitcher, W. S., Wilson, J. J., Baldock, J. W., Taylor, W. P., McCourt, W., and Snelling, N. J., 1981: The geology of the western Cordillera of northern Peru. Inst. Geol. Sci., Overseas Mem. *5*, 143 p.

Coney, P. J., 1971: Structural evolution of the Cordillera Huayhuash, Andes of Peru. Geol. Soc. America Bull. *82*, 1863-1883.

Cross, T. A., and Pilger, R. H., Jr., 1982: Controls of subduction geometry, location of magmatic arcs, and tectonics of arc and back-arc regions. Geol. Soc. America Bull. *93*, 545-562.

Dahlstrom, C.D.A., 1970: Structural geology in the eastern margin of the Canadian Rocky Mountains. Bull. Can. Petroleum. Geology *18*, 332-406.

Dalmayrac, B., Lancelot, J. R., and Leyreloup, A., 1977: Two-billion-year granulites in the Late Precambrian metamorphic basement along the southern Peruvian coast. Science *198*, 49-51.

Dalmayrac, B., Laubacher, G., and Marocco, R., 1980: Géologie des Andes péruviennes: Caractères généraux de l'évolution géologique des Andes péruviennes. ORSTOM Trav. Doc. *122*, 501 p.

De Jong, K. A., 1974: Melange (olistostrome) near Lake Titicaca, Peru. Am. Assoc. Petroleum Geologists Bull. *58*, 729-741.

Dewey, J. F., 1980: Episodicity, sequence, and style at convergent plate boundaries. *In*: Strangway, D. W. (ed), The Continental Crust and its Mineral Deposits. Geol. Assoc. Canada Spec. Paper *20*, 553-573.

Douglas, J. A., 1920: Geological sections through the Andes of Peru and Bolivia. II, From the port of Mollendo to the Inambari River. Geol. Soc. London Quart. Jour. *76*, 1-61.

Feininger, T., 1982: The metamorphic ''basement'' of Ecuador. Geol. Soc. America Bull. *82*, 3325-3346.

James, D. E., 1971: A plate tectonic model for the evolution of the central Andes. Geol. Soc. America Bull. *82*, 3325-3346.

James, D. E., Brooks, C., and Cuyubamba, A., 1975: Early evolution of the central Andean volcanic arc. Carnegie Inst. Washington, Yearbook *74*, 247-250.

Janjou, D., Bourgois, J., Mégard, F., and Sornay, J., 1981: Rapports paléogéographiques et structuraux entre Cordillères occidentale et orientale des Andes nord péruviennes: Les écailles du Marañon. Soc. géol. France Bull., Sér. 7, *23*, 697-705.

Jenks, W. F., 1948: Geología de la hoja de Arequipa al 200,000. Perú Inst. Geol. Bol. *9*, 104 p.

Jordan, T. E., Isacks, B. L., Allmendinger, R. W., Brewer, J. A., Ramos, V. A., and Ando, C. J., 1983: Andean tectonics related to the geometry of subducted Nazca plate. Geol. Soc. America Bull. *94*, 341-361.

Kontak, D. J., Clark, A. H., Farrar, E., and Strong, D. F., 1985: The rift-associated Permo-Triassic magmatism of the Eastern Cordillera: A precursor to the Andean orogeny. *In*: Pitcher, W. S. et al. (eds), Magmatism at a Plate Edge: The Peruvian Andes. Blackie, Glasgow, 36-44.

Kulm, L. D., Prince, R. A., French, W., Johnson, S., and Masias, A., 1981: Crustal structure and tectonics of the central Peru continental margin and trench: Lithostratigraphy, biostratigraphy, and tectonic history. Geol. Soc. America Mem. *154*, 445-468.

Lancelot, J. R., Laubacher, G., Marocco, R., and Renaud, U., 1978: U/Pb radiochronology of two granitic plutons from the Eastern Cordillera (Peru)—extent of Permian magmatic activity and consequences. Geol. Rundschau *67*, 236-243.

Laubacher, G., 1978: Géologie des Andes péruviennes: Géologie de la Cordillère orientale et de l'altiplano au nord et au nord-ouest du Lac Titicaca (Pérou). ORSTOM Trav. Doc. *95*, 217 p.

Laurent, H., and Pardo, A., 1975: Ensayo de interpretación del basamento del Nororiente peruano. Soc. Geol. Perú Bol. *48*, 25-48.

Lavenu, A., 1978: Néotectonique des sédiments plio-quaternaires du nord de l'Altiplano Bolivien (régions de la Paz, Ayo-Ayo, Umale). ORSTOM Cah., Sér. géol., *10*, 115-126.

Lefèvre, C., 1979: Un exemple de volcanisme de marge active dans les Andes du Pérou (Sud) du Miocène à l'actuel. Sc. D. thesis (unpublished), Univ. Montpellier.

Loughman, D. L., and Hallam, A., 1982: A facies analysis of the Pucara group (Norian to Toarcian carbonates, organic-rich shale and phosphate) of central and northern Peru. Sed. Geology *32*, 161-194.

McKee, E. H., and Noble, D. C., 1982: Miocene volcanism and deformation in the western Cordillera and high plateaus of south-central Peru. Geol. Soc. America Bull. *93*, 657-622.

Marocco, R., 1978: Géologie des Andes péruviennes: Un segment E-W de la chaîne des Andes péruviennes: La déflexion d'Abancay. ORSTOM Trav. Doc. *94*, 195 p.

Marocco, R., and García-Zabaleta, F., 1974: Estudio geológico de la región entre Cuzco y Machu Picchu. Bull. Inst. français, Etudes Andines, III, *2*, 1-27; Ing. Geol. *16*, 9-35.

Martinez, C., 1980: Structure et évolution de la chaîne hercynienne et de la chaîne andine dans le nord de la Cordillère des Andes de Bolivie. ORSTOM Trav. Doc. *119*, 352 p.

Mégard, F., 1967: Commentaire d'une coupe schématique à travers les Andes centrales de Pérou. Rév. Géog. phys. Géol. dyn., Sér. 2, *9*, 335-345.

Mégard, F., 1973: Etude géologique d'une transversale des Andes au niveau du Pérou central. Sc. D. thesis (unpublished), Univ. Montpellier.

Mégard, F., 1978: Etude géologique des Andes du Pérou central. ORSTOM Mem. *86*, 310 p.

Mégard, F., 1984: The Andean orogenic period and its major structures in central and northern Peru. Geol. Soc. London Jour. *141*, 893-900.

Mégard, F., Dalmayrac, B., Laubacher, G., Marocco, R., Martinez, C., Paredes, J., and Tomasi, P., 1971: La chaîne hercynienne au Pérou et en Bolivie: Premiers résultats. ORSTOM Cah., Sér. géol., *3*, 5-43.

Mégard, F., Marocco, R., Vicente, J. C., and Mégard-Galli, J., 1983: Découverte d'une discordance angulaire tardi-hercynienne (Permien moyen) dans les Andes du Pérou central. Acad. Sci. Paris C. R., Sér. II, *296*, 1267-1270.

Mégard, F., Noble, D. C., McKee, E. H., and Bellon, H., 1984: Multiple pulses of Neogene compressive deformation in the Ayacucho intermontane basin, Andes of central Peru. Geol. Soc. America Bull. *95*, 1108-1117.

Mégard, F., and Philip, H., 1976: Plio-Quaternary tectono-magmatic zonation and plate tectonics in the central Andes. Earth Planet. Sci. Lett. *33*, 231-238.

Myers, J. S., 1974: Cretaceous stratigraphy and structure, western Andes of Peru between latitudes 10°-10°30′. Am. Assoc. Petroleum Geologists Bull. *58*, 474-487.

Myers, J. S., 1975a: Vertical and crustal movements of the Andes in Peru. Nature *254*, 672-674.

Myers, J. S., 1975b: Cauldron subsidence and fluidization: Mechanisms of intrusion of the Coastal Batholith of Peru into its own volcanic ejecta. Geol. Soc. America Bull. *86*, 1209-1220.

Newell, N. D., 1949: Geology of the Lake Titicaca region, Peru and Bolivia. Geol. Soc. America Mem. *36*, 111 p.

Noble, D. C., Bowman, H. R., Hebert, A. J., Silberman, M. L., Heropoulos, C. E., Fabbi, B. F., and Hedge, C. E., 1975: Chemical and isotopic constraints on the origin of low-silica latite and andesite from the Andes of central Peru. Geology *3*, 501-504.

Noble, D. C., Farrar, E., and Cobbing, E. J., 1979: The Nazca group of south-central Peru: Age, source, and regional volcanic and tectonic significance. Earth Planet. Sci. Lett. *45*, 80-86.

Noble, D. C., and McKee, E. H., 1982: Nevado Portugueza volcanic center, central Peru: A Pliocene central volcanic-collapse caldera complex with associated silver mineralization. Econ. Geology *77*, 1893-1900.

Noble, D. C., McKee, E. H., Eyzaguirre, V. R., and Marocco, R., 1984: Age and regional tectonic and metallogenetic implications of igneous activity and mineralization in the Andahuaylas-Yaurí belt of southern Peru. Econ. Geology *79*, 172-176.

Noble, D. C., McKee, E. H., Farrar, E., and Petersen, U., 1974: Episodic Cenozoic volcanism and tectonism in the Andes of Peru. Earth Planet. Sci. Lett. *21*, 213-220.

Noble, D. C., McKee, E. H., and Mégard, F., 1979: Early Tertiary "Incaic" tectonism, uplift, and volcanic activity, Andes of central Peru. Geol. Soc. America Bull. *90*, 903-907.

Noble, D. C., Sébrier, M., Mégard, F., and McKee, E. H., 1985: Demonstration of two pulses of Paleogene deformation in the Andes of Peru. Earth Planet. Sci. Lett. *73*, 345-349.

Noble, D. C., Silberman, M. L., Mégard, F., and Bowman, H. R., 1978: Comendite (peralkaline rhyolite) and basalt in the Mitu Group, Peru: Evidence for Permian-Triassic lithospheric extension in the central Andes. U.S. Geol. Surv. Jour. Res. *6*, 453-457.

Noble, D. C., Vogel, T. A., Peterson, P. S., Landis, G. A., Grant, N. K., Jezek, P. A., and McKee, E. H., 1984: Rare-element–enriched, S-type ash-flow tuffs containing phenocrysts of muscovite, andalusite, and sillimanite, southeastern Peru. Geology *12*, 35-39.

Pardo, A., 1982: Características estructurales de la faja subandina del Norte del Perú. Com. Simp. "Exploración petrolera en las cuencas subandinas de Venezuela, Colombia, Ecuador, y Perú," Bogotá.

Pitcher, W. S., 1978: The anatomy of a batholith. Geol. Soc. London Jour. *135*, 157-182.

Pitcher, W. S., Atherton, M. P., Cobbing, E. J., and Beckinsale, R. D. (eds), 1985: Magmatism at a Plate Edge: The Peruvian Andes. Blackie, Glasgow, and Halsted Press, New York, 328 p.

Pitcher, W. S., and Bussell, M. A., 1977: Structural control of batholithic emplacement in Peru: A review. Geol. Soc. London Jour. *133*, 249-256.

Price, R. A., 1981: The Cordilleran foreland thrust and fold belt in the southern Canadian Rocky Mountains. *In*: McClay, K. R., and Price, N. J. (eds), Thrust and Nappe Tectonics. Geol. Soc. London Spec. Publ. *9*, 427-448.

Reyes, L., 1980: Geología de los cuadrángulos de Cajamarca, San Marcos y Cajabamba. Perú Inst. Geol. Min. Metal. Bol. *31*, 67 p.

Rodríguez, A., 1979: Las estructuras salinas en la Faja Subandina. Soc. Geol. Perú Bol. *62*, 141-159.

Rodríguez, A., and Chalco, A., 1975: Cuenca Huallaga, reseña geológica y posibilidades petrolíferas. Soc. Geol. Perú Bol. *45*, 187-212.

Roeder, D., 1982: Geodynamic model of the Subandean zone in Alto Beni area, Bolivia. Com. Simp. "Exploración petrolera en las cuencas subandinas de Venezuela, Colombia, Ecuador, y Perú," Bogotá.

Romani, M., 1982: Géologie de la région minière Uchucchacua-Hacienda Otuto, Pérou. 3rd cycle thesis (unpublished), Univ. Grenoble.

Sébrier, M., Mercier, J. L., Mégard, F., Laubacher, G., and Carey-Gailhardis, E., 1985: Quaternary normal and reverse faulting and the state of stress in the central Andes of south Peru. Tectonics *4*, 739-780.

Shackleton, R. M., Ries, A. C., Coward, M. P., and Cobbold, P. R., 1979: Structure, metamorphism and geochronology of the Arequipa massif of coastal Peru. Geol. Soc. London Jour. *136*, 195-214.

Soulas, J. P., 1977: Les phases tectoniques andines du Tertiaire supérieur, résultats d'une transversale Pisco-Ayacucho (Pérou central). Acad. Sci. Paris C. R. (D) *284*, 2207-2210.

Soulas, J. P., and Mégard, F., ms., in prep.: Longitudinal strike-slip faults and sigmoidal fault and fold systems, Andes of northern and central Peru.

Stauder, W., 1973: Mechanism and spatial distribution of Chilean earthquakes with relation to subduction of the oceanic plate. Jour. Geophys. Res. *78*, 5033-5061.

Steinmann, G., 1921: Geologie von Peru. Karl Winter, Heidelberg, 448 p.

Stewart, J. W., Evernden, J. F., and Snelling, N. J., 1974: Age determinations from Andean Peru: A reconnaissance survey. Geol. Soc. America Bull. *85*, 1107-1116.

Suárez, G., Molnar, P., and Burchfiel, B. C., 1983: Seismicity, fault plane solutions, depth of faulting, and active tectonics of the Andes of Peru, Ecuador, and southern Colombia. Jour. Geophys. Res. *88*, 10,403-10,428.

Thornburg, T. M., and Kulm, L. D., 1981: Sedimentary basins of Peru continental margin: Structure, stratigraphy, and Cenozoic tectonics from 6°S to 15°S latitude. Geol. Soc. America Mem. *154*, 393-422.

Uyeda, S., 1982: Subduction zones: An introduction to comparative subductology. Tectonophysics *81*, 133-159.

Vicente, J. C., 1981: Elementos de la estratigrafía mesozoica surperuana. *In*: Volkheimer, W., and Musacchio, E. A. (eds), Cuencas sedimentarias del Jurásico y Cretácico de América del sur. Comité sud. Amer. del Jur. y Cret. *1*, 319-351.

Vicente, J. C., Beaudoin, B., Chavez, A., and Léon, I., 1982: La cuenca de Arequipa (Sur-Perú) durante el Jurásico-Cretácico inferior. Vto Congreso latinoamericano de Geología, Buenos Aires, Actas *1*, 121-153.

Vicente, J. C., Sequeiros, F., Valdivia, M. A., and Zavala, J., 1979: El sobre-escurrimiento de Chincha-Lluta: Elementos del accidente mayor andino al NW de Arequipa. Soc. Geol. Perú Bol. *61*, 67-99.

Webb, S., 1976: The volcanic envelope of the Coastal Batholith in Lima and Ancash, Peru. Ph.D. thesis (unpublished), Univ. Liverpool.

Wilson, J. J., 1963: Cretaceous stratigraphy of Central Andes of Peru. Am. Assoc. Petroleum. Geologists Bull. *47*, 1-34.

Wilson, J. J., Reyes, L., and Garayar, J., 1967: Geología de los cuadrángulos de Mollebamba, Tayabamba, Pomabamba y Huarí. Perú Serv. Geol. Min. Bol. *16*, 95 p.

Chapter 11

THE KINEMATIC PUZZLE OF THE NEOGENE NORTHERN ANDES

H. P. LAUBSCHER

Universität Basel

North of Ecuador, the South American Andes split into a number of diverging branches which terminate on approaching the Caribbean coast (Figure 11-1). Plate-tectonically, these branches are a sort of extended triple-junction domain between the South American plate (SA), the Caribbean plate (Ca), and the complex northern end of the Nazca plate (Na) (Figure 11-2; cf. Beck et al. 1975, Lonsdale and Klitgord 1978, Case and Holcombe 1980). This paper deals mostly with tectonics on land, and particularly with the Cordillera Oriental of Colombia, the Santander massif, the Sierra de Perijá, the Sierra Nevada de Santa Marta, and the Mérida Andes of Venezuela (Figure 11-2). These high mountain ranges, culminating in the Sierra Nevada de Santa Marta—at an altitude of 5,800 m—with Precambrian in its core

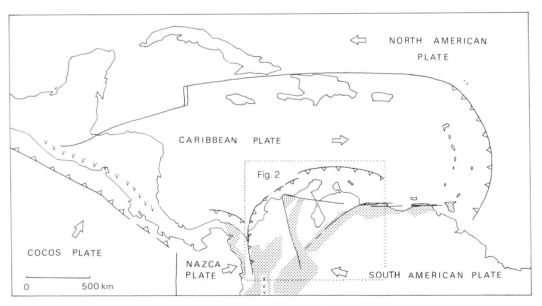

FIGURE 11-1. Plate-tectonic situation of the South America-Caribbean-Nazca triple-junction domain. Shaded areas: Branches of the Neogene northern Andes. Lines with triangles: Subduction zones. V-pattern: Volcanic chains. (After Plafker 1976.)

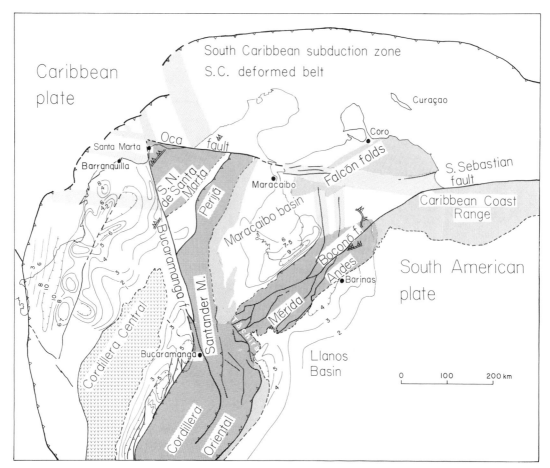

FIGURE 11-2. Generalized tectonic map of the South American-Caribbean-Nazca triple-junction domain. Only a selection of the main faults (heavy lines; thrusts with triangles) is shown. Light shading: Postulated diffuse strike-slip zones. Medium shading: Marginal mountain ranges and some foothill belts. Darkest shading: Intracontinental Andean ranges. V-pattern: Cordillera Central (high elevations are volcanic). Thin contours: Approximate base of Cretaceous (km) in mostly Neogene depressions. Double lines with double triangles: Laramide metamorphic front at the Boconó, Bucaramanga, and Oca faults. (Mainly after Martin 1978 and Case and Holcombe 1980.)

(Tschanz et al. 1974), and their flanking depressions are essentially of Neogene age, although the Neogene motions are superposed on a multiply deformed terrain with a long deformational history (Case and Holcombe 1980, Kellogg and Bonini 1982). It is with the Neogene movements that this paper is principally concerned. Older deformations are of interest for two reasons: they suggest that the location of Andean branches is co-determined by inherited structure (Renz 1959), and they provide a frame of reference for estimating Neogene motion (Tschanz et al. 1974, Bellizzia et al. 1976, Pümpin 1979). The terrain involved is vast; large parts of it are difficult of access and have been explored only cursorily, and other parts are hidden below the sea or alluvial plains. Though much information is known to the oil companies, little has been published. One very useful recent compilation, partly of unpublished material, is the map by Case and Holcombe (1980; cf. Martin 1978).

The style of deformation is quite different from that of the Alps and appears somewhat exotic to people accustomed to thinking in Alpine terms such as nappes, although kinematic consistency requires more thrusting than is generally assumed. An illustration is provided in the companion paper by Meier, Schwander, and Laubscher (this volume). The kinematics of these mountain ranges is controversial (e.g., Pennington 1981, Schubert 1981), but there are some quantitative data, such as dissection of preexisting structures and crustal compression, which provide means for developing models of an internally consistent field of motion. The most informative preexisting structure for estimating strike-slip components is the Cretaceous-early Eocene (Laramide) system of Caribbean nappes (Figure 11-2; cf. MacDonald et al. 1971). These nappes are particularly important, since strike-slip evidently plays a major role at the southern margin of the Caribbean plate (e.g., Bowin 1976, Figure 16; Kafka and Weidner 1981). Compressive (normal) components can be estimated on the basis of crustal shortening (Figures 11-3, 11-4). The resulting quantities can then be used to provide an overall kinematic framework for more detailed deformation, such as that in Táchira (see Meier, Schwander, and Laubscher, this volume).

QUANTIFICATION OF MOTION AT THE BLOCK BOUNDARIES OF THE MARACAIBO SYSTEM

A first, cursory inspection of the structure of northwestern South America reveals the existence of a mosaic of comparatively stable blocks surrounded by zones of motion (Figure 11-2). The most prominent of these blocks and the one conforming most closely with the ideal of internal rigidity is the Maracaibo block, and for that reason the entire system of blocks occupying the domain of the triple plate junction SA-Ca-Na is named the "Maracaibo block mosaic." Deformation in this system is unevenly distributed. Maximum concentration is along regional fault systems that are very prominent topographically as well as geologically; these

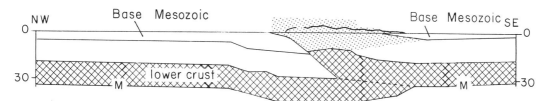

FIGURE 11-3. A crustal section from Maracaibo to Barinas (see Figure 11-2) through the central Mérida Andes based on gravity data. (From Bonini et al. 1977, Kellogg and Bonini 1982.) A relative high in the gravity trough is interpreted as obducted lower crust, reminiscent of the interpretation of the Ivrea gravity high (seismically constrained) in the western Alps (Berckhemer 1968). Although this interpretation is not constrained by seismic information, it is roughly balanced kinematically; the squeezed-out upper crust (that part of the upper crust above the line connecting the top of the basement in the two foreland troughs—stippled) would result from shortening of about 30 km, as would the thickening of the entire crust. For a different view, see Hospers and van Wijnen (1959) and Shagam (1975, Figure 15); Shagam's section, however, seems incompatible with the results of detailed mapping in Táchira (Meier, Schwander, and Laubscher, this volume).

W 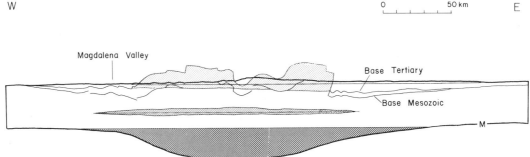 E

FIGURE 11-4. A crustal section through the Cordillera Oriental of Colombia, south of Bucaramanga (see Figure 11-2). (From Irving 1971, simplified; base of crust added.) Light shading: Mass squeezed up; reference is base of Tertiary. Intermediate shading: Putative mid-crustal detachment zone. Heavy shading: Estimated lower crust squeezed down; reference is M. From material balance estimates for the squeezed-up mass, we conclude that there must have been compression exceeding 100 km, which is far more than that suggested by folds and thrusts on geological cross sections (Julivert 1970, Irving 1971, Gansser 1973). We surmise that particularly the marginal thrusts are vastly underestimated. Depression of M would be kinematically compatible (materially balanced).

faults are usually accompanied by somewhat more extensive belts into which they branch out, and which take up compressional deformation. The interior of the blocks is never entirely rigid but invariably manifests some deformation. The concept of "block mosaic" is thus an idealization. Consequently, the model (Figures 11-7 to 11-9, discussed in more detail below), is a drastic simplification.

Exact stratigraphic allocation of phases of movement is feasible only locally. Dissection of the Laramide system strictly defines the Andean system (consisting of the various branches of present-day Andean mountain ranges) only as post-early Eocene; however, the Oligocene and even the Miocene are generally deformed to the extent that the Andean system may be termed essentially Neogene (Andean orogeny s.s., cf. Mencher et al. 1953, Macellari 1982); it is still active today. This can be shown conclusively for Táchira (Meier, Schwander, and Laubscher, this volume) and for both margins of the Mérida Andes and the Cordillera Oriental. According to Kellogg and Bonini (1982), it is also true for the Sierra de Perijá and the Sierra Nevada de Santa Marta, and, according to Case and Holcombe (1980), for the South Caribbean deformed belt and subduction zone. Indeed, as will be demonstrated below, the kinematic relations between the parts of the block mosaic are such that it would be strange if they were not contemporaneous. Consequently, stratigraphic evidence in the south around Táchira constitutes strong evidence for the time of much of the deformation in the north around the Sierra Nevada de Santa Marta. Earlier deformations are present, but they seem to belong to different systems, although they have been partly reactivated by the Andean orogeny s.s.

A rough quantification of strike-slip motion can be obtained across the Boconó fault of the Mérida Andes (Rod 1956, Bellizzia et al. 1976, Pümpin 1979) and the border faults of the Sierra Nevada de Santa Marta (Tschanz et al. 1974; see also Figure 11-2). When assessing the significance of the apparent displacements, we must keep in mind that exact strike-slip displacements can be figured out only if subsequent tilting and erosion of the preexisting structures are known. For feebly inclined structures such as nappes, the margin of error is considerable, and for the case at hand, only orders of magnitude are indicated. They are between 50 and slightly over 100 km, and round figures of 100 km have been assumed for both the Oca and Bucaramanga faults and 60 km for the Boconó fault. There are other estimates, but they are based on less satisfactory correlations (see, e.g., Féo-Codecido 1972). As for quantification of compression, the gravimetric profile of Wittke (in Bonini et al. 1977 and Kellogg and Bonini 1982) across the central Mérida Andes (Figure 11-3) permits a measurement of crustal thickening, which suggests about 30 km of compression. The northern part of the Cordillera Oriental of Colombia (Figure 11-4) has a similar structural relief but is much wider, and as it was the site of accumulation of thick Mesozoic and early Tertiary sediments, these play a large role in its composition. A rough estimate of crustal compression normal to the strike yields more than 100 km. As there are dextral en échelon patterns on both sides of this mountain range (Case and Holcombe 1980), a strike-slip component should probably be added, but at this time it may not be quantified. Quantities for the shortening in the Sierra de Perijá, the Sierra Nevada de Santa Marta, and the South Caribbean subduction zone may be obtained from Kellogg and Bonini (1982); they are discussed in some detail on pp. 217, 219, and 220.

THE TRANSPORT FIELD OF THE MARACAIBO BLOCK MOSAIC

The estimates given above can be used as fundamental quantities for modeling the transport field of the idealized block mosaic (see Figures 11-7 to 11-9, below, for the location of and symbols for the blocks). If we begin with the Mérida Andes, the Ma/SA boundary, the simplest orderly procedure for describing the other block boundaries is to move from there to a diagonal southeast-northwest section through the Sierra Nevada de Santa Marta, beginning in the southeast; then to the boundaries west of the Bucaramanga fault from north to south; and finally to those north of the Oca fault, beginning in the west.

Ma/SA Boundary

Straightforward interpretation of published crustal sections and maps yields the figures given in the previous section: 60 km dextral slip and 30 km compression. These and other figures used in this paper, though supported by some, have been doubted by others (see, e.g., Shagam 1975, Kellogg and Bonini 1982). Rather than enter into a lengthy evaluation, it is our intention here to find out whether they are mutually compatible within a kinematic model. For the Ma/SA boundary, Kellogg and Bonini (1982) play down the importance of strike-slip, which, moreover, they want to restrict to the past 10,000 years. However, simultaneity and transpression throughout the last 10 Ma conform better with the data and particularly with the results of detailed mapping in Táchira (Meier, Schwander, and Laubscher, this volume).

Ma/SM Boundary (Sierra de Perijá)

Data on compression in the Sierra de Perijá are given by Kellogg and Bonini (1982, particularly their Figures 4 and 6). The evidence is best for the Cerrejón thrust at the northwest margin (see Figure 11-5 of this paper), with an estimated 25 km of shortening. The other thrusts,

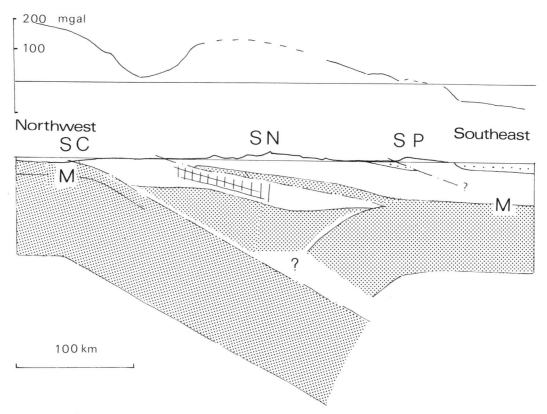

FIGURE 11-5. Crustal structure of the Sierra Nevada de Santa Marta (SN), the Sierra de Perijá (SP), and the South Caribbean subduction zone (SC), modified after Kellogg and Bonini (1982, their Figures 9 and 12). M: Mohorovičić discontinuity. The upper figure is the Bouguer gravity profile; the solid part of the curve is measured. To account for the large positive anomaly in the Sierra Nevada de Santa Marta, there must be either a high of the M-discontinuity at normal crustal thickness (Kellogg and Bonini 1982) with virtually no crustal compression, or an obduction of dense lower crust to upper mantle which involves two superposed layers of less than normal crustal thickness with a shortening of 160 km as required by kinematics (Figure 11-7). Both crusts indeed are anomalous (Tschanz et al. 1974, Duque-Caro 1978). Corresponding shortening of the lithosphere must be disharmonic, perhaps in the "flaking" mode advocated by Laubscher and Bernoulli (1982) for the Ivrea body in the southern Alps. The Benioff zone of the South Caribbean subduction is shortened in accordance with the kinematic models (Figures 11-6 and 11-7), but it would seem to interfere with lithospheric deformation under the Sierra Nevada even then. I suspect that this Benioff zone does not portray conditions under the Barranquilla block, which according to Figure 11-6 is expected to be severely dissected. No densities are assigned for quantitative modeling of gravity. Gravity data without seismic or other (e.g., kinematic) constraints are ambiguous in such complex structures.

and particularly the interpretation of the Tigre fault (near the southwest margin in the northeast part of the Sierra de Perijá; cf. Figure 11-2) as a Miocene thrust, are doubtful. In contrast to the Cerrejón fault, the Tigre fault is a straight feature on the surface, indicative of a steep fault rather than a low-angle thrust, and sediment distribution on the two flanks would favor a Paleogene rather than a Neogene age (cf. Mencher 1963, his Figure 5). Kellogg and Bonini's interpretation (1982, their Figure 6) that the entire Sierra de Perijá is due to a stack of ramps issuing from a flat décollement surface at 8 km depth that also passes under the Maracaibo basin, is difficult to believe in view of the latter's rigidity. To let such basin-internal features as the La Paz structure originate similarly from that décollement plane does not take into account its history and overall character as a transpressive "flower structure" (P. Stalder, personal communication). Because the Sierra de Perijá is a more modest mountain than the Mérida Andes, we would tend to assign it less compression, say 20 km, and this would be the order of magnitude of Cerrejón thrusting. As to the direction of this translation, the fact that it interferes only slightly with the Oca fault would suggest that it is almost parallel to that fault.

Ba/SM Boundary

This boundary is one of the most puzzling, as it includes the isolated Santa Marta massif (Gansser 1955, Tschanz et al. 1974) as well as the intersection of the large Oca and Bucaramanga strike-slip faults. Intersections of simultaneously active complementary shears are complex kinematic problems everywhere and give rise to bewildering tectonic complications; look at the intersection of the San Andreas and Garlock faults in the Transverse Ranges of California, for example. An experimental illustration of some of the problems involved is the sandbox experiment by Horsfield (1980). To explore the Oca-Bucaramanga intersection, I have, in Figure 11-6, constructed a cut-paper model. This is done by tracing the map-view block boundaries on semitransparent paper, cutting the paper along these boundaries, and performing the plane kinematics in a series of steps. Each step is recorded by first fixing the blocks in their new position and then dry-copying against a dark background: displacements are clearly visible on such a copy, with shortenings as lighter bands, and extensions as darker bands. The individual records (copies) are then assembled into a sequence of kinematic steps which allows one to see clearly the kinematics that led from a simple beginning to a complex result.

We first note that intersecting complementary shears without additional major complications are possible only in a static fracture pattern prior to finite motion (Figure 11-6a). The point of intersection at this stage is a quadruple junction of blocks 1 to 4. When finite motion begins, there is considerable freedom for the behavior of the individual blocks (cf. Laubscher 1965, Figure 37), depending on the distribution of resisting forces at their boundaries, including the base. If only block 2 is free to move, then blocks 1, 3, and 4 are the stable surroundings against which block 2 moves obliquely by transpression on the boundary faults. This choice is usually made for the Santa Marta massif, as by Kellogg and Bonini (1982, their Figure 2). At the other extreme of the range of possibilities is the one shown in Figure 11-6b; here, motion along the intersecting faults is pure strike-slip. This case requires block 1 to move by the same amount as block 2 with respect to blocks 3 and 4, the latter being pushed sideways in the process. The compression is now concentrated at the tips of blocks 1 and 2 where one overrides the other; the original quadruple point at the intersection of the shears has degenerated into the diamond-shaped area of compression with two triple junctions, 1-2-3 and 1-2-4. Which extreme or intermediate choice has been made by nature can be determined only by observation, which shows for the northwestern foreland of the Santa Marta block, simplified to "Barranquilla block" in this paper, an intriguing similarity to our Figure 11-6b, at least in

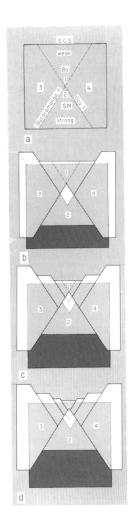

FIGURE 11-6. A simplified cut-paper model of the kinematic interaction between the Santa Marta (SM) and Barranquilla (Ba) blocks. White: Double layer of paper (sink of mass transport; shortening). Black: Gap (source of mass transport). As this is a simple model with general applicability to the kinematics at intersecting shears (quadruple junction of four blocks), block numbers rather than names are used. To facilitate comparison to northwestern South America, however, names for some blocks and block boundaries are given as well in Figure 6a.

6a: Static rupture. The configuration before the onset of motion is essentially that of a quadruple junction (blocks 1-4) at intersecting complementary shears. SCS: South Caribbean subduction zone.

6b: First kinematic phase (about 60 km on both shears, in nature including dextral drag in the accretionary wedge). At the intersection, the faults of block 1 are displaced, and a diamond-shaped transverse range (white) emerges, the continental block 2 overriding the marginal block 1. The quadruple junction degenerates into two triple junctions, 1-2-3 and 1-2-4.

6c: Second kinematic phase (about 30 km on both shears; in nature Bucaramanga fault phases I and II are not separable and together produce about 90 km, including sinistral drag in the accretionary wedge). The development in nature is not exactly symmetrical, thus agreeing with the experiment by Horsfield (1980). However, for the cut-paper model, a symmetrical solution has been chosen.

6d: Third kinematic phase (Oca and Bucaramanga faults, about 10 km).

Notice the decrease in shortening in the South Caribbean subduction zone and the piling up of masses under the tip of the Santa Marta block. These masses apparently are more widely distributed in nature. For a rough scale, the final displacement on the strike-slip boundaries 2-3 (Bucaramanga fault) and 2-4 (Oca fault) is 100 km.

the behavior of the South Caribbean subduction zone (see Case and Holcombe 1980; Kellogg and Bonini 1982, their Figure 6). This similarity would seem to warrant the search for further evidence in the Barranquilla block of the path followed by nature. For a trial, let us take the offsets of the South Caribbean subduction zone at face value, assigning them to distributed shear zones with directions of the Oca and Bucaramanga faults, as in Figure 11-2. These fault zones are not on the maps of Case and Holcombe (1980) or Kellogg and Bonini (1982), but as they would have to be found in the subsurface and information on this is scarce, their existence cannot be considered disproved on present evidence. On the contrary, Case and Holcombe's map (cf. Krause 1971, Shepard 1973) shows large-scale structural and topographic irregularities along the trends shown in Figure 11-2. If we accept them for the time being, a remarkable kinematic story unfolds, one that is strikingly similar to Horsfield's (1980) experiment (upper half of his Figure 3), and which is reconstructed in Figure 11-6. Briefly, the kinematics of this model involves three steps that account for the inflections of the South Caribbean subduction zone. While the original shears remained active all the time on the boundaries of the Santa Marta block, they were used for only 60 km out of a total of 100 km in the Barranquilla block. Thereafter, they were replaced by new faults more in line with the Oca and Bucaramanga faults—in agreement with Horsfield's experiment (1980). These new faults in turn were replaced by a third set of faults after a mere 30 km of slip. Though nature is not as symmetrical as the model, the similarity is close enough to be taken seriously. Most remarkable are the quantitative agreements: the 100 km of displacement of the Barranquilla block boundaries with respect to the Oca and Bucaramanga faults is that required by strike-slip on the two faults as estimated independently, using the amount of displacement of the Laramide metamorphic front (Figure 11-2). Moreover, compression is concentrated in the Santa Marta massif rather than distributed all along the Oca and Bucaramanga faults, as would be required by the Kellogg and Bonini (1982) scenario; the Santander massif along the southern part of the Bucaramanga fault is separated from the Santa Marta massif by the Cesar Valley low and seems to be connected kinematically with the Sierra de Perijá rather than with the Santa Marta massif.

The Transverse Ranges of California have been mentioned in connection with kinematic complications arising at intersecting shears. According to the model, Figure 11-6, the Sierra Nevada de Santa Marta is in some respects the equivalent of the Transverse Ranges; whenever dextral strike-slip faults such as the San Andreas and Oca faults are offset sinistrally, continued motion along them must give rise to "Transverse Ranges." The same applies, of course, to sinistral faults offset dextrally. This phenomenon, then, is the counterpart of pull-aparts which arise when dextral faults are offset dextrally (or sinistral faults sinistrally), as observed along the Oca fault further east in Falcón and at its junction with the San Sebastián fault (see pp. 221, 225).

One of the many important questions still to be answered concerns the different behavior of the Barranquilla and Santa Marta blocks: the latter, the "strong" one, seems to have imposed its boundaries on the former, the "weak" one. In a vague way this is not surprising, as the Santa Marta block is composed at least partly of pre-Mesozoic continental fragments welded together by Mesozoic intrusions (Tschanz et al. 1974), the Barranquilla block mostly of Mesozoic and younger accretionary material (Duque-Caro 1978). Another interesting aspect of our model (Figure 11-6) is that a part of the compression of the South Caribbean subduction zone in the Barranquilla block domain is transferred to the Sierra Nevada de Santa Marta, and shortening across the South Caribbean subduction zone is increased outside the Barranquilla block by the equivalent of the lateral push of blocks 3 and 4.

A related major problem is the distribution of compression in cross section through the Sierra Nevada de Santa Marta, which is constrained by the large positive gravity anomaly there (cf. Kellogg and Bonini 1982, particularly their Figures 3 and 9). To achieve a match with observed gravity, Kellogg and Bonini modeled mass distribution below the Santa Marta massif in a simple way which, however, is at variance with the inferred strike-slips of 100 km on both the Oca and Bucaramanga faults. Our model in Figure 11-6 implies doubling for up to 160 km of that part of the crust that is not subducted or an equivalent distribution of the masses piled up in the transverse range of the Santa Marta massif. A comparison of Figures 11-2 and 11-6 of this paper shows that this amount is equivalent to doubling of at least the upper crust under the entire Sierra Nevada. The scenario presented in Figure 2 of Kellogg and Bonini (1982) aggravates the material balance problem. It is certainly not resolved by the paltry 10 km of thrusting shown in their Figure 9. The problem would be easier to solve if there were a large negative Bouguer anomaly indicative of crustal thickening. It would seem that the positive anomaly, together with large crustal compression, would require thinner than normal continental crust, with lower crust and possibly some upper mantle material obducted at the base of the upper plate, in analogy with models of the Ivrea body in the western Alps (Berckhemer 1968, see also Laubscher and Bernoulli 1982). Stretching of the Santa Marta crust did indeed occur in the Mesozoic (see Tschanz et al. 1974; for the Barranquilla crust see Duque-Caro 1978). The puzzling high of the M-discontinuity under the Sierra Nevada might thus possibly be modeled into some sort of root. Whatever the final solution, kinematics and density distribution should be compatible. For the overall block kinematics model, Figure 11-7, the dissection of the Barranquilla block performed in Figure 11-6 is ignored, and all the strike-slip motion is assigned to the putative original continuations of the Oca and Bucaramanga faults.

BA/CA BOUNDARY

This boundary corresponds to the South Caribbean deformed belt or subduction zone (Case 1975, Talwani et al. 1982), modified according to Figure 11-6 by the interference with the Bucaramanga and Oca strike-slip systems. Compression is at a minimum at the center and increases stepwise at the strike-slip zones. Kellogg and Bonini (1982, Figure 12) estimated the length of the corresponding Benioff zone at about 400 km and the actual amount of subduction at about 200 km for the last 10 Ma by projecting earthquake hypocenters onto a profile passing through the apex of the Santa Marta block. This is a rather risky procedure in view of the complex kinematics, but it may give a measure of the full amount of compression across the subduction zone outside the Barranquilla block. According to our models, Figures 11-6 and 11-9, the 200 km of convergence between Ca and CC, or between Ca and Cu, should be reduced for the boundary Ca/Ba, the amount depending on movement along the Oca and Bucaramanga-Santa Marta faults. For the assumptions of Figure 11-7, Ca/Ba convergence turns out to be 100 km.

BA/CC, CA/CC, CC/SM, CC/SA, CC/CO, and CO/SA BOUNDARIES

We now take up the block boundaries in a southerly direction on the west side of the Bucaramanga fault. As shown by Case and Holcombe (1980), there is no significant Neogene movement between the broad and complex band of deformation that continues the South Caribbean subduction zone and the deformed belt southward through the Sinu valley (south-southwest of Barranquilla; see Figure 11-2; Duque-Caro 1978, Théry 1980), and the Magdalena depression (southwest of Bucaramanga, Figure 11-2), which is the western foredeep of the Cordillera Oriental (see map by Case and Holcombe 1980). The mighty Cordillera Central is essentially a chain of young volcanoes (Irving 1971, Gansser 1973). For simplicity we unite this entire terrain between the South Caribbean deformed belt and the western front of the Cordillera Oriental into the Central Colombia block CC. Its boundary displacements are already fixed by our previous assumptions: Ba/CC (West Oca fault zone) = 100 km dextral strike-slip; Ca/CC (South Caribbean deformed belt) = 200 km; CC/SM (Bucaramanga fault) = 100 km sinistral strike-slip. The Santander block (Sa) is introduced between CC and Ma only to separate strike-slip along the Bucaramanga fault and compression in the southern continuation of the Sierra de Perijá. Since these two quantities are already fixed, they determine the vectors shown in Figure 11-7. Likewise the CC/Co vector is determined by slip on the Bucaramanga fault. We are not even free to choose a value for the Co/SA vector, the southeastern part of the Cordillera Oriental; the vector addition SA/Ma + Ma/Sa + Sa/Co fixes motion Co/SA at about 80 km with a direction slightly north of east, very similar to that of Ma/SA. This results in a total compression of Cordillera Oriental (CC/Co + Co/SA) of about 130 km, consonant with the independent estimate of crustal compression (Figure 11-4). The Na/CC convergence is assumed to be 700 km east-west, or 7 cm/year for 10 Ma (Kellogg and Bonini 1982; cf. Pennington 1981).

Ba/Cu, Ca/Cu, Cu/SM, Cu/Ma, Cu/La, and Ca/SA Boundaries

Except for Cu/La, displacements at the block boundaries along the Oca fault are completely implicit in previous assignations and need no extensive discussion. However, some points that can easily be overlooked merit consideration. Along the Oca fault, Ba/SM is 100 km dextral in map view (for the profile cf. Figures 11-5 and 11-6), as is Cu/SM, but Cu/Ma is 120 km dextral strike-slip as 20 km compression along the Sierra de Perijá are added (Figure 11-7). The most perplexing boundary is Cu/La. It is characterized by a maze of compressional features, dextral strike-slip faults, and pull-aparts, the exact dating of which is difficult except that they are Miocene to Recent. In this maze the Oca fault gets lost to the extent that its existence has been doubted (González de Juana 1980; cf. Shagam 1975, p. 381). The kinematic situation, so far as I have been able to gather from the existing literature and from maps, aerial photographs, and SLAR pictures, is characterized by simultaneous northward movement of the Lara block along sinistral faults such as the Valera fault (Habicht 1960) that branch off from the Andes and separate La from Ma; by compression across the northern boundary of the block because of this motion; and by dextral motion along a somewhat distributed Oca fault zone that roughly coincides with this northern boundary. The Oca fault zone seems here to be staggered dextrally, which would account for the pull-aparts.

Qualitatively, the bewildering superposition of compression, strike-slip, and pull-apart is implied in our block-mosaic kinematics, although the quantities require further investigation.

With La our model terminates, as the complex partition of the Caribbean margin east of Cu is not investigated (see Case and Holcombe 1980). The overall relation Ca/SA is dextrally convergent, and on the eastern margin of Cu there should follow considerable pull-apart, but the distribution (Kafka and Weidner 1981) of these quantities is a puzzle in its own right; possibly it is also tractable by means of a block-mosaic kinematic model.

A SYNOPSIS OF THE BLOCK-MOSAIC KINEMATICS OF NORTHWESTERN SOUTH AMERICA

The overall kinematics of the Maracaibo block mosaic is represented in Figure 11-8 as a field of translational vectors with respect to SA, obtained by addition of boundary vectors, and in

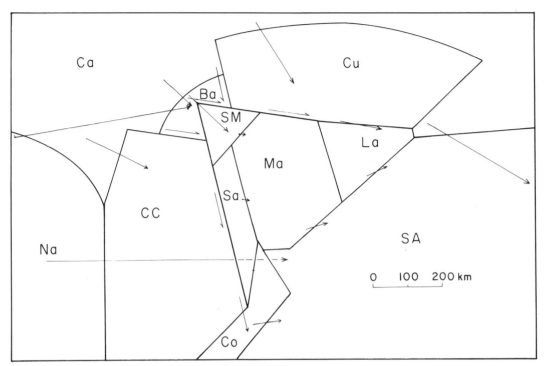

FIGURE 11-7. Relative motions (translations; for possible rotations see MacDonald and Opdyke 1972) across block boundaries of the Maracaibo block mosaic (simplified). *Plates*: SA: South American plate. Ca: Caribbean plate. Na: Nazca plate. *Blocks*: Ma: Maracaibo. SM: Santa Marta. Ba: Barranquilla. CC: Central Colombia. Sa: Santander massif. Co: Cordillera Oriental. Cu: Curaçao (internal Neogene structures in the Gulf of Venezuela are neglected). La: Lara.

Figure 11-9 as a simplified cut-paper model (for the technique, see Figure 11-6). The translational vector field shown in Figure 11-8 may be viewed as composed of the three plate translations Sa (here considered fixed), Ca, and NA, at whose triple junction the mosaic developed, and the mediating field of the mosaic (''extended triple junction''). It is seen that the Caribbean plate dragged along dextrally and compressed, in a northwest-southeast direction, the northwestern tip of South America, while the Nazca plate compressed it in an east-west direction. Mass escape of the compressed domains was lateral as well as up; the two large strike-slip faults pushed the blocks that adjoin them (Cu and CC) sideways over the South Caribbean subduction zone.

The role played by the blocks and their boundaries becomes even more transparent in the cut-paper model shown in Figure 11-9. Comparison with Figure 11-8 shows that, while the essential kinematic quantities are retained, the movement system has been further simplified and divided into two subsystems. As they play different roles, these subsystems have been developed separately and have then been superposed for the total kinematics.

Subsystem 1 (Figure 11-9b) is composed of the Oca and Bucaramanga strike-slip boundaries and those block boundaries in which they terminate: the South Caribbean subduction zone in the north and west and the northwestern margin of the Cordillera Oriental in the south.

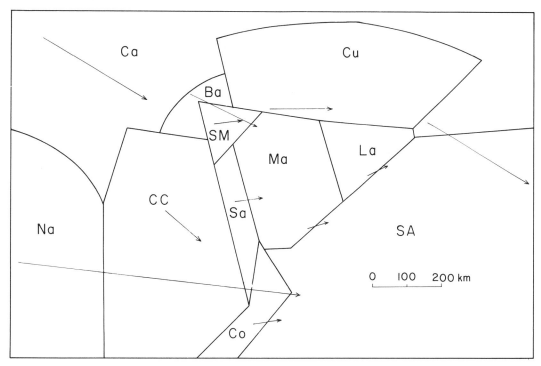

FIGURE 11-8. Translational vectors of blocks with respect to SA, obtained by adding boundary translations.

These boundaries constitute a self-contained subsystem with overall northwest-southeast compression of the triple-junction domain and its simultaneous northeast-southwest extension attended by shove over the South Caribbean subduction zone. Subsystem 2 (Figure 11-9c) consists of the boundaries of the South America and the Maracaibo blocks with approximately east-west compression portraying Nazca-South America convergence. This model thus makes clear that the Cordillera Oriental is composed of the confluence of two kinematically different belts or, viewed the other way, that its branches (the Mérida Andes, the Santander massif, and the Bucaramanga fault) all play different kinematic roles.

At this juncture we should mention the relation between fields of finite block motions and fields of static stress. In a nonhomogeneous medium the relation between the two can be complex, depending on the distribution of resisting forces. As these are not generally known, it would seem wiser at this time not to make any inferences about the stress field and the boundary forces that produced it.

This caveat is also valid for such compressional (or at least partly compressional) features as the Mérida Andes. In a straightforward application of the Mohr-Coulomb criterion they might be regarded as due to a northwest-southeast–directed main principal compressive stress, as Kellogg and Bonini (1982) have interpreted them. Indeed, in a homogeneous medium this conclusion would be difficult to escape. However, as the Andes occupy an inherited inhomogeneity (Early Cretaceous trough, possibly Jurassic grabens), the Mohr-Coulomb criterion does not so simply apply, and transpression is possible from the very beginning of Neogene deformation.

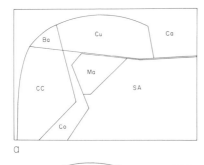

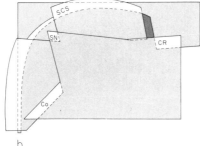

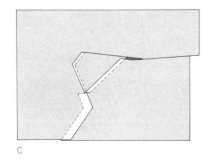

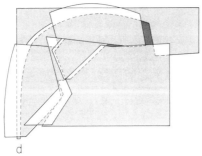

FIGURE 11-9. A cut-paper model of the simplified Maracaibo block mosaic (La is eliminated). For a rough scale, the strike-slip displacements at SN (Figure 11-9b) are 100 km. White: Compression. Black: Extension.

9a: System before the onset of motion.

9b and 9c: The two subsystems can be treated separately, since they seem to be largely decoupled. Subsystem 1 (b) undergoes northwest-southeast shortening with northeast-southwest extension achieved by complementary strike-slip on the Oca and Bucaramanga faults. CC and Cu are shoved over the South Caribbean subduction zone (SCS). Shortening in this zone is drastically reduced in this figure (cf. Figure 11-7) in order to maintain legibility. Subsystem 2 (c) is essentially due to the easterly translation of Na and Ca. The large amount of Na/SA convergence (Figure 11-8) at the Pacific margin (Na/CC; see Figure 11-7) is not included, as it does not directly interfere with the intracontinental Maracaibo block kinematics.

9d: Superposition of the two subsystems.

CONCLUSIONS

Although our detailed knowledge of the northern Andes is inadequate, the large-scale Neo-gene features can be assembled in a quantitative kinematic block-mosaic model that reveals in an internally consistent way the relations between strike-slip faulting, compression, and elevation. The quantitative information falls into two general categories. The first is crustal thickening, interpreted as due to horizontal compression; estimates should preferably be based on geophysical data but can be expanded by geomorphological indications of uplift together with mapped folding and thrusting. The second concerns strike-slip components, as implied by the disruption of pre-Neogene boundaries. There is considerable uncertainty about the exact quantities involved, although the overall kinematic consistency lends plausibility to the order-of-magnitude estimates. Northwestern South America in the Neogene represented an extended triple-junction domain of the South America, Caribbean, and Nazca plates, with dextral transpression between the South American and Caribbean plates superposed on east-west compression between the Nazca and South American plates. In particular, the complementary Oca and Bucaramanga strike-slip faults, associated with southeast-northwest convergence, produced some northeast-southwest expansion that accommodates the diverging arc of the South Caribbean or Curaçao deformed belt; at their intersection, the peculiarly triangular Sierra Nevada de Santa Marta developed as a sort of "Transverse Range."

Kinematic balance of the block mosaic is established by estimating relative translation vectors between neighboring blocks (neglecting rotations) and varying them in such a way as to achieve an internally consistent transport field dominated by what is considered the best quality information: the strike-slip along the Oca and Bucaramanga faults and the dextral transpression across the Venezuelan Andes. Such puzzling features as the Sierra Nevada de Santa Marta, the Táchira gap, the undulations of the South Caribbean deformed belt, the interference of the Oca fault with the Falcón ranges, and the pull-aparts of the Golfo Triste region, where the Oca fault zone joins the San Sebastián fault (Figure 11-2), all find a plausible, if large-scale, explanation in the model. Although many more interesting features are not accounted for, the model seems promising in its present state and would appear to offer a good base for expansion and refinement. Similar techniques are probably applicable to other three-dimensionally complex areas.

ACKNOWLEDGMENTS

This article benefited from the Táchira mapping project supported by MARAVEN, Caracas, and from discussions with a number of geologists, particularly P. Stalder.

REFERENCES

Beck, R. H., Lehner, P., Diebold, P., Bakker, G., and Doust, H., 1975: New geophysical data on key problems of global tectonics. 9th World Petroleum Congress, Tokyo 1975, Proc. *2*, 3-17.

Bellizzia, A., Pimentel, N., and Bajo, R., 1976: Mapa geológico estructural de Venezuela, 1:500,000. Venezuela Ministerio de Minas e Hidrocarburos, Caracas.

Berckhemer, H., 1968: Topographie des "Ivrea-Körpers" abegeleitet aus seismischen und gravimetrischen Daten. Schweiz. Min. Petr. Mitt. *48*, 235-246.

Bonini, W., Pimstein de Gaete, C., and Graterol, V., 1977: Mapa de anomalías gravimétricas de Bouguer de la parte norte de Venezuela y areas vecinas, 1:1,000,000. Venezuela Ministerio de Energía y Minas, Caracas.

Bowin, C., 1976: Caribbean gravity field and plate tectonics. Geol. Soc. America Spec. Paper *169*, 79 p.

Case, J. E., 1975: Geophysical studies in the Caribbean Sea. *In*: Nairn, A.E.M., and Stehli, F. G. (eds), The Ocean Basins and Margins, Vol. 3, The Gulf of Mexico and the Caribbean. Plenum, New York, 107-180.

Case, J. E., and Holcombe, T. L., 1980: Geologic-tectonic map of the Caribbean region, 1:2,500,000. U.S. Geol. Survey Misc. Invs. Ser., Map I-1100.

Duque-Caro, H., 1978: Major structural elements of northwestern Colombia. Am. Assoc. Petroleum Geologists Mem. *29*, 329-351.

Féo-Codecido, G., 1972: Breves ideas sobre la estructura de la falla de Oca, Venezuela. 6th Conf. Geol. del Caribe, Margarita 1971, Venezuela, Trans. *6*, 184-190.

Gansser, A., 1955: Ein Beitrag zur Geologie und Petrographic der Sierra Nevada de Santa Marta (Kolumbien, Südamerika). Schweiz. Min. Mitt. *35*, 209-279.

Gansser, A., 1973: Facts and theories on the Andes. Geol. Soc. London Quart. Jour. *129*, 93-131.

González de Juana, C., 1980: Geología de Venezuela y de sus cuencas petrolíferas. Tomo I. Caracas.

Habicht, K., 1960: La sección de El Baño, Serranía de Trujillo, estado Lara. Venezuela, Dir. Geol., Bol. Geol., Publ. Esp. *3*, 192-207 (Memoria, III Congreso Geol. Venez., Caracas).

Horsfield, W. T., 1980: Contemporaneous movement along crossing conjugate normal faults. Jour. Struct. Geology *2*, 305-310.

Hospers, J., and Van Wijnen, J. C., 1959: The gravity field of the Venezuelan Andes and adjacent basins. K. Nederl. Akad. Wet., Afd. Natuurk., Verh. *23*, 1, 95 p.

Irving, E. M., 1971: La evolución estructural de los Andes mas septentrionales de Colombia. Colombia, Serv. Geol. Nac., Bol. Geol. *19*, 2, 89 p.

Julivert, M., 1970: Cover and basement tectonics in the Cordillera Oriental of Colombia, South America, and a comparison with some other folded chains. Geol. Soc. America Bull. *81*, 3623-3646.

Kafka, A. L., and Weidner, D. J., 1981: Earthquake focal mechanisms and tectonic processes along the southern boundary of the Caribbean plate. Jour. Geophys. Res. *86*, 2877-2888.

Kellogg, J. N., and Bonini, W. E., 1982: Subduction of the Caribbean plate and basement uplifts in the overriding South American plate. Tectonics *1*, 251-276.

Krause, D. C., 1971: Bathymetry, geomagnetism, and tectonics of the Caribbean Sea north of Colombia. Geol. Soc. America Mem. *130*, 35-54.

Laubscher, H. P., 1965: Ein kinematisches Modell der Jurafaltung. Eclogae geol. Helvetiae *58*, 231-318.

Laubscher, H. P., and Bernoulli, D., 1982: History and deformation of the Alps. *In*: Hsü, K. J. (ed), Mountain Building Processes. Academic Press, London, 169-180.

Lonsdale, P., and Klitgord, K. D., 1978: Structure and tectonic history of the eastern Panama Basin. Geol. Soc. America Bull. *89*, 981-999.

MacDonald, W. D., Doolan, B. L., and Cordani, U. G., 1971: Cretaceous-early Tertiary metamorphic K-Ar age values from the south Caribbean. Geol. Soc. America Bull. *82*, 1381-1388.

MacDonald, W. D., and Opdyke, N. D., 1972: Tectonic rotations suggested by paleomagnetic results from northern Colombia, South America. Jour. Geophys. Res. *77*, 5720-5730.

Macellari, C. E., 1982: El Mio-Plioceno de la depresión del Táchira (Andes venezolanos): Distribución paleogeográfica e implicaciones tectónicas. GEOS (Caracas) *27*, 3-14.

Martin F., C., 1978: Mapa Tectónico, Norte de América del Sur. Venezuela Ministerio de Energía y Minas y Comisión Presidencial de Geología y Sismología, Caracas.

Meier, B., Schwander, M., and Laubscher, H. P., 1987 (this volume): The tectonics of Táchira: A sample of north Andean tectonics. *In*: Schaer, J.-P., and Rodgers, J. (eds), The Anatomy of Mountain Ranges. Princeton University Press, Princeton, N.J., 229-237.

Mencher, E., 1963: Tectonic history of Venezuela. Am. Assoc. Petroleum Geologists Mem. *2*, 73-87.

Mencher, E., Fichter, H. J., Renz, H. H., Wallis, W. E., Patterson, S. M., and Robie, R. H., 1953: Geology of Venezuela and its oil fields. Am. Assoc. Petroleum Geologists Bull. *37*, 690-777.

Pennington, W. D., 1981: Subduction of the eastern Panama Basin and seismotectonics of northwestern South America. Jour. Geophys. Res. *86*, 10,753-10,770.

Plafker, G., 1976: Tectonic aspects of the Guatemala earthquake of 4 February 1976. Science *193*, 1201-1208.

Pümpin, V., 1979: The structural setting of northwestern Venezuela. Exploration Bull. *183*.

Renz, O., 1959: Estratigrafía del Cretáceo en Venezuela occidental. Venezuela, Dir. Geol., Bol. Geol. *5*, 3-48.

Rod, E., 1956: Strike-slip faults of northern Venezuela. Am. Assoc. Petroleum Geologists Bull. *40*, 457-476.

Schubert, C., 1980: Late Cenozoic pull-apart basins, Boconó fault zone, Venezuelan Andes. Jour. Struct. Geology *2*, 463-468.

Schubert, C., 1981: Are the Venezuelan fault systems part of the southern Caribbean plate boundary? Geol. Rundschau *70*, 542-551.

Shagam, R., 1975: The northern termination of the Andes. *In*: Nairn, A.E.M., and Stehli, F. G. (eds), The Ocean Basins and Margins, Vol. 3, The Gulf of Mexico and the Caribbean. Plenum, New York, 325-420.

Shepard, F. P., 1973: Sea floor off Magdalena delta and Santa Marta area, Colombia. Geol. Soc. America Bull. *84*, 1955-1972.

Talwani, M., Stoffa, P., Buhl, P., Windisch, C., and Diebold, J. B., 1982: Seismic multichannel towed arrays in the exploration of the oceanic crust. Tectonophysics *81*, 273-300.

Théry, J. M., 1980: Evolution géotectonique de l'Occident colombien; nouvelles données. Cent. Rech. Explor.-Prod. Elf-Aquitaine Bull. *4*, 2, 649-660.

Tschanz, C. M., Marvin, R. F., Cruz, B. J., Mehnert, H. H., and Cebula, G. T., 1974: Geologic evolution of the Sierra Nevada de Santa Marta, northeastern Colombia. Geol. Soc. America Bull. *85*, 273-284.

Chapter 12

THE TECTONICS OF TÁCHIRA: A SAMPLE OF NORTH ANDEAN TECTONICS

B. MEIER, M. SCHWANDER, AND H. P. LAUBSCHER

Universität Basel

Táchira is the connecting link between the Mérida Andes, the Cordillera Oriental of Colombia, and the Santander massif (see Laubscher, this volume, Figure 11-2). Consequently, it is ideally located for illustrating in detail what the general block-mosaic kinematics of northern South America, as presented by Laubscher (this volume), implies on a regional scale. Indeed, the article by Laubscher is the result of an attempt to establish a regional frame for the puzzling tectonics of Táchira. The dating of motions in northern South America is somewhat controversial; because the dating of motions in Táchira is crucial for understanding this problem, it will be reviewed in some detail.

The map compiled by Case and Holcombe (1980), which in a reduced form contains the information given in Bellizzia et al. (1976) and Martin (1978), among other sources, shows the essence of Táchira's structural position and the problems involved. Some of these problems are particularly obvious: (1) Táchira is a pronounced structural depression relative to the Andean branches it connects and (2) it contains a bewildering network of structures, usually severely faulted, that contains all the elements of the surrounding Andean branches.

The time of deformation also can be inferred from the map of Case and Holcombe (1980), at least roughly. The "upper Tertiary" in the northern part is fully involved in the deformation. In the southern part, Andean structures submerge under the Quaternary plain of the Llanos (Figure 12-1 and inset) at what is called in this paper the "Santo Domingo reentrant" (Bellizzia et al. 1976; cf. Laubscher, this volume, Figure 11-2, southwest of the word "Mérida"). Seismic work (Maraven files) show here an important intra-Miocene unconformity, the structures below containing the trends of the Mérida Andes which thus are partly older than in the north but still late Tertiary; these older structures subsequently have been selectively reactivated to some extent, but mostly they were tilted into the foredeep of the Cordillera Oriental of Colombia (see Laubscher, this volume, Figure 11-2), which is consequently very young also; even Quaternary strata are involved in the deformation (Bellizzia et al. 1976). A late Tertiary age is obvious also for the Mérida Andes. Their foredeeps shown on the map by Case and Holcombe (1980; cf. Laubscher, this volume, Figure 11-2) developed from the Miocene up (González de Juana 1980, Macellari 1982); possibly a first structural unrest began in the Oligocene (Mencher et al. 1953). Quaternary terraces are tilted, and the Boconó fault has an appreciable Quaternary to Recent component (Rod 1956, Schubert and

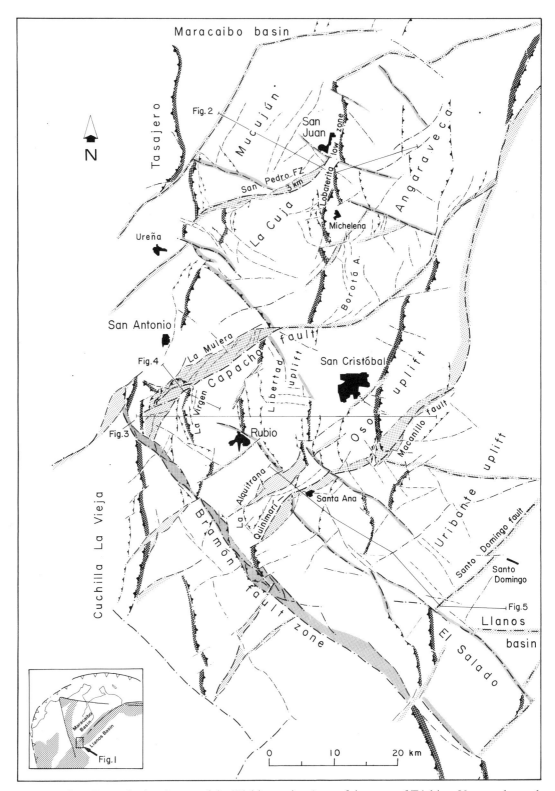

FIGURE 12-1. Tectonic sketch map of the Táchira region (part of the state of Táchira, Venezuela, and surroundings). Dark coarse stipple: Almost purely compressional system, approximately north-south.

Laredo 1979, Schubert 1980). Neogene structural relief between the northern foredeep and the high Andes is on the order of 15 km. It has overwhelmed preexisting Eocene structures in the southeastern foredeep (Barinas Llanos), also shown on the map by Case and Holcombe (1980). These structures were studied in detail by the oil companies (e.g., private reports by Laubscher and others between 1948 and 1958 in Corpoven files; Smith 1962). Their relief is mostly on the order of 100 m and does not exceed 1 km; it is negligible with respect to the Neogene structures. Their structural pattern is unrelated to the Neogene one; they are not fore-runners of the Andes but rather foreland structures of the Late Cretaceous-Eocene Laramide orogeny, which in this area was discordantly overwhelmed and dissected by the Neogene Andean pattern (map by Bellizzia et al. 1976). These Laramide structures disappear westward and do not seem to be of any significance in Táchira. On the other hand, the Mérida Andes follow approximately the Early Cretaceous Uribante trough (Renz 1959a), which, after long quiescence, they appear to have reactivated by dextral transpression (Rod 1956; cf. Laubscher, this volume). Similarly, the Cordillera Oriental south of Táchira now occupies the site of an even more pronounced Jurassic-Cretaceous trough (Bürgl 1967), and its eastern foredeep, at least in the northern part, is essentially late Tertiary (seismic and well data, Maraven files; cf. Bellizzia et al. 1976). Inasmuch as these foredeeps are largely the response to loading by the rising mountains, they are late Tertiary also, that is, coeval with the Mérida Andes. This conclusion agrees with a widely accepted view (e.g., Bürgl 1967, Kellogg and Bonini 1982, Macellari 1982).

In this paper we discuss the structural style and kinematics of Táchira revealed by mapping done by Meier and Schwander between 1978 and 1981 and supplemented by information contained in both published (Renz 1959b, Trump and Salvador 1964) and private reports (particularly those in the files of Maraven, Caracas), and by the study of aerial photographs and SLAR pictures. Inasmuch as the tectonics of Táchira reflect, on a smaller scale, the structure of the surrounding Andes, it should provide insight into Andean deformation that is usually difficult to come by.

THE TECTONIC STYLE OF TÁCHIRA

The tectonic style of Táchira is illustrated by the simplified map (Figure 12-1) and the cross sections (Figures 12-2–12-5) presented in this article. The quality of the information used to construct these sections varies.

There are numerous excellent exposures between extensive areas of poor outcrops. As topographical relief is much smaller than structural relief, the deeper part of the cross sections

Dark fine stipple: Sinistrally transpressive system, approximately northwest-southeast. Light stipple: Dextrally transpressive system, approximately southwest-northeast. The shading is applied only to the more important trends. Individual faults are shown by dash-dot lines, the least important ones by thin dashes. The thickness of the lines indicates the degree of importance of a fault. Faults with a predominant thrust component are marked by triangles or barbs. In addition to the more important faults shown, there is ubiquitous bending and folding of strata within the deformational belts. (From Meier 1983 and Schwander 1984; cf. Renz 1959b and Trump and Salvador 1964.)

FIG. 12-2

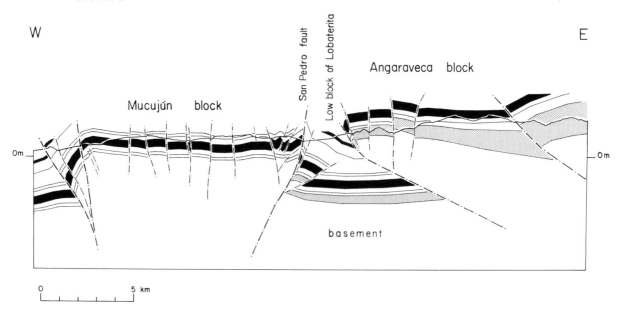

W   E

Mucujún block

San Pedro fault

Low block of Lobaterita

Angaraveca block

Om   Om

basement

0   5 km

FIG. 12-3

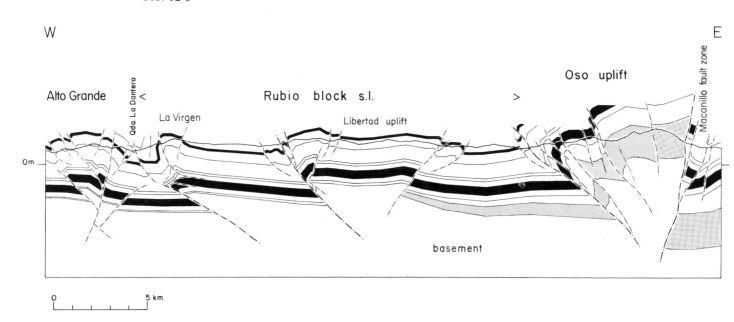

W   E

Alto Grande

Qda. La Dantera

La Virgen

Rubio block s.l.

Libertad uplift

Oso uplift

Macanillo fault zone

Om

basement

0   5 km

FIGURES 12-2 to 12-5. Typical cross sections of Táchira structures. For locations, see Figure 12-1. (From Meier 1983 and Schwander 1984.)

FIG. 12-4

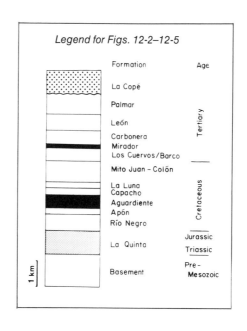

Legend for Figs. 12-2—12-5

Formation	Age
La Copé	
Palmar	
León	
Carbonera	Tertiary
Mirador	
Los Cuervos/Barco	
Mito Juan – Colón	
La Luna	
Capacho	
Aguardiente	Cretaceous
Apón	
Río Negro	
La Quinta	Jurassic
	Triassic
Basement	Pre-Mesozoic

1 km

FIG. 12-5

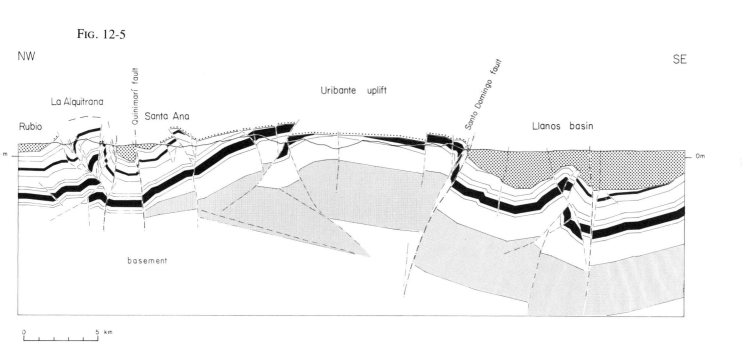

had to be extrapolated by following some rules that were based on direct field observation, analysis of the mapping results, and general considerations of material balance:

1. All deformation occurred in the brittle domain, faulting as well as folding. Basement at present is shallower than 4000 m below the surface, and it probably was never much deeper. The whole sedimentary sequence was deformed at temperatures hardly exceeding 100°C. Consequently, a brittle style is maintained throughout the cross sections.

2. Folding is important, and several of the shallow folds we were able to map are so narrow that décollement on one or several incompetent horizons is required, particularly on the shales of the Apón, the Capacho (Seboruco member), and the Colón formations. The equivalent disharmonic compression at greater depths is shown mostly as thrusts or reverse faults. Such thrusts are also required to raise the extensive high blocks between the zones of deformation.

3. Northeast and southeast trends are transpressive, and their complex fold-fault patterns are interpreted as ''flower structures'' (Harding and Lowell 1979) with folds, thrusts, and reverse faults accommodating the compressive components, and the ''stem of the flower'' permitting strike slip. It goes without saying that the specific structures at depth are quite uncertain, although the style is believed to be realistic, particularly as similar structures can be studied on the surface in several places.

A rough sort of kinematic consistency and material balance, as well as a rheological style that is compatible with observations and the depth of deformation, are believed achievable by following these three rules.

The cross sections presented here, chosen from some 30 constructed sections, illustrate a few typical features: blocks and block boundaries with faults, folds, thrusts, disharmonies, flower structures, and the intra-Miocene unconformity (base of La Copé formation).

Figure 12-2 shows two major uplifted blocks with a complex low belt in between. The western boundary of the Mucujún block is in Colombia and is not known in detail. It is constructed here according to information in private reports, supplemented by aerial photographs, without re-mapping by the authors. The structural style (flexures and steep faults) conforms to that found in areas that have actually been mapped. The eastern boundary as seen on the cross section actually consists of the confluence of two important zones of deformation: the east-northeast trending San Pedro fault zone, which is dextrally transpressive and bounds the Mucujún block in the south, and the north-south striking fault zone of the Lobaterita low, which is filled with complexly deformed Tertiary sediments and separates the Mucujún block from the Angaraveca block (Figure 12-1). These blocks are dissected internally by minor faults. It would appear that the bordering fault zones of the Mucujún block converge somewhere in the middle crust. This is true for elevated blocks generally, whereas, as a rule, they diverge under depressed blocks. Deep structure of the Lobaterita low belt is uncertain, as there is considerable disharmonic deformation in the surficial Tertiary sediments.

Figure 12-3 passes through the Rubio block s.l., the central low block of Táchira, which is subdivided by a comparatively narrow, complexly bounded, uplifted wedge. The western end of the section is at the very complex junction of the sinistral and dextral transpressive belts of the Bramón and Capacho fault zones. The east-northeast trending Alto Grande-La Mulera structure is interpreted as a dextrally convergent flower structure (cf. Figure 12-4), interfering with a thrust trending north-south, whereas the La Virgen structure is a narrow fold trending north-south, disharmonically developed above the Aguardiente formation in the Upper Cretaceous–Tertiary part of the section. The deeper part of the La Virgen structure is interpreted as an almost blind thrust that helps raise the Rubio block with respect to the Quebrada la Dantera low zone. This low zone, like the Lobaterita low zone of Figure 12-2, has a somewhat uncertain deep structure because of interference and overlap of the Alto Grande and La Virgen structures. The Oso uplift east of San Cristóbal terminates the Táchira depression.

Figure 12-4 is an attempt to interpret the confusing structure at the junction of the Capacho dextrally transpressive belt, immediately north of its maximum structural relief at Cerro Rangel, and the La Virgen fold. The very narrow slice of basement and Lower Cretaceous at Cerro Rangel is bounded on all sides by a complex network of faults and separates the very low segments of the San Antonio and Quebrada la Dantera depressions. It is interpreted as a flower structure with the compressive component at depth accommodated by "wedge into split-apart" structures often discernible on seismic sections (cf. Harding and Lowell 1979, Figure 6; Jones 1982; Laubscher and Bernoulli 1982).

Figure 12-5 shows a section in southeastern Táchira where an intra-Miocene unconformity beneath the La Copé formation is found both at the surface and in seismic sections in the Llanos of the Santo Domingo reentrant. This unconformity disappears toward San Antonio (but cf. Macellari 1982). The evidence here is clearly for a Neogene age of virtually all the Andean deformation. It also indicates a shift of activity along the Andean northeast trend from the southeast (Llanos) to the northwest. The intra-Miocene unconformity in some structures, such as the Uribante uplift, cuts down into the Early Cretaceous Rio Negro formation, which is equivalent to a relief of up to 2 km. Intensity of re-deformation of the pre-unconformity structures generally increases to the northwest, as in the La Alquitrana structure and the Oso uplift.

These patterns all fit in a kinematic scheme of east-west compression (and perhaps some north-south extension) generally consonant with the regional picture (see Laubscher, this volume) but in detail possibly suggestive of some temporal variations in the relative direction of block movements. The cross sections permit a rough estimate of compressions across block boundaries, although material balance considerations, which are so helpful in cross sections of cylindrical structures, are weakened in such obviously three-dimensional patterns. Compression usually turns out to be on the order of 1 to 4 km, and the total amount across the whole of Táchira is less than 10 km; in some places it is probably as little as 6 km.

This modest amount of shortening is reflected in modest topographical elevations; Táchira forms a depression between the high Andean branches where crustal compression is roughly estimated to range from 30 km to over 100 km (Laubscher, this volume). On the other hand, this reduction in compression calls for an explanation within the scheme of regional kinematics. Examination of the transport vectors (as given by Laubscher, this volume, Figures 11-7 to 11-9) of Maracaibo mosaic blocks that join in Táchira suggests a translation of the Maracaibo block slightly north of east with respect to South America. This implies that little of the compressional pile-up that would result in high mountains would occur along a roughly east-west trending Táchira link between the Andean branches (cf. Laubscher, this volume, Figure 11-9). This solution, though inescapable as far as we can see, is not immediately obvious in the bewildering maze of structures shown in Figure 12-1. The structures trending east-west, like the east-west trending segment of the Capacho fault (Figure 12-1,

Meier 1983), are not conspicuous, although they are doubtless important. The Mérida Andes terminate against the Santo Domingo reentrant of the Llanos basin (Laubscher, this volume, Figure 12-2) by plunging under it along a zone trending east-west that is difficult to interpret except as a zone of distributed dextral shear (Schwander 1984). More difficult to estimate are the amounts of strike-slip. There is no known preexisting frame of tectonic, facies, or igneous configurations that might be used. Tentatively, it can be assumed that the Borotá anticline on the north side and the Libertad uplift on the south side of the Capacho fault were nucleated as one and the same structure at an early stage of deformation and were later cut off and displaced by the fault. This assumption results in a rough estimate of about 10 km dextral displacement. Considering that there are perhaps 7 to 8 such dextral transpressive zones, a total amount of 60 km dextral strike-slip as inferred for the Boconó fault at the northeast end of the Mérida Andes, would not appear excessive.

CONCLUSIONS

Táchira, a link between branches of the Andes, is composed of a sub-mosaic of the Maracaibo block mosaic, with mostly transpressive block boundaries, dextral for the northeast-trending, sinistral for the southeast-trending zones of deformation. This system is supplemented by the more purely compressive north-south trends. The blocks, though at different elevations, are hardly deformed internally, and this requires variable crustal thickening underneath, presumably below shallow thrusts converging in the middle crust, with disharmonic deformation of the lower crust. Deformation is brittle, since observed sediments were never buried below about 4 km. Compression is by folding and thrusting, often disharmonic, and the complex combinations of both with steep faults in the transpressive zones are interpreted as "flower structures." The total amount of east-west compression across Táchira seems not to exceed 10 km, which is compatible with the east-west trend of the southwest end of the Mérida Andes southwest of San Cristóbal (Figure 12-1) and the structural low of the "Táchira gap." Strike-slip is more difficult to estimate, but a total of 60 km of dextral shear would not seem excessive.

ACKNOWLEDGMENTS

We gratefully acknowledge MARAVEN, Caracas, for its financial and logistical support of the Táchira mapping project. H. Krause, V. Pümpin, and P. Rowland helped logistically and otherwise, and the authors benefited from many discussions, particularly with A. Mozetic, V. Pümpin, O. Renz, P. Stalder, and C. White. A. Bellizzia of Dirección de Geología, Ministerio de Energía, Venezuela, also supported the project in various ways. A. Useche was helpful in introducing us to some aspects of the geology of Táchira.

REFERENCES

Bellizzia, A., Pimentel, N., and Bajo, R., 1976: Mapa geológico estructural de Venezuela, 1:500,000. Venezuela Ministerio de Minas e Hidrocarburos, Caracas.

Bürgl, H., 1967: The orogenesis in the Andean system of Colombia. Tectonophysics 4, 429-443.

Case, J. E., and Holcombe, T. L., 1980: Geologic-tectonic map of the Caribbean region, 1:2,500,000. U.S. Geol. Survey Misc. Invs. Ser., Map I-1100.

González de Juana, C., 1980: Geología de Venezuela y de sus cuencas petrolíferas. Tomo I. Caracas.

Harding, T. P., and Lowell, J. D., 1979: Structural styles, their plate-tectonic habitats, and hydrocarbon traps in petroleum provinces. Am. Assoc. Petroleum Geologists Bull. 63, 1016-1058.

Jones, P. B., 1982: Oil and gas beneath east-dipping underthrust faults in the Alberta foothills. In: Powers, R. B. (ed), Geologic Studies of the Cordilleran Thrust Belt. Rocky Mountain Assoc. Geologists, Denver, Colo., 61-74.

Kellogg, J. N., and Bonini, W. E., 1982: Subduction of the Caribbean plate and basement uplifts in the overriding South American plate. Tectonics 1, 251-276.

Laubscher, H. P., and Bernoulli, D., 1982: History and deformation of the Alps. In: Hsü, K. J. (ed), Mountain Building Processes. Academic Press, London, 169-180.

Macellari, C. E., 1982: El Mio-Plioceno de la depresión del Táchira (Andes venezolanos): Distribución paleogeográfica e implicaciones tectónicas. GEOS (Caracas) 27, 3-14.

Martin F., C., 1978: Mapa Tectónico, Norte de América del Sur. Venezuela Ministerio de Energía y Minas y Comisión Presidencial de Geología y Sismología, Caracas.

Meier, B. P., ms., 1983: Lithostratigraphie und Blocktektonik im nördlichen Teil der Táchira-Senke (W. Venezuela). Diss., Univ. Basel.

Mencher, E., Fichter, H. J., Renz, H. H., Wallis, W. E., Patterson, S. M., and Robie, R. H., 1953: Geology of Venezuela and its oil fields. Am. Assoc. Petroleum Geologists Bull. 37, 690-777.

Renz, O., 1959a: Estratigrafía del Cretáceo en Venezuela occidental. Venezuela, Dir. Geol., Bol. Geol. 5, 3-48.

Renz, O., 1959b: Guia de la Excursión C-7 Andes surocidentales, sección de Santo Domingo a San Antonio (estado Táchira). Venezuela, Dir. Geol., Publ. Esp. 3, 87-91 (Memoria, III Congreso Geol. Venez., Caracas).

Rod, E., 1956: Strike-slip faults of northern Venezuela. Am. Assoc. Petroleum Geologists Bull. 40, 457-476.

Schubert, C., 1980: Late Cenozoic pull-apart basins, Boconó fault zone, Venezuelan Andes. Jour. Struct. Geology 2, 463-468.

Schubert, C., and Laredo, M., 1979: Late Pleistocene and Holocene faulting in Lake Valencia basin, north-central Venezuela. Geology 7, 289-292.

Schwander, M., ms., 1984: Die südliche Táchira-Depression (W. Venezuela): Lithostratigraphie, intramiocäne Diskordanz und tektonische Entwicklung im Neogen. Diss., Univ. Basel.

Smith, F. D. (coordinator), 1962: Mapa geológico-tectónico del norte de Venezuela. Primero Congreso Venezolano de Petróleo, Caracas.

Trump, G. W., and Salvador, A., 1964: Guidebook to the geology of western Táchira. Asociación Venezolano de Geología, Minería y Petróleo, Caracas.

Part V

NORTH AMERICA

Chapter 13

THE APPALACHIAN GEOSYNCLINE

JOHN RODGERS

Yale University

THE CONCEPT OF A GEOSYNCLINE

It was with reference to the Appalachian mountains of eastern North America (Figure 13-1) and the stratigraphic sequence of which they are built that the concept of the *geosynclinal* or *geosyncline* was first developed. Using the then relatively new concept of faunal succession, the paleontologist James Hall had played a large part in establishing (1836–1842) the Paleozoic sequence in the state of New York as typical for eastern North America (up to the Carboniferous), and thereafter he was repeatedly consulted on stratigraphic questions by the geologists of the Appalachian states of Pennsylvania, Virginia, and Tennessee and by those of the midwestern states from Ohio to Iowa. In the course of journeys to these regions, he was struck to observe that the strongly folded Paleozoic strata in the linear Appalachian mountain belt are much thicker than the flat-lying Paleozoic strata in the broad mid-western lowland, although their fossils showed him that they cover the same time span; indeed the strata in the east-west outcrop belt across New York State, with which he was most familiar, display the transition from thick to thin. In a presidential address to the recently established American Association for the Advancement of Science in 1857, he proposed a genetic connection between the great thickness and the strong deformation in the Appalachian belt, suggesting that merely the subsidence of the trough in which the thick deposits were accommodated caused the folding (Hall 1883). A few years later, J. D. Dana (1873), dissatisfied by what he called ''a theory for the origin of mountains with the origin of mountains left out'' (Dana 1866, p. 210), proposed to explain both the trough and the folding in turn by the operation of tangential forces (related to the earth's contraction during cooling), and he named the trough of subsidence a ''geosynclinal.'' The further vicissitudes of the geosynclinal idea have been discussed many times and need not detain us here; rather, I wish to explore the question, What was the Appalachian geosyncline like while it existed? If we had visited the Appalachian region while the thick sediments were being deposited but before they were deformed, what kind of geography would we have seen?

I cannot hope to discuss the whole history of the Appalachian geosyncline in a single lecture or article or to cover its entire extent—20° of latitude by 30° of longitude; if placed on a map of Europe at the same latitude, the Appalachians would reach from Warsaw to Casablanca. Hence I must confine my discussion to a specific time and a specific cross section, and I have chosen the Cambrian and Early to Middle Ordovician, up to the first major orogenic

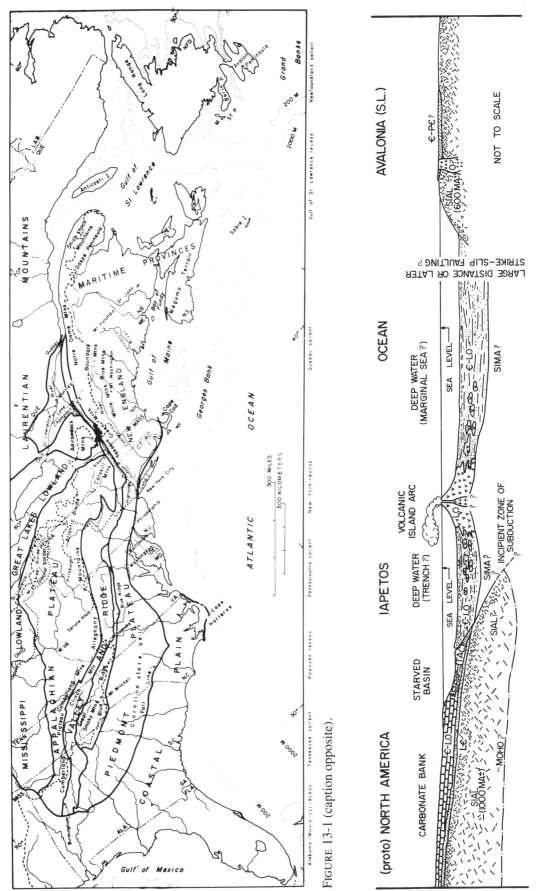

FIGURE 13-1 (caption opposite).

FIGURE 13-2 (caption opposite).

FIGURE 13-1. Location map of Appalachian Mountains, eastern North America. Abbreviations for states, provinces, and so forth: ALA. (Alabama); CT. (Connecticut); DEL. (Delaware); FLA. (Florida); GA. (Georgia); KY. (Kentucky); LAB. (Labrador); M. (Miquelon); MASS. (Massachusetts); MD. (Maryland); ME. (Maine); MISS. (Mississippi); N.B. (New Brunswick); N.C. (North Carolina); NFD. (Newfoundland); N.H. (New Hampshire); N.J. (New Jersey); N.S. (Nova Scotia); N.Y. (New York); OHIO (Ohio); ONT. (Ontario); P.E.I. (Prince Edward Island); PENNA. (Pennsylvania); QUE. (Québec); R.I. (Rhode Island); S.C. (South Carolina); ST.P. (Saint-Pierre); TENN. (Tennessee); VA. (Virginia); VT. (Vermont); W.VA. (West Virginia). (From *Mountain Building Processes*, K. J. Hsü, ed, Academic Press, London and New York, 1983. Reproduced with permission of the publisher.)

FIGURE 13-2. Schematic cross section of southern New England as it might have appeared during late Early and early Middle Ordovician time, at about the beginning of the Taconic orogeny. P€: Precambrian. €: Cambrian (L€: Lower Cambrian). O: Ordovician (LO: Lower Ordovician). TAC.: Taconic sequence (materials of future Taconic klippen).

climax, and the cross section from Hall's (and my) New York State across southern New England to the present Atlantic Ocean. Figure 13-2 shows, schematically, what I think the cross section looked like toward the end of that time; Figure 13-3 shows, more exactly, how I interpret its present-day structure.

In the original conception of Hall and Dana, the geosyncline coincided more or less with the belt of folded but unmetamorphosed rocks in the present-day Valley and Ridge province and its northern continuation in the Hudson, Champlain, and Saint Lawrence Valleys; New England and the Piedmont to the east lay outside it, and indeed for a long time their metamorphic rocks were considered mostly Precambrian (see the Geological Map of the United States of 1933), probably the eroded roots of a complementary geanticline (which came to be called Appalachia) that supplied terrigenous sediments to the geosyncline throughout its history.

If, in accordance with the principle that "the present is the key to the past," we look for a comparable situation in the world today, where would we find a linear belt of thick but essentially shallow marine or indeed partly continental sediments supplied by the erosion of a parallel linear uplifted belt? The Ganges Plain at the foot of the Himalaya comes to mind, or the Persian Gulf and Mesopotamia besides the Zagros, or possibly the Po Valley and the head of the Adriatic Sea at the southern foot of the Alps. But the simple picture of a geosynclinal-geanticlinal pair has been revised substantially in our own century, beginning with the ideas of Haug (1900), who also rejected the view that all geosynclinal sediments are shallow-water deposits; these ideas were brought to North America especially by Bucher (1933) and Kay (in the 1940s; see Kay 1951). At about the same time it became evident that the metamorphic rocks in large parts of New England, probably in most of it, are not Precambrian but Paleozoic, and that they were derived from very thick piles of sediments and volcanics, as thick as or thicker than those in the unmetamorphosed belt. Thus the former deposits are the real core of the geosyncline, the latter only a marginal part. But, because of pervasive, commonly high-grade and, in many places, repeated metamorphism, the history and paleogeography represented in the core belt, and hence the original nature of the geosyncline itself, are difficult to decipher.

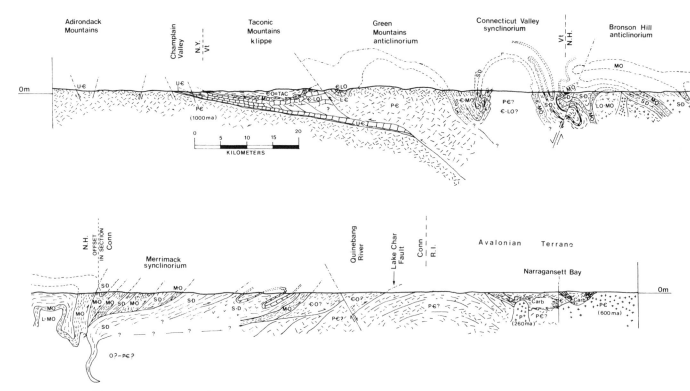

FIGURE 13-3. Generalized but to scale cross section of southern New England as it appears today. PЄ: Precambrian. Є: Cambrian (LЄ: Lower Cambrian. UЄ: Upper Cambrian). O: Ordovician (LO: Lower Ordovician. MO: Middle Ordovician). S: Silurian. D: Devonian. Carb: Carboniferous. P: Permian. TAC: Taconic klippen. Om: Sea level.

CARBONATE BANK

Upper Cambrian and Lower Ordovician strata are spread over the whole central lowland of the United States as far west as Minnesota and Texas, where they are interrupted along a belt that has been called the ''continental backbone'' (though certainly it has no great relief). The basal layers of the Upper Cambrian–Lower Ordovician sequence are largely clean quartz sandstone and rest unconformably upon Precambrian basement rocks whose deformation and metamorphism range in age from 1,000 to 3,500 Ma. Except for a little shale in the Upper Cambrian, the rest is carbonate, mainly dolostone; outside the Appalachian region it is 100 or 200 m thick. Approaching the Appalachian Mountains, however, the thickness increases rapidly, partly by actual thickening of the strata of each epoch, partly by the addition of older and older strata beneath. Thus the basal sandstone becomes older (and also thicker) eastward until, where it is exposed at the eastern ridge of the Valley and Ridge province, it is Lower Cambrian or even uppermost Precambrian and ranges from 30 meters to several kilometers thick; clean sandstone forms only the uppermost part of the thicker sequences, the rest being dirty quartz sandstone, siltstone, arkose, and arkosic conglomerate. Here (as in the central lowland) the rest of the sequence, which ranges from low in the Lower Cambrian to the top of the Lower Ordovician, is mostly carbonate—largely dolomitic but more and more calcitic to the east—and it reaches a total thickness between 2 and 3 km. Quartz sandstone layers (representing temporary regressions) are intercalated in this carbonate sequence near the top of the Lower Cambrian and the base of the Upper Cambrian (locally also near the Cambrian-Ordovician boundary) merging westward into the transgressive basal sandstone unit; between these two sandstone units in the southern Appalachians are considerable bodies of clay shale. But *all* these terrigenous materials clearly came from the west or northwest, for they coarsen in that direction; all are derived from erosion of the "continental backbone" or other then-exposed parts of the Precambrian craton, none from any landmass to the east within the Appalachian belt.

The terrigenous and especially the carbonate strata in this thick pile show abundant evidence of deposition in very shallow water, such as abundant cross-bedding (in both sandstone and carbonate), local scour surfaces, ripple marks, and even mud-cracks; there are also flat-pebble conglomerates in the carbonate strata. The fossils confirm this evidence and show that the water was marine. Algae were widespread and locally formed veritable bioherms with stromatolitic structures; trilobites, brachiopods, and gastropods were present from the first, and cephalopods and pelecypods appeared as they evolved.

These carbonate strata and the included terrigenous rocks thus record a long period of deposition in very shallow water (perhaps 80 Ma) and reach a thickness measured in kilometers, implying slow subsidence with which deposition kept pace. One might ask whether such a situation is probable; the answer is yes, for several areas in the world today display carbonate banks with long lives and equally great or even greater thicknesses. Along the west coast of Florida and on the Bahama Banks, around the Yucatán Peninsula and between the coast of Nicaragua and the island of Jamaica, shallow-water carbonate sediments are now accumulating, and deep wells show that in the first two areas at least they have been accumulating since the Early Cretaceous, more than 100 Ma, and now have thicknesses up to 5 km. A similar area is the Sahul shelf along the north coast of Australia, facing the eastern islands of Indonesia; carbonate deposition may have started there in the Late Jurassic.

We may conclude therefore that during the Cambrian and Early Ordovician the eastern part of North America supported a vast carbonate bank, reaching at least from (present) Alabama to Newfoundland (and indeed extending to the northern and western margins of the continent, now in the Innuitian and Cordilleran Ranges). Carbonate began to accumulate first along the borders of the continent as it then existed and spread more and more widely across it, though never covering it entirely. Only along the borders, however, was the subsidence, and hence the deposition, large. If we can judge by the disposition of present-day carbonate banks, most of the North American continent was then in the tropics, as indeed paleomagnetic data indicate. The immense extent of this early Paleozoic bank has no parallel, however, in the present world.

STARVED (TACONIC) SEQUENCE

But what lay to the east of the carbonate bank? Already much of the bank's eastern edge is now included within the metamorphic belt east of the Valley and Ridge province (furnishing the commercial marbles of Vermont, North Carolina, Georgia, and Alabama), so that the

rocks that originally lay still farther east are now largely metamorphic and difficult to decipher. Luckily for us, however, considerable bodies of the rocks were displaced from their
original positions east of the carbonate bank to lie on top of it, where they now form the Taconic klippen; as they were displaced during the Ordovician before any Paleozoic metamorphism could affect them, they still preserve, despite their considerable deformation, enough
fossils for accurate dating and enough sedimentary features to tell us in what environments
they were deposited. (Both their stratigraphy and their tectonic position were the subject of
long and furious debate; I believe the matter is now settled, but to give the evidence would
take an article as long as this one; see Bird et al. 1963; Zen 1967; Rodgers 1970, pp. 75-90.
In several respects, the Taconic klippen resemble the Préalpes and related klippen in the western Alps.) These rocks were probably deposited on the east side of or a little eastward from
Precambrian basement rocks that have since been uplifted to form the Green Mountains and
similar anticlinoria in western New England, which bound the Valley and Ridge province
there on the east.

The sequence of strata in these displaced rock bodies or klippen is called the Taconic
sequence, from the Taconic Mountains on the New England–New York border, which they
underlie. The Taconic sequence is known by its fossils to range in age from Early Cambrian
to early Middle Ordovician, that is, exactly the time range of the eastern part of the carbonate
bank, but it is almost entirely different. To begin with, it is considerably thinner, barely 1 km.
The lower Lower Cambrian strata at the bottom (their base is not exposed) are somewhat like
the dirtier sediments of the same age that underlie the carbonate strata at the eastern edge of
the Valley and Ridge province, but the rest, from the level at which the carbonate bank to the
west became established up to the Middle Ordovician, is mostly shale, slate, and argillite,
some silty, some limy, and it forms a ''comprehensive series'' in the European sense—the
result of slow deposition of fine mud over a long period of time. In places, the siltier beds
appear to be distal turbidites. The fossils also are different from those in the carbonate-bank
sequence. Graptolites dominate in the Ordovician, and dendroids are present in the Cambrian;
other fossils include algae, small sponges and worms, inarticulate brachiopods, conularia,
and phyllocarids, and also trilobites but of different genera from those in the carbonate strata.
Many appear to represent floating rather than bottom-dwelling organisms, as first clearly recognized by Ruedemann (1934).

Other rock types are present as layers intercalated in the dominantly argillitic sequence.
Chert or highly siliceous argillite is significant in the Ordovician part; some of it contains considerable feldspar (now albite), and it may be of volcanic provenance, but outcrops of volcanic rock (all basaltic) are rare and small. Layers of clean quartz sandstone, with either
quartz or ankerite as cement, are intercalated at certain levels, notably in the lower Upper
Cambrian, exactly when such quartz sandstone spread widest on the bank. Perhaps the most
instructive of these intercalated layers are those made of limestone-breccia or -conglomerate
(in places associated with or grading laterally into limestone turbidites), which occur all along
the western side of the Taconic klippen at all levels from Lower Cambrian to Middle Ordovician. They resemble some of the Jurassic limestone-breccias in the western Alps. They consist of fragments of carbonate, many though not all recognizable as pieces of the carbonate-
bank sequence, including such characteristic rock types as cross-bedded calcarenite, flat-peb-

ble conglomerate, and algal stromatolites; some fragments contain the typical shelly fauna of the bank, generally of the same age as or only slightly older than fossils in the enclosing argillite. The matrix is variably argillite, carbonate mud, quartz sand, or a mixture. The most spectacular exposure of these layers is at Cow Head on the west coast of Newfoundland (Kindle and Whittington 1958, 1959; Hubert et al. 1977), where an Upper Cambrian to Middle Ordovician sequence of limy shale and muddy limestone 300 m thick includes numerous layers meters to tens of meters thick of limestone-breccia containing blocks from a meter to tens of meters across. Evidently the relatively small area where the sequence was deposited received limestone boulders repeatedly over a period of perhaps 70 million years. The source of these boulders must have been the margin of the carbonate bank, which could not have been much more than a few kilometers away for much of the time. The limestone-breccia and conglomerate layers elsewhere could have been deposited somewhat farther from the source area, but they all tell the same story—from Newfoundland all the way to Pennsylvania the western margin of the basin in which the Taconic sequence was deposited was adjacent to the eastern margin of the carbonate bank.

It is not difficult to match such a geography in the world today around the margins of the modern carbonate banks mentioned above (p. 245) as analogues of the early Paleozoic bank in the Appalachians. The oversteepened margins of the Great Bahama Bank stand 1 or 2 km above the adjacent sea floor; in places the upper slopes are practically vertical for hundreds of meters, and the lower slopes have gradients of 50%. Hence they are capable of supplying great amounts of coarse carbonate debris to the adjacent basins. Furthermore, cores on the bottom of the Gulf of Mexico between Florida and Yucatán cut turbidite layers made entirely of carbonate fragments, intercalated with turbidites made of siliceous-fragmental sediment from the Mississippi and other rivers.

The steep slopes of modern carbonate reefs depend on the ability of modern hexacorals and calcareous algae to thrive in the breaker zone at the reef edge, and we do not know whether the early Paleozoic organisms could have done so. In the case of the exceptional Cow Head deposit, perhaps it was down-to-basin normal faulting that kept the bank edge steep to vertical for so long.

The basin in which the Taconic sequence was deposited probably began in the Early Cambrian as a simple extension of the continental margin, but as soon as the carbonate bank was established most clastic sediment was prevented from reaching the basin; only fine mud arrived to accumulate slowly through the rest of the Cambrian and Early Ordovician, interrupted from time to time by quartz sand that succeeded in spreading across the bank to its margin (or in coming down submarine canyons in that margin), by carbonate debris washed by storms off the bank itself, and apparently in the Early Ordovician at least by fine-grained volcanic material from some other source. Throughout this period, therefore, the Taconic basin was a "starved" basin. Finally, in the early Middle Ordovician, coarse clastic sediment began to arrive in large amounts, although at that time, as shown by the fossils, the bank to the west was still receiving only carbonate sediments; similar clastic sediment arrived there only later in the Middle Ordovician. Evidently a new source of sediment had appeared, outboard of both bank and starved basin. The presence in the clastic sediment of volcanic feldspar, chromite, and serpentine fragments tells us what the new source area was like.

THICK, PARTLY VOLCANIC, DEEP-WATER SEQUENCE (TRENCH?)

The next sequence of rocks eastward is exposed on the east flank of the Green Mountains anticlinorium, or rather in the Connecticut Valley synclinorium beyond, probably separated from the anticlinorium by a major thrust fault or zone of thrust faults representing the belt from which the Taconic klippen were expelled. Indeed, the whole sequence may be cut up by large thrust faults. Moreover, all the rocks have been metamorphosed in the deeper greenschist or amphibolite facies, so that the environment and history of deposition are difficult to determine.

The sequence is very thick, but just how thick is not clear. A succession of stratigraphic units has been worked out that can be recognized across Vermont and Massachusetts into Connecticut. In north-central Vermont this succession was reported to measure about 30 km across the strike (White and Jahns 1950, pp. 190-191), a "thickness" that is evidently the result of much reduplication by isoclinal folding and thrusting, whereas in places in southeastern Vermont the same succession is reduced to 2 or 3 km, manifestly by thinning on the flanks of diapir-like gneiss domes (Thompson, in Billings et al. 1952, pp. 17-18, 20). Clearly the original thickness lay somewhere between these two extremes; in any case it was undoubtedly much greater than that of the Taconic sequence and probably somewhat greater than that of the carbonate bank.

The age is equally difficult to be sure of. The upper third or so contains a few fossils that may be Silurian and traces northward into lower-grade rocks in Quebec that are dated as Silurian and Early Devonian; hence we can leave this part out of account. A unit near the middle can apparently be traced northward into strata in Quebec containing Middle Ordovician graptolites, and a unit farther down has been followed, but with a good deal of uncertainty, across the north-plunging end of the Green Mountains anticlinorium into lower Upper Cambrian strata on the west. We used to think the sequence extended down to the unconformity on the Precambrian basement of the anticlinorium, but the major thrust fault mentioned above intervenes. Nevertheless, our best guess is that the sequence, like both the carbonate-bank and Taconic sequences, ranges from the Lower Cambrian through the Middle Ordovician (the Silurian probably lies disconformably or unconformably above).

Despite the metamorphism, it is clear that this sequence differs greatly from the other two. Originally the dominant rock was probably quartzose graywacke; flysch-like alternations of more quartzose (and feldspathic) and more argillitic (now micaceous) layers are common in much of the succession, and graded bedding has been recognized in a few places. Units of clay shale are now represented by two-mica and garnet-mica schist, some graphitic (black shale). Carbonate layers as such are rare (except in the overlying Silurian), though scattered calc-silicate layers record the former presence of impure carbonate-bearing strata; relatively pure quartzite occurs in just that part of the section that may correlate with lower Upper Cambrian. The chief novelty is the abundance of metavolcanics, largely mafic and now greenstone, greenschist, or amphibolite depending on grade; they are present almost throughout though especially prominent in the presumed Middle Ordovician, where they are associated with more restricted but locally thick bodies of felsic metavolcanics. Another novel rock type in the Ordovician part of the sequence is "coticule," relatively thin layers of quartz-spessartine rock interlayered with generally graphitic two-mica aluminum-silicate schist and probably representing manganiferous chert in black shale. (More ordinary chert would now be difficult to tell from metaquartzite.)

The possible flysch, the graded bedding, the manganiferous chert, coupled with the probably large thickness of clastics and the rarity of carbonate, suggest a deep-water basin into which much clastic debris was brought. In view of the intense folding the strata have undergone, the basin could have been very wide (perhaps hundreds of kilometers instead of the present few tens). The metavolcanics show that a volcanic source was available nearby, but the relatively high quartz content of most of the metasedimentary rocks argues for a continental source as well. All that quartz could hardly have crossed the carbonate bank and the starved basin; if it came from North America at all, it must have found a by-pass of which we have no other knowledge. In any case, the Taconic "basin" may well have been simply the continental slope or rise beside this deep-water basin, starved because coarse sediment was prevented from entering across either the carbonate bank to the west or the deep basin to the east.

A further question is whether the crust beneath the basin was continental or oceanic. When it was thought that the thick sequence rested unconformably on the Precambrian basement of the Green Mountains anticlinorium, at least its western margin could be assumed to be continental, but after the discovery of the major thrust fault zone between, this assumption would apply only to the Taconic sequence, and even that is not certain. Along and near the thrust fault zone, in the lower part of the thick sequence, are scattered pods of ultramafic rock; traced into northern Vermont and adjacent Quebec they coalesce into large bodies that, in Quebec at least, are now considered ophiolitic, providing an argument for an oceanic crust. On the other hand, the trough of the Connecticut Valley synclinorium from central Vermont southward is diversified by a line or two of gneiss domes, which rose like diapirs (late in the Acadian orogeny; i.e., Early to Middle Devonian). The cores of some of these domes expose only felsic metavolcanics, presumably belonging to the lower Paleozoic succession, but some expose gneisses underlying the full succession of units and generally assigned (as on the Geologic Map of Vermont, Doll et al. 1961) to the Precambrian basement, although as far as I know isotope geochronology has not confirmed such an age. To turn from positive to negative arguments, shortening of the Connecticut Valley synclinorium has been extreme, involving much recumbent folding followed by steeper isoclinal folding before the rise of the gneiss domes; if the strata so shortened rested on a continental crust, that crust must have been shortened and thickened proportionally and should be evident somewhere. The gneiss domes in question hardly seem adequate to represent it. If the crust were oceanic, it could have been returned to the mantle. In sum, my guess (model, in the current jargon) is that the sequence we have been discussing was deposited on oceanic crust, in part, at least, in a deep-sea trench west of a volcanic island arc, but the question must certainly remain open for the present.

MIDDLE ORDOVICIAN OROGENIC EVENTS

In discussing the carbonate bank, we considered only the strata up through the Lower Ordovician. In most places these strata (and locally the very base of the Middle Ordovician, the

White Rock stage) are truncated above by a disconformity, recording a period of subaerial exposure during which a pronounced karst developed in the carbonate rocks below. The relief on the erosion surface locally reached at least 50 m (Bridge 1955); the disconformity is not universal, however, and in a few places deposition was apparently continuous. In early Middle Ordovician time the sea returned and covered the whole bank again, and carbonate deposition resumed; in the central part of the continent it continued throughout the Middle Ordovician. (Especially to the east, the basal layers above the unconformity include chert gravel, chert and quartz sand and silt, and mud, much of it red.) Because of the appearance of many new kinds of shelly animals, the character of the limestone (dolostone is now less common) changed considerably, but the sea remained fairly shallow.

Along the outboard eastern side of the bank, however, shortly after the return of the sea, black mud appeared from the east, and it spread westward through the rest of the Ordovician, fairly continuously though with some sudden bursts, until by Late Ordovician time it had reached the present location of the upper Mississippi River, nearly 1,500 km to the west. Behind it, silt and sand appeared and likewise spread westward, reaching approximately to the present east end of Lake Ontario by the Late Ordovician; the rate of deposition also increased. The sand is not pure quartz sand but contains many fragments of slate, argillite, chert, and feldspar, and some of chromite and serpentine. Many of the sand layers are obvious turbidites, and distal turbidites reach into the eastern part of the mud facies. Evidently, the depth of water increased, at least on the eastern part of the carbonate bank. As time went on, the rate of deposition outstripped subsidence, the seaway filled up, and red deltaic sediments spread out over the former carbonate bank as far as western New York State.

Naturally, the spread of mud killed off most of the shelly organisms that had been building up the carbonate bank; the fossils found in the mud and sand layers are largely graptolites and the other planktonic types mentioned above (p. 246). Indeed, it was his study of the graptolites that enabled Ruedemann (1908, 1912) to refute the prevailing layer-cake conception of the stratigraphy (sandstone over shale over limestone) and demonstrate the transgressive nature of all the facies, perhaps the first such demonstration in North American stratigraphy.

As noted above (p. 247), similar sand had arrived in the starved basin (ending its starvation) *before* black mud reached the carbonate bank; thus the transgressive facies can be traced backward into the basin sequence. In a number of areas from Newfoundland to Pennsylvania, great masses of the Taconic sequence now arrived on top of the carbonate bank to form the Taconic klippen. Many of us believe that these masses arrived by gravity on the sea floor while deposition was still going on, as shown by olistostromal breccias (''Wildflysch breccias'' or ''blocks-in-shale'') in the mud and sand underlying and surrounding them. In effect, the klippen themselves would be gigantic olistoliths, tens of kilometers on a side. (For a contrary view see Rowley and Kidd 1981; for discussion, see Rodgers 1982, Rowley and Kidd 1982.)

The source of this great influx of material must be sought in rising highlands farther east. We have already mentioned (p. 249) the ultramafic rocks near the zone of thrust faults bounding the deep-water basin on the west; the chromite and serpentine fragments in the sand show that such rocks had been lifted above sea level and were being eroded. Although in the Taconic region of western New England and eastern New York any such rocks have been removed by erosion, they remain farther north; the ophiolitic suites of Thetford (southeastern

Quebec) and the Bay of Islands (western Newfoundland) form thrust sheets above other klip-pen of the Taconic sequence. Thus oceanic crust was emplaced on top of the klippen, above the associated sand and mud and the underlying carbonate bank and its continental basement. The final emplacement took place within a relatively short time in the Middle Ordovician, bracketed within two or three graptolite zones in Newfoundland (but the time span is *not* the same in Newfoundland as in New England or Quebec; see Rodgers 1971).

There can be no doubt that the emplacement of the ophiolitic complexes over Taconic klippen and continental crust was a case of obduction; the North American continent ap-proached and attempted to go down an east-dipping subduction zone that was probably lo-cated, at least at first, in or along the west side of the deep-water basin or trench, as shown by the ultramafics there, and part of the oceanic crust of that basin was thrust up over the conti-nental margin. All the rest of the phenomena follow logically. Even the widespread discon-formity on the carbonate bank and the subsequent differential subsidence of the bank's eastern margin are natural consequences (Jacobi 1981). The sum of all these related events is what is called the Taconic orogeny, first recognized in the form of an angular unconformity at Becraft Mountain in the Hudson Valley by H. D. Rogers in 1837.

Can we match such an arrangement in the present world? The northwestern margin of the Australian continent, bearing the carbonate bank of the Sahul shelf, is today approaching the east end of the north-dipping subduction zone south of the Sunda Islands of Indonesia, and seismic profiles show that the carbonate layers bend down along the south side of the Timor trench. The island of Timor itself would correspond to the rising highlands that supplied sed-iments to the trench and onto the downbent margin of the bank; olistostromes with enormous olistoliths are characteristic of the geology of Timor and may well be present also to the south in the trench. In due course the rocks now accumulating in the trench will be pushed up onto the edge of the Australian continent, perhaps accompanied by obduced oceanic crust.

VOLCANIC SEQUENCE (ISLAND ARC?)

East of the Connecticut Valley synclinorium, mainly east of the Connecticut River, comes the Bronson Hill anticlinorium, which is not conventional like the Green Mountains anticlinor-ium with its Precambrian basement core but consists of irregular strings of gneiss domes. The rocks are metamorphic, mostly amphibolite facies but locally shallower, but a distinctive stra-tigraphic succession is present, and sparse fossils date its upper part as Silurian and Early De-vonian. Indeed it was in this belt that Marland Billings (1937, 1950), working out from a long-known but largely ignored fossil locality, first demonstrated the fallacy of assigning most of the metamorphic rocks of New England to the Precambrian.

We are concerned with the lower, less well-dated part of the sequence, beneath the un-conformable Silurian. The upper strata in this part are largely graphitic mica schist but contain lenses of metavolcanics; long-range correlations to fossil localities in Maine and Quebec date them as Middle Ordovician. Lower down, the rocks are either clearly metavolcanics, both mafic and felsic, or simply felsic gneisses of mainly igneous chemistry, some clearly plutonic and some ambiguous. Isotope geochronology dates these rocks also as Ordovician, probably Early or Middle. Only in a few domes do lower rocks appear, also mainly metaigneous gneisses but including more clearly metasedimentary rocks; some of the gneisses give latest Precambrian isotopic ages (considerably younger than any ages from the Precambrian basement of the Green Mountains farther west). No rocks datable as Cambrian have been found.

The thickness of the Ordovician sequence is not easy to determine, because of intense and repeated recumbent folding followed by the rise of the gneiss domes, but it appears to be a good deal less than that of the corresponding, less volcanic "trench" sequence. Both above and below the well-established unconformity beneath the Silurian, which truncates granodiorite intrusions as well as stratigraphic units, other unconformities have been reported or postulated by several workers, but there has been a dismaying lack of consistency as to their position in the succession, suggesting perhaps that the individual unconformities are local though they may be important in their localities. The presence of these unconformities helps to explain the lesser thickness of the sequence.

The dominant characteristic of this sequence is the abundance of calc-alkaline volcanics, at least in the Ordovician. (In Massachusetts and New Hampshire, volcanics are absent in the Silurian but reappear in the Devonian; northeastward along strike in Maine, volcanics are common in the Silurian as well.) Several workers have therefore suggested that the Bronson Hill belt represents a volcanic island arc, or its roots, perhaps related genetically to the east-dipping subduction zone responsible for the deep-sea trench to the west and, ultimately, for the obduction of oceanic crust onto North America (p. 251). The laterally inconsistent unconformities in the belt would thus find a natural explanation; individual volcanic edifices would have built up into islands and then been eroded to disappearance, but there need have been no coordination from edifice to edifice. The peculiar character of the anticlinorium, formed of discontinuous strings of separated domes, might also be reflecting the lateral discontinuity of the various edifices along the volcanic chain. In any case, the Bronson Hill volcanic belt provides a perfect source area for the volcanic material in the adjacent basin sequences, at least as far as the Ordovician is concerned; in the absence of any known Cambrian rocks in the belt we cannot say the same for volcanics in the Cambrian strata to the west, if indeed they are correctly dated.

ANOTHER DEEP-WATER BASIN SEQUENCE

Beyond the volcanic Bronson Hill belt lies the Merrimack synclinorium, the highest-grade metamorphic belt in southern New England, mostly in deep amphibolite facies but attaining granulite facies over considerable areas. In view of this metamorphism and at least two phases of recumbent folding, one of them associated to the south and southeast with intense slicing probably along thrust faults, the succession is poorly known. For some time, all the strata were assigned to the Lower Devonian, but they are now believed to reach down at least to the Middle Ordovician. There is no evidence of a post-Ordovician unconformity in the sequence, except in places along the western margin next to the Bronson Hill belt. The base is unknown except that an enigmatic gneiss body on the east flank in southeastern New Hampshire has given a very late Precambrian isotopic date. The thickness is indeterminate but must be large.

Common rock types, interlayered on all scales, are quartzose metagraywacke, commonly layered and not infrequently graded; more feldspathic, perhaps volcanogenic, meta-graywacke; graphitic and sulfidic schist; and calc-silicate rock, locally in thick units. Meta-volcanics are present here and there but are not dominant, coticule occurs locally, but clean quartzite and marble are rare.

All these characteristics suggest another, probably wide, deep-water basin. The same question arises—whether the crust beneath was oceanic or continental—and evidence is even sparser. If continental, it was not part of the North American Precambrian basement but latest Precambrian, related rather to the basement present in the next belt to the east. If oceanic, no obvious fragments remain; ultramafic rocks are rare in the synclinorium (and not common on either side). Again, negative arguments persuade me it was oceanic, but others are not convinced.

CONTINENTAL FRAGMENT—"AVALONIA"

The eastern margin of the Merrimack synclinorium reaches into a belt where faulting plays a larger role than to the west (although early Mesozoic faulting is important in and beside the Mesozoic graben of the southern Connecticut Valley, sited on the west flank of the Bronson Hill anticlinorium, and is present farther west). Some of the faults, characterized by silification and open-work breccias, presumably date from the late Paleozoic or early Mesozoic, but some, characterized instead by commonly thick zones of mylonite, formed late in the Acadian orogeny; they are the ones mentioned above (p. 252) as slicing up the strata in the southeastern part of the synclinorium. A particularly well marked and continuous fault of this group, the Honey Hill–Lake Char–Bloody Bluff fault, separates the Merrimack synclinorium from the next terrane to the southeast, called the Avalonian terrane because it has many of the characteristics of and is on strike with the terranes that underlie southern New Brunswick, the northern half of Nova Scotia, St. Pierre et Miquelon, and the Burin and Avalon Peninsulas of southeastern Newfoundland. In New England, this terrane underlies eastern and southeastern Massachusetts, Rhode Island, and a narrow strip in eastern Connecticut.

In the eastern part of this terrane, fossiliferous, little metamorphosed though commonly severely deformed Cambrian strata crop out in several separate areas, resting unconformably on a complex of metasedimentary, metavolcanic, and especially metaplutonic rocks that give latest Precambrian isotopic ages. Ordovician strata have not been found (though Ordovician alkalic intrusives are known). In the western part of the terrane, the late Precambrian metamorphism has been overprinted by Paleozoic metamorphism (of several ages), but no Cambrian or Ordovician rocks have been recognized.

The Cambrian strata have nothing in common with those in the three western sequences we have described; even their faunas are unrelated to standard North American faunas and resemble rather those in southern Britain, northern France, Spain, and the Baltic region. The rocks are largely dark silty shale but include some sandstone (mainly at the base) and some carbonate; probably they were deposited in moderately shallow water, not in the tropics but in a cool temperate climate. They are not particularly thick, though associated Lower Ordovician strata in parts of Atlantic Canada do attain a considerable thickness; in general, they appear to represent "starved" sedimentation. They are sometimes described as the deposits of a platform, but their later history of deformation belies this description. In any case, there is no doubt that they rest on a continental or sialic "basement," though the latter may have been created only shortly beforehand. Whether it formed part of a large continent, a continental peninsula, or one or several continental islands (like Madagascar, for example) is not yet clear. If the deep-water basins to the west were in fact oceanic, then one might think of the Avalonian terrane as marking the southeastern shore of that ocean, but there may well have been more ocean still farther southeast, for the Avalonian terrane in northern Nova Scotia is adjoined on the south by another synclinorium containing a thick Lower Ordovician and probably Cambrian clastic succession (Meguma group), again probably deposited in deep water. We have not yet traversed the entire Appalachian geosyncline.

GEOGRAPHIC RELATIONS OF THE DIFFERENT BELTS

We have hitherto proceeded as though the geographic order and position of the belts we have been describing were the same in the early Paleozoic as they are today. But more and more it is becoming evident that different belts and terranes in mountain chains have had quite different relations in the past; especially in the North American Cordillera, such ideas have led to the recognition of many "suspect" and "exotic" terranes, the former perhaps and the latter certainly derived from regions far away from their present locations relative to the main body of the continent. The evidence for long-distance migration of such terranes is partly faunal—evidence of geographic isolation and of a different climate—and partly paleomagnetic—from the determination of magnetic paleolatitudes. We must therefore ask which of the Appalachian belts can have moved very little, which might have moved a good distance, and which almost certainly did move far.

There can be no doubt that the carbonate bank and the underlying sandstones have been an integral part of the North American continent since they were deposited, for they can be seen to rest on continental basement from Newfoundland to Tennessee. The eastern part of the bank was involved in the folding and thrusting of the Valley and Ridge province, and its eastern edge was uplifted and thrust a considerable distance westward along with the (clearly North American) basement rocks of the Green Mountains and related anticlinoria; these parts have certainly been transported tens of kilometers toward the continental interior, but even before being transported they lay upon or very close to the edge of the continent as it then existed.

The Taconic sequence we have attributed to the continental slope and rise adjacent to the eastern margin of the carbonate bank, and the presence of fragments of the bank sequence in the limestone breccias along the western side of the klippen confirms this attribution. Nothing proves that the rock masses now forming the various klippen originally lay directly outboard of those parts of the bank they now overlie; they could have been brought in obliquely, with a large component of strike slip parallel to the continental margin, but they cannot be properly exotic, and there is no positive evidence for strike slip either.

We have already pointed out that the strata deposited in the western deep-water basin have been vastly shortened; there is no adequate measure of the shortening, and hence of the original width of the basin (they now form a belt only tens of kilometers wide), but nothing prevents its having had a width of several hundreds of kilometers, perhaps as wide as the present marginal seas along the east coast of Asia (at its widest the Philippine Sea reaches 2,000 kilometers). Again, the transport could have involved any amount of strike slip, but evidence is lacking; on the other hand, by Middle Ordovician time ocean crust was being obducted onto North America, furnishing chromite and serpentine fragments to the sediments first in the Taconic sequence and then over the carbonate bank and the continent itself. Thus there is no reason to postulate a truly exotic provenance, and the deep-water basin was probably adjacent to some part of eastern North America, though not necessarily exactly opposite its present position.

The position of the volcanic island arc is another matter. The present smooth curve of the Bronson Hill and related anticlinoria from coastal Connecticut to northeastern Maine is certainly an artifact, produced when (in the Devonian?) the belt finally smashed up against North America, molding itself like the other transported belts against the southern side of the Quebec reentrant (north side of the New York promontory) (Rodgers 1975). The original trend of the arc could have taken almost any direction. As just mentioned, the width of the sea between the arc and the continent may well have been equal to that of one of the marginal seas of eastern Asia—hundreds of kilometers or even a thousand or two. The presence in a few of the Bronson Hill domes of a non–North American basement, whose isotopic age and some of whose metasedimentary rocks resemble instead those in the southwestern part of the Avalonian terrane, suggests indeed that the Bronson Hill arc may have originally started to grow at the margin of Avalonia, wherever that was at the time, and then migrated away over the east-dipping subduction zone, leaving the eastern deep-water basin behind as a new marginal sea. Here then we may well be dealing with a truly exotic terrane. We do not know whether the Bronson Hill belt approached and perhaps collided with North America during the Taconic orogeny—certainly an attractive hypothesis—but if it did, it must then have migrated away again to provide the basin in which the Silurian and Lower Devonian strata of the western deep-water sequence were deposited, before again being driven against North America in the Acadian orogeny. Speculation about the provenance of the eastern deep-water basin would follow the same lines.

Avalonia, finally, is a prime example of a terrane that was probably exotic, both because of its non–North American fauna, probably belonging to a quite different climatic belt, at least during the Cambrian and Early Ordovician, and because of reported paleomagnetic latitudes that place it far down in the southern hemisphere. Taken at face value, however, the paleomagnetic results raise serious problems, for they seem to demand that Avalonia did not arrive in its present position relative to North America until Middle or Late Carboniferous, if not Permian, time. Yet Acadian (and even Taconic?) metamorphism overprints its western margin, as though it was already nearly in place; indeed, an Early Devonian arrival would provide a splendid motor for the Acadian orogeny. Moreover, at least one paleomagnetist, formerly committed to a Carboniferous arrival, now has new data permitting if not demanding arrival in the Devonian. Proponents of a late arrival have sought a Carboniferous (or Permian) suture line, and have suggested it might follow the Honey Hill–Lake Char–Bloody Bluff fault (or the

Clinton-Newbury fault or some other branch of the system) in southern New England and comparable faults in eastern Maine and New Brunswick (Norumbega fault) and in Newfoundland (Dover-Hermitage fault). Unfortunately, the geological evidence is not favorable to such late suturing along these faults (for example, the Dover-Hermitage fault was clearly sealed by the Early Carboniferous Ackley bathylith). Attempts to find the suture elsewhere have been quite *ad hoc*, disregarding the geological evidence. Avalonia may be and probably is exotic, but its time of arrival is still uncertain; Devonian seems at least as likely as Carboniferous.

WHAT WAS THE APPALACHIAN GEOSYNCLINE LIKE?

In sum, if we had taken a cruise, starting at the eastern shore of North America during the Early Ordovician and sailing directly out to sea (Figure 13-2), we would first have crossed the carbonate bank and the starved slope and rise beyond it, and then entered a large deep-water sea, perhaps a marginal sea, perhaps a true ocean. (By Middle Ordovician time, deformation was producing islands in this sea.) Whether we would then have encountered the Bronson Hill island arc is not so clear, but perhaps it was already well on its way to colliding with North America, the (east-dipping) subduction zone associated with it eating up the deep-water sea in the process. Having passed the arc and entered the marginal sea behind it, we simply do not know whether we should have looked for Avalonia dead ahead or hundreds or thousands of kilometers off to the right; if the latter, a direct course might instead have brought us to the Baltic-Russian continent. All this time, in my view, we would have been sailing over the Appalachian geosyncline.

Furthermore, an Early Ordovician journey like this would not have been all that different from a present-day journey into eastern Indonesia, sailing northward from the northwest coast of Australia over the Sahul shelf and its continental slope, across the Timor trench, around one end or the other of the island of Timor, then through the volcanic chain of the Lesser Sunda Islands into the Banda Sea. Beyond, off to the right, would then lie western New Guinea (the Vogelkop or Birdhead Peninsula), a continental promontory that quite possibly is presently undergoing left-handed strike-slip motion that may ultimately bring it into position across our traverse.

At the start of this article, I asked, What was the Appalachian geosyncline like while it existed?, and my answer is eastern Indonesia. The idea is not new of course; I learned it from Olaf Holtedahl in 1946, who ended a lecture on the early Paleozoic "Caledonian" geosyncline of Scandinavia by comparing it to the present-day East Indies, and it goes back to Emile Argand's great work, *La Tectonique de l'Asie* (1922).

ACKNOWLEDGMENTS

I am grateful to the many geologists in many places where I have presented this article in lecture form, for their fruitful discussions of the ideas it contains. Figure 2 was drafted by Carol Ann Phelps and Figure 3 by Peter Sisterson and Craig Dietsch, in each case after originals drawn by Michel Guélat.

REFERENCES

Argand, Emile, 1922: La tectonique de l'Asie. 13th Internat. Geol. Congr., Brussels 1922, Comptes rendus, 171-372.

Billings, M. P., 1937: Regional metamorphism of the Littleton-Moosilauke area, New Hampshire. Geol. Soc. America Bull. *48*, 463-565.

Billings, M. P., 1950: Stratigraphy and the study of metamorphic rocks. Geol. Soc. America Bull. *61*, 435-447.

Billings, M. P., Rodgers, John, and Thompson, J. B., Jr., 1952: Geology of the Appalachian highlands of east-central New York, southern Vermont, and southern New Hampshire. 65th Ann. Mtg., Geol. Soc. America, Boston 1952, Gdbk. (Guidebook for field trips in New England), 1-71.

Bird, J. M., Zen, E-an, Berry, W.B.N., and Potter, D. B., 1963: Stratigraphy, structure, sedimentation and paleontology of the southern Taconic region, eastern New York. Albany, N.Y. 76th Ann. Mtg., Geol. Soc. America, New York 1963, Gdbk. Trip 3, 67 p.

Bridge, Josiah, 1955: Disconformity between Lower and Middle Ordovician series at Douglas Lake, Tennessee. Geol. Soc. America Bull. *66*, 725-730.

Bucher, W. H., 1933: The deformation of the earth's crust. Princeton Univ. Press, 518 p. Reprinted 1957 (Hafner, New York).

Dana, J. D., 1866: Observations on the origin of some of the earth's features. Am. Jour. Sci., 2nd ser., *42*, 205-211, 252-253.

Dana, J. D., 1873: On some results of the earth's contraction from cooling, including a discussion of the origin of mountains, and the nature of the earth's interior. Am. Jour. Sci., 3rd ser., *5*, 423-443, 474-475; *6*, 6-14, 104-115, 161-172, 304, 381-382.

Doll, C. G., Cady, W. M., Thompson, J. B., Jr., and Billings, M. P., 1961: Geologic Map of Vermont. Vermont Geol. Survey.

Hall, James, 1883: Contributions to the geological history of the North American continent. 31st Ann. Mtg., Am. Assoc. Adv. Sci. Proc., Montreal 1882, 29-69. (Abstr. in Canadian Naturalist and Geologist [1857] *2*, 284-286.) *See also* Natural History of New York (1859), Div. 6, Paleontology, *3*, 1-86.

Haug, Emile, 1900: Les géosynclinaux et les aires continentales. Soc. géol. France Bull., 3me sér., *28*, 617-711.

Hubert, J. F., Suchecki, R. K., and Callahan, R.K.M., 1977: The Cow Head breccia: Sedimentology of the Cambro-Ordovician continental margin, Newfoundland. Soc. Econ. Pal. Min. Spec. Publ. *25*, 125-154.

Jacobi, R. D., 1981: Peripheral bulge—a causal mechanism for the Lower/Middle Ordovician disconformity along the western margin of the northern Appalachians. Earth Planet. Sci. Lett. *56*, 245-251.

Kay, Marshall, 1951: North American geosynclines. Geol. Soc. America Mem. *48*, 143 p.

Kindle, C. H., and Whittington, H. B., 1958: Stratigraphy of the Cow Head region, western Newfoundland. Geol. Soc. America Bull. *69*, 315-342.

Kindle, C. H., and Whittington, H. B., 1959: Some stratigraphic problems of the Cow Head area in western Newfoundland. New York Acad. Sci. Trans., Ser. 2, *22*, 7-18.

Rodgers, John, 1970: The tectonics of the Appalachians. Wiley-Interscience, New York, 271 p.

Rodgers, John, 1971: The Taconic orogeny. Geol. Soc. America Bull. *82*, 1141-1177.

Rodgers, John, 1975: Appalachian salients and recesses (abstr.). Geol. Soc. America Abstr. Programs *7*, 111-112.

Rodgers, John, 1982: Stratigraphic relationships and detrital composition of the medial Ordovician flysch of western New England: Implications for the tectonic evolution of the Taconic Orogeny: A discussion. Jour. Geology *90*, 219-222.

Rowley, D. B., and Kidd, W.S.F., 1981: Stratigraphic relationships and detrital composition of the medial Ordovician flysch of western New England: Implications for the tectonic evolution of the Taconic Orogeny. Jour. Geology *89*, 199-218.

Rowley, D. B., and Kidd, W. S. F., 1982: Stratigraphic relationships and detrital composition of the medial Ordovician flysch of western New England: Implications for the tectonic evolution of the Taconic Orogeny: A reply to Rodgers. Jour. Geology *90*, 223-226.

Ruedemann, Rudolf, 1908: Graptolites of New York. Part 2, Graptolites of the higher beds. New York State Mus. Mem. *11*, 583 p.

Ruedemann, Rudolf, 1912: The Lower Siluric shales of the Mohawk valley. New York State Mus. Bull. *162*, 151 p.

Ruedemann, Rudolf, 1934: Paleozoic plankton of North America. Geol. Soc. America Mem. *2*, 141 p.

White, W. S., and Jahns, R. H., 1950: Structure of central and east-central Vermont. Jour. Geology *58*, 179-220.

Zen, E-an, 1967: Time and space relationships of the Taconic allochthon and autochthon. Geol. Soc. America Spec. Paper *97*, 107 p.

Part VI

ASIA

Chapter 14

COMPARATIVE STUDIES ON PROFILES ACROSS THE NORTHWEST HIMALAYAS

W. FRANK

Geologisches Institut, Vienna

A. BAUD

Institut et Musée de Géologie, Lausanne

K. HONEGGER

V. TROMMSDORFF

Institut für Mineralogie und Petrographie, Zürich

This study deals with profiles in Ladakh, Kashmir, and the Kulu area (Figure 14-1) of the Northwest Himalayas. Considerable fieldwork in this region in the past ten years (Frank et al. 1973; Gupta and Kumar 1975; Frank, Gansser, and Trommsdorff 1977; Frank, Thöni, and Purtscheller 1977; Bassoullet, Colchen, et al. 1978; Sharma and Kumar 1978; Srikantia et al. 1978; Fuchs 1979, 1981, 1982a, 1982b; Bassoullet, Colchen, Marcoux, and Mascle 1980; Gaetani et al. 1980; Baud, Arn, et al. 1982; Honegger et al. 1982; Bassoullet et al. 1983; Baud et al. 1983; Honegger 1983; Trommsdorff et al. 1983) has led to a number of attempts to reconstruct its geological history and evolution. In this paper we shall try to summarize some of the present knowledge, focusing particularly on the crystalline nappes of the High Himalayas. We will discuss only briefly the geological features in the neighborhood of the Indus suture zone and will refer to the papers on this topic by Andrieux et al. (1977); Gansser (1977, 1980); Bassoullet, Colchen, et al. (1978); Brookfield and Reynolds (1981); Baud, Arn, et al. (1982); Honegger et al. (1982); and Trommsdorff et al. (1983). We start our review with a brief survey of the different geological units, beginning in the south. For orientation, see the map in Figure 14-1 and the profiles in Figure 14-2.

STRATIGRAPHY OF THE LESSER HIMALAYAS
(units 2 to 4, profiles B, D, and E)

The tectonic units below the large crustal wedge of the Himalayan crystalline nappe represent the Proterozoic sedimentary cover of the Indian continental basement. Their oldest parts show facies variations parallel to the strike direction of the much younger Himalayan fold belt. The uniform, dark-colored Simla slate formation is the oldest sequence in the south and is derived from the Indian shield (Valdiya 1970). In the north, the Chail-Berinag slates (several kilometers thick) interfinger with massive quartzites of deltaic origin. The Chail-Berinag forma-

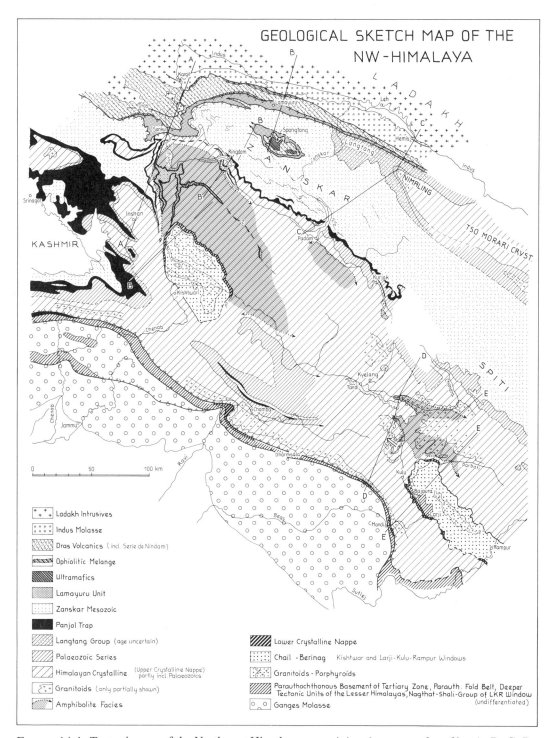

GEOLOGICAL SKETCH MAP OF THE
NW-HIMALAYA

Ladakh Intrusives

Indus Molasse

Dras Volcanics (incl. Serie de Nindam)

Ophiolitic Melange

Ultramafics

Lamayuru Unit

Zanskar Mesozoic

Panjal Trap

Langtang Group (age uncertain)

Palaeozoic Series

Himalayan Crystalline (Upper Crystalline Nappe) partly incl. Palaeozoics

Granitoids (only partially shown)

Amphibolite Facies

Lower Crystalline Nappe

Chail - Berinag Kishtwar and Larji-Kulu-Rampur Windows

Granitoids - Porphyroids

Parauthochthonous Basement of Tertiary Zone, Parauth. Fold Belt, Deeper Tectonic Units of the Lesser Himalayas, Nagthat-Shali-Group of LKR Window (undifferentiated)

Ganges Molasse

FIGURE 14-1. Tectonic map of the Northwest Himalayas containing the traces of profiles A, B, C, D, and E, as shown in Figure 14-3.

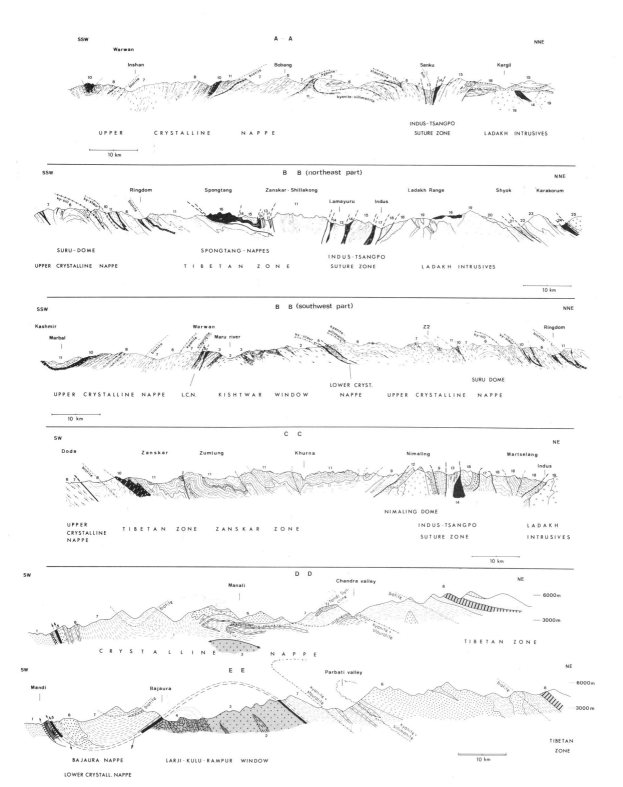

FIGURE 14-2. Profiles across the Northwest Himalayas: A-A from Honegger (1983), B-B in part after Trommsdorff et al. (1983), C-C after Baud, Arn, et al. (1982), and D-D and E-E after Frank et al. (1973). For profile traces, see Figure 14-1. 1: Siwaliks or Ganges molasse. 2-4: Lesser Himalayan units (2: porphyroids, granitoids. 3: Chail-Berinag. 4: Nagthat-Shali Group). 5: Lower crystalline nappe. 6: Granitoids of the crystalline nappe. 7: Metasediments of the crystalline nappe. 8: Paleozoic. 9: Langtang Group. 10: Panjal Trap. 11: Zanskar Mesozoic. 12: Crystalline of the Tsomorari (Nimaling) unit. 13: Lamayuru Formation. 14: Ophiolitic Mélange. 15: Dras unit. 16: Ultramafics. 17: Indus flysch. 18: Indus molasse. 19: Ladakh batholiths. 20: Khardung-La volcanics. 21-25: Shyok and Karakorum Formations (21: Upper Carboniferous to Lower Permian platform-type limestones. 22: Shyok intrusives. 23: Garnet/kyanite mica schists. 24: Meta-basalts and serpentinites. 25: Karakorum batholith). Metamorphic isograds are indicated where mapped.

tion contains a few typical diabases and also contemporaneous quartz porphyries and granitic rocks. Some distinctive quartz porphyries (blue quartz and large rounded K-feldspars) have been found in the Kishtwar window, in the Kulu area, in Kumaon, and in western Nepal 88 km along strike to the east. They yielded an age of 1,900 Ma in the Kulu (Sainj) area. Their melts evolved already in a fairly old continental basement.

The younger sediments resting above both the Simla slate and the Chail formation represent a shallow-water sequence starting with red quartzites (Nagthat, Khaira), followed by multicolored carbonates (Blaini), and capped by massive stromatolitic carbonates (Shali, Krol). Local, basinal shales are often intercalated in the deeper part of the shallow-water sequence, which is up to 1800 m thick. The famous Blaini tillites occur in some places (e.g., Simla) on top of Simla slates. There is a long discussion in the literature (see Frank, Gansser, and Trommsdorff 1977) about whether the age of the whole sequence might reach into the Paleozoic, and whether the Blaini tillite corresponds to the Permocarboniferous Talchir tillites. Transgression of the Lower Cambrian Hazira formation (Mostler 1980) on the same sequence in the Abbottobad area (Pakistan), including identical stromatolite limestones and the Tanakki tillite, clearly shows the Proterozoic age of the typical Lesser Himalayan series. There are some striking similarities with parts of the Sinian system in China. Locally, the Jurassic-Cretaceous Tals and the Paleocene-Miocene beds transgress on the Proterozoic.

TECTONICS OF THE PARAUTOCHTHONOUS UNIT
(part of unit 4, southwestern ends of profiles D and E)

The Simla slate–Shali sequence forms the lowermost "Parautochthonous Unit" of the Himalayas, overthrust along the Main Boundary fault upon the Tertiary zone (Fuchs 1981). Its transport distance is usually considered of minor importance, only a few tens of kilometers. On Figure 14-1, north of the Sutlej River, only the narrow (Par-)"Autochthonous" fold belt (Wadia 1934) and the Shali limestone above the Tertiary zone belong to this unit, typically developed between the Sutlej and western Nepal. The metamorphic grade of this unit is very weak, mostly anchizonal reaching partly to shallowest greenschist facies. It should be mentioned that the stratigraphy of the "Parauthochthonous fold belt" is sometimes uncertain, and that the Proterozoic-Permocarboniferous series known in Kashmir are sometimes included in this unit (e.g., Fuchs 1975, his Figure 2).

CHAIL NAPPE
(units 2 and 3, profiles D and E)

The Chail nappe(s) (Fuchs 1967) represent(s) a characteristic middle tectonic unit below the crystalline nappes which have the Main Central thrust at their base all along the Himalayas. Muscovite/phengite–chlorite schists are typical rocks, biotite is rare, and garnet is only locally present. The highest horizons in the root zone just below the crystalline nappe may also be affected by the low-grade part of the reversed metamorphism.

Partly because of some lithologic similarities between the uniform Proterozoic-Paleozoic slate formation in the southern part of the crystalline nappe and the Chail series, Fuchs (1981) has regarded the southern front of the Kashmir and Chamba basins as part of the Chail nappe. But from field evidence we know that the early Paleozoic granites of Mandi and their counterparts in the frontal zone to the northwest are intruded in the southern part of the crystalline nappe. We realize, of course, that narrow remnants of the Chail nappe exist below the Main Central thrust (e.g., at Gulabgarh; see Fuchs 1975, his Figure 2); they are not separated in Figure 14-1.

LOWER CRYSTALLINE NAPPE
(unit 5, profiles B, D, and E)

The lower crystalline nappe (Fuchs and Frank 1970) represents a sequence only 10 to 100 m thick of peculiar lithologies traceable in many places immediately at the base of the main (upper) crystalline nappe. In the Kulu area at Bajaura (Thöni 1977) carbonaceous schist and marble, acid volcanics, and granitic rocks are typical members, all highly deformed, with an obvious south-vergent simple shear component. It is tentatively suggested that this unit represents a rock sequence (age uncertain) in the deeper levels of the southern crystalline nappe, which, because of its special lithology, became later the "décollement horizon" and was dragged below the main mass of the crystalline nappe. The unit does not necessarily represent a separate paleogeographic unit.

CRYSTALLINE NAPPE (units 6 and 7, profiles A, B, C, D, and E), TIBETAN ZONE (units 8 to 12, profiles A, B, and C), AND THEIR RELATIONSHIP

The crystalline nappe of the High Himalayas comprises Precambrian to, at the youngest, Cambrian rocks of pelitic to arenitic composition that were intruded about 500 Ma ago by a distinct suite of granitic rocks with high $^{87}Rb/^{86}Sr$ ratios. They are often associated with migmatization and probably with a low-pressure (pre-Alpine) metamorphism. In the area southeast of Padam (see Figure 14-1), for example, sillimanite gneisses and migmatites are cut by pre-Alpine leucogranites. During the Alpine (Main Himalayan) orogenesis, these rocks were metamorphosed at medium grade. This metamorphism dies out gradually in the overlying late Proterozoic to early Paleozoic rocks, which are generally thought of as the lower part of the Tibetan zone. This feature is typical all along the Himalayas.

Tectonically, most authors regard the crystalline nappe and Tibetan zone as lower and upper parts of one large continental wedge; they recognize considerable internal deformation in between but no large-scale thrust faults (Gansser 1964; Frank, Gansser, and Trommsdorff 1977; Fuchs 1982a). One of us (A.B.) believes, however, that there is an important thrust plane between them (Baud, Gaetani, et al. 1982, 1984; Baud 1983).

The stratigraphic sequence at the base of the Tibetan zone is often continuous from the late Proterozoic far into the Phanerozoic with only minor gaps and unconformities. In a wider sense (Gansser 1964, Fuchs 1981) the Tibetan zone comprises Tethyan platform sediments along the northern continental margin of the Indian Peninsula throughout most of the Phanerozoic. Figure 14-3 is a simplified sketch, after Fuchs (1982b), of the lithologic-stratigraphic situation in the Zanskar area for the Mesozoic and early Tertiary. In the early Paleozoic to Devonian there are large-scale facies variations: carbonate sequences several thousands of meters thick in western Nepal; pronounced clastic sedimentation in Spiti, Ladakh, and Kashmir; and uniform argillaceous rocks in the Chamba basin. The early Paleozoic sequence in southeastern Zanskar is equivalent to that of Spiti (Fuchs 1982a). Carboniferous sediments show evidence of a humid, moderate climate; they are predominantly argillaceous rocks with intercalated coarse clastic rocks (Po series) and, locally, carbonates at their base (Lipak series). These Carboniferous sequences are rather uniform all along the Himalayas. Tillites similar to the Talchir tillites on the Indian continent have been described (Wadia 1934). In the Upper Carboniferous, volcanic activity began in the Kashmir area and produced several hundred meters of intermediate to acid pyroclastics and welded tuffs interfingering with sediments—the Agglomeratic slate (Pareek 1976). It is overlain by plateau-type alkaline basaltic flows—the Panjal Trap. The climax of volcanism was in Lower Permian times, and it lasted in places until the Middle Triassic.

The thickness of the whole volcanic suite reaches 2,500 m in the Kashmir basin and decreases toward the east and west. The Panjal Trap forms the only well-defined stratigraphic marker that can be easily recognized in the high-grade metamorphosed sequences of the High Himalayas in the Suru and Zanskar regions. (For details on the lithology, mineralogy, and chemistry of the Panjal Trap, see Pareek 1976, Bhat and Zainuddin 1978, Honegger et al. 1982.) The Panjal Trap is only local when compared with the size of the whole Himalaya belt. Its easternmost recorded outcrops are at Spiti (Fuchs, personal communication). Otherwise, the Permian strata are quite thin (see Fuchs 1982a).

Conditions along the whole continental shelf were comparatively homogeneous during the Mesozoic. Carbonate sedimentation had already started locally in the Permian and had become widespread in the Triassic, thus producing a uniform and stable shelf area with cyclic sedimentation of carbonates, marls, and, locally, shales and quartzites (Lilang Group, Rhetian quartzite series). The climax of carbonate sedimentation is marked by the well-bedded, uniform Kioto limestone. A break in sedimentation during the early Late Jurassic was caused by a rapidly subsiding platform (at the onset of continental drift of peninsula India) and produced a sequence of dark shales (Spiti shales) and sandstones. The sedimentation during Lower to Middle Cretaceous times is characterized by rapid changes in lithology and local variations in thickness and facies, which may be correlated with the final breaking up of the Gondwana continent. The Cretaceous ferruginous Giumal sandstone (disappearing toward the north) and the pelagic variegated Chikkim (Shillakong) limestone are typical members of this sequence. In the Upper Cretaceous, sedimentation again became uniform and produced quite

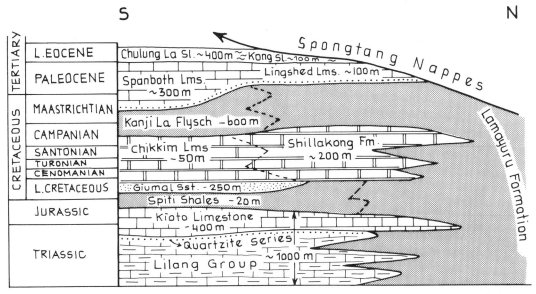

FIGURE 14-3. Simplified stratigraphic scheme of the Mesozoic-Tertiary sequence of the Zanskar area (after Fuchs 1982b).

thick homogeneous slates and, locally, sandstones (Kangi-La area) with no obvious lithologic differences from the slate sedimentation that lasted during the whole Mesozoic along the continental margin—the Lamayuru unit in the sense of Fuchs (1981, 1982b), discussed below. A prominent fossil-rich limestone layer (Spanboth and especially Lingshed limestone) marks the end of marine sedimentation in the early Tertiary (Gaetani et al. 1980, 1983). Red slates of continental affinity (Chulung La) represent the final sedimentary stage.

In this description we can present only the general and large-scale aspects of the whole Tibetan zone. Considerable work on the stratigraphy, lithology, and facies variations of the suite has led to numerous local subdivisions (including tectonic divisions of the zone; see, for example, Baud, Gaetani, et al. 1982, 1984; Fuchs 1982b; Bassoullet et al. 1983), which we omit in this account.

Several authors accept the idea that a tectonic separation exists between the Tibetan zone s.s. (Fuchs 1982b) and the northern Zanskar unit (Fuchs 1982b) or Zanskar zone (Baud, Arn, et al. 1982), as outlined in profile C in Figure 14-2. The amount of shortening along this line, its regional extent, and its significance are still being disputed.

In Kashmir and at the base of the Tibetan zone in west Ladakh, a pronounced unconformity is reported at the base of the Agglomeratic slate and the Panjal Trap, respectively (Srikantia and Bhargava 1978, Fuchs 1979). Locally at Karcha Gompa near Padam, isoclinal folds with an amplitude of 100 m occur below the unconformity. Thus it can be interpreted as a sedimentary unconformity without major tectonic overprinting. Baud, Gaetani, et al. (1982, 1984) have interpreted this unconformity as tectonic (thrust plane of the Zangla nappe, omitted in the figures). The deformation below this unconformity, which is correlated with a slaty cleavage, is obviously connected to the well-defined 500 Ma event in the crystalline nappe and is very widespread and common along strike in the whole mountain belt.

At present there is insufficient evidence of an independent Hercynian orogenic event in the crystalline basement. Recently, Baud, Gaetani, et al. (1982, 1984) published the interpretation that the crystalline nappe and the Tibetan zone s.l. (partly "nappes du Haut Himalaya" of Baud, Arn, et al. 1982, 1983) are in tectonic contact in the Ladakh and Lahul regions. Although there are local Alpine shear zones in the low-grade part of the crystalline nappe east of Padam and tectonic repetition and deformation in the early Paleozoic sequence near Kurjiak, a fundamental tectonic separation between the crystalline and Tibetan zones is strongly contested by the other authors of this review. Of course further investigation in the disputed area is needed.

LAMAYURU UNIT
(unit 13, profiles A, B, and C)

The Lamayuru unit was first described by Frank, Gansser, and Trommsdorff (1977). Fossils prove that it extends from the Middle Triassic to Dogger, and it may even reach the Cretaceous. Its lithology has been described by Bassoullet, Colchen, Marcoux and Mascle (1980). The major part of the Lamayuru unit consists of monotonous, sometimes silty slates. Only minor parts exhibit a true flysch character. Lenses of limestones are intercalated. They have different lithologies and origins and have only in part been interpreted as olistoliths (Bassoullet, Colchen, Marcoux, and Mascle 1980). Other lenses may represent local stratigraphic intercalations of their counterparts in the Tibetan platform. In part Lamayuru unit rocks have strong affinities to the calcareous Bündnerschiefer of the Alps. Obviously, this unit represents continental-slope sediments adjacent to the Tibetan platform. In some regions—for example, southwest of Lamayuru, in the Spongtang area—lenses of serpentinite separate Lamayuru rocks from those of the Tibetan platform, but the regional extent of this tectonic separation is not known well enough to outline clearly a so-called "Lamayuru nappe" (Bassoullet, Colchen, Juteau, et al. 1980). The difficulty is that Upper Cretaceous shales east of Spongtang and indistinguishable from those of the true Lamayuru unit are intercalated in stratigraphic continuity within the Tibetan zone. One of the authors (A.B.) is, however, in favor of a Lamayuru nappe hypothesis.

DRAS VOLCANIC UNIT
(unit 15, profiles A and B)

The Dras unit constitutes a complex volcanic to volcanosedimentary sequence comprising basaltic and dacitic flows, pillow lavas, doleritic sills, and pyroclastics. Intercalations of radiolarian cherts of Callovian to Tithonian age and local inclusions of Albian to Cenomanian

Orbitolina limestones set minimum limits for the time span of volcanic activity (Honegger et al. 1982). Accompanying, steeply dipping ophiolitic mélange zones on both sides have thrust contacts against the Dras unit. Toward the east, between profiles B and C of Figure 14-2, the Dras suite grades into flyschoid volcanosedimentary rocks (Frank, Gansser, and Trommsdorff 1977, p. 101) with basaltic and dacitic fragments—the Nindam formation (Bassoullet, Colchen, et al. 1978). On the basis of its tectonic position, its chemical and petrographic characteristics, and its sediment association, the Dras unit is interpreted as representing an Upper Jurassic to Cretaceous island-arc association in front of the Eurasian continent (Dietrich et al. 1983). Oceanic basalts adjacent to and within the Dras unit (Honegger et al. 1982) may be interpreted as transitional from the island-arc rocks into oceanic crust.

TRANS-HIMALAYAN PLUTONS
(unit 19, profiles A, B, and C)

The batholithic belt of the Trans-Himalayan plutons or Ladakh intrusives, 30 to 50 km wide and over 2,500 km long, comprises in Ladakh rocks ranging from olivine norite and gabbro with local cumulate structures through diorites and tonalites to granodiorites and granites. Chemically, these rocks represent a typical calc-alkaline suite. They intrude Dras volcanics and mélange in the Kargil area (Frank, Gansser, and Trommsdorff 1977). To the east near Mt. Kailas in Tibet, the Trans-Himalayan plutonic rocks intrude Upper Jurassic sediments (Heim and Gansser 1939). The Ladakh intrusives are overlain by Upper Cretaceous molasse deposits. Radiometrically, they yield ages from 103 Ma (concordant zircons in diorite, Kargil area) to 39 Ma (Rb/Sr on Kailas granites) (Frank, in Honegger et al. 1982).

SPONGTANG OPHIOLITE NAPPES
(units 14 to 16, profile B)

In the Spongtang area a huge slab of unserpentinized peridotites (5) rests upon an edifice that contains, from bottom to top, (1) Mesozoic platform sediments (Tibetan zone); (2) Mesozoic sediments of a continental margin, the Lamayuru "flysch" formation, separated locally from (1) by serpentinite lenses and containing sills and lavas of alkaline chemistry; (3) an ophiolitic mélange zone; and (4) volcanic and volcanosedimentary rocks of the Dras (Nindam) formation. Units (3) to (5) form the Spongtang *ophiolite* nappes. Together with wedges of the continental slope sediments (Lamayuru unit), they override in their southwestern part Paleocene to Eocene Nummulite limestones (Gaetani et al. 1980, Fuchs 1982b) that form the top of the Tibetan-zone sequence. The emplacement of the ophiolite nappes thus postdates these sediments. Apart from the generally accepted fact that units (1) to (5) represent, from a paleogeographic standpoint, successively more northerly units, quite different opinions about the tectonic nature and emplacement of the edifice have been published (Frank, Gansser, and Trommsdorff 1977; Fuchs 1977, 1982b; Trommsdorff et al. 1983). Two possible tectonic models are discussed below. Remarkable similarities in the general tectonic picture between these ophiolite nappes and those in the Amlang La region in Tibet (Heim und Gansser 1939) indicate that the nappes represent a kinematic event of regional significance.

INDUS MOLASSE
(unit 18, profile A, B, and C)

The Eocene and post-Eocene Indus molasse is transgressive on Trans-Himalayan plutons near Kargil and in the Upper Indus region (Frank, Gansser, and Trommsdorff 1977), and on Dras volcanics south of Kargil (Fuchs 1981). Its conglomeratic parts strongly resemble the molasse deposits described by Heim and Gansser (1939) from the Mt. Kailas region in Tibet. In the Upper Indus area, however, distinct lateral variations in the facies of these deposits are recognized (Frank, Gansser, and Trommsdorff 1977) and have repeatedly led to a confusion of molasse and flysch deposits. Paleogeographically more southern parts of the molasse have been involved in late north-directed tectonics, for example, at Hemis Gompa (Frank, Gansser, and Trommsdorff 1977). The more southern molasse units have been described along several profiles by Pal et al. (1978), Baud, Arn, et al. (1982), and Brookfield and Andrews-Speed (1984).

METAMORPHISM AND DEFORMATION IN THE
SURU-ZANSKAR REGION

Except for eo-Himalayan exotic blocks with glaucophane and omphacite occurring along thrust contacts of the Dras unit (Frank, Gansser, and Trommsdorff 1977), no apparent discontinuities are found in the pattern of regional metamorphic zones north of the Suru dome. Metamorphism increases from zeolite grade (molasse) through prehnite-pumpellyite and greenschist facies (Dras unit; see Honegger 1983) to deep amphibolite facies in the High Himalayas. Around the major intrusive masses of the Trans-Himalayan batholiths, contact metamorphism has been recognized (Frank, Gansser, and Trommsdorff 1977).

In the Suru-Zanskar region of the High Himalayas (profiles A and B), a metamorphic belt has been recognized and mapped (Frank, Thöni, and Purtscheller 1977, Honegger et al. 1982). This belt is of Barrovian type and roughly follows the southeast-trending structural Suru dome. In the deepest-seated part of the dome center (sillimanite + K-spar grade), crystallization began prekinematically and then continued during and outlasted deformation, as can be read from inclusion and zoning patterns in garnet. As one moves from the center of the dome to its more external parts (amphibolite to greenschist facies), the apparent beginning of crystallization becomes synkinematic and finally postkinematic (Honegger 1983). Toward the south, the metamorphic belt is thrust over lower-grade metamorphic rocks of the Lesser Himalayas (Kishtwar window, profile B; see Honegger 1983). This situation can be compared with that along the Kulu (profile E; see Frank et al. 1973, Thöni 1977), where a huge recumbent fold with a high-grade, plastic core and a lower-grade, more rigid mantle has been thrust southward. This kind of deformation is one way to produce reversed metamorphic zoning patterns.

In other parts of the High Himalayas farther to the east, however, the mechanism for producing reversed metamorphism is synmetamorphic isograd deformation by large-scale south-vergent shearing at the base of the crystalline nappe (Frank et al. 1973, Le Fort 1975, Pecher 1978).

EVOLUTION OF HIMALAYAN DEFORMATION AND METAMORPHISM

The present northern border of the Indian subcontinent contains a Precambrian basement overlain by Proterozoic slate and quartzite formations and finally by Proterozoic shelf carbonate platforms. It formed a stable block rimmed by clastic Proterozoic to early Paleozoic deposits, which about 500 Ma ago became an orogenic belt with widespread generation of granites and probably low-pressure metamorphism and deformation. For three of the authors (W.F., K.H., and V.T.), all available information favors the interpretation that this old tectonic belt had gradational tectonic and metamorphic contacts.

Sedimentation continued during this event and by the Mesozoic the area had become a stable shelf. That the original orientation of the north rim of India in the Triassic must have been north-south can be deduced from the sedimentary facies, which indicates a subtropical climate in Kashmir and a more moderate climate in the eastern Himalayas (L. Krystyn, personal communication). This orientation is in good agreement with a counterclockwise rotation of India of about 70°, as deduced by Klootwijk et al. (1979) from paleomagnetic data. According to Klootwijk (1980), India separated from Gondwanaland in the Jurassic. In the Cretaceous, probably as a result of the consumption of oceanic crust, the Dras volcanics and Ladakh intrusives were produced (their chemical and structural evolution is discussed in Honegger et al. 1982 and Trommsdorff et al. 1983). India collided with Asia in the Oligocene, when a rapid drop in the drift rates is evident (Klootwijk 1980). From this time to the present the further movement and the counterclockwise rotation of India mentioned above have led to enormous crustal shortening in the Himalayan mountain belt.

The resulting internal structure of the crystalline nappes and the Tibetan zone can be simply characterized as follows: Just south of the suture zone itself, steep isoclinal folds of kilometer size are exposed in the impressive gorges of the Zanskar area. This very tight compression with enormous shortening (approximately 75%) becomes gradually more open to the south, and instead of numerous isoclinal folds fewer and larger (and recumbent) folds reaching several kilometers in thickness are the typical structures. The internal structure of the outer part of the crystalline nappe south of the Kishtwar and Larji-Kulu-Rampur windows is a fairly simple open syncline caused by a late shortening process.

Whereas the typical shortening features in the north are fold systems parallel to the strike directions—and sometimes also north-vergent (e.g., the Tandi syncline), depending on the local distribution of rigid blocks—the dominant deformation in the southern part of the crystalline nappe is thrusting associated with stretching lineations in the direction of (southwest-vergent) shear ''flow.'' The thrust sheets become younger toward India. The style of deformation in this area resembles that of the Caledonian mountain belts in northern Europe; reversed isograds are one of its peculiarities.

Only one clear regional nappe system exists, the ophiolite nappes in the Spongtang area; their evolution was discussed by Fuchs (1979) and by Trommsdorff et al. (1983). Their emplacement was post-Eocene, and obviously the rigid mass of peridotites on top is one reason for the difference in tectonic style between the nappes and their strongly (disharmonically) folded basement. After a short field trip he took in 1982, one of the authors (W.F.) would argue that folding is really the main shortening process in this area also, and that most of the tectonic subdivisions found in a detailed investigation by Baud, Arn, et al. (1982) disappear in continuous fold systems. We suggest two main reasons why obvious nappe structures, similar to the nappes in the northern Limestone Alps, are missing. The most important is the absence of a massive and rigid dolomite layer in the Mesozoic. The other reason may be elevated temperatures (up to beginning greenschist facies) during deformation, so that the well-stratified limestone deposits behaved more plastically.

A special situation is found in Kashmir, where the Suru Mountains trend across the strike of the main Himalayas. Here, the crystalline basement plus the Mesozoic rocks have undergone large-scale folding and metamorphism together. Even in the Mesozoic units middle amphibolite facies is reached, and the style of mountain building strongly resembles that of the classic belt of the Lepontine Alps. This situation, however, is unique in the Himalayas and may have to do with the early collision of the spur of India with Asia.

In the crystalline rocks, the pre-Alpine metamorphic mineral assemblages have usually been overprinted and obliterated by the Alpine metamorphism. The limited available geochronological data suggest that the temperatures may already have been quite high around 30 to 35 Ma ago and may have remained high in the high-grade part of the Suru area until 11 to 12 Ma ago. The end of metamorphism at the southern front of the crystalline nappe took place during a complicated interaction of deformation and erosion and especially during thrusting on the Lesser Himalayan units. This thrusting can be clearly correlated with the pronounced onset of the middle Miocene Siwalik molasse deposition around 16 Ma ago. At present new thrust planes are developing below the Ganges plain as one result of continuing shortening in the whole region of approximately 3 cm per year (Molnar et al. 1977).

ACKNOWLEDGMENTS

The authors thank A. Gansser for constructive discussions, A. Staubli for assistance with the illustrations, and John Rodgers and E. H. Perkins for constructively correcting the English in the manuscript.

REFERENCES

Andrieux, J., Brunel, M., and Shah, S. K., 1977: La suture de l'Indus au Ladakh (Inde). Acad. Sci. Paris C. R., Sér. D, *284*, 2327-2330.

Bassoullet, J. P., Bellier, J. P., Colchen, M., Marcoux, J., and Mascle, G., 1978: Découverte de Crétacé supérieur calcaire pélagique dans le Zanskar (Himalaya du Ladakh). Soc. géol. France Bull., Sér. 7, *20*, 961-964.

Bassoullet, J. P., Colchen, M., Juteau, T., Marcoux, J., and Mascle, G., 1980: L'édifice des nappes du Zanskar (Ladakh, Himalaya). Acad. Sci. Paris C. R., Sér. D, *290*, 389-392.

Bassoullet, J. P., Colchen, M., Juteau, T., Marcoux, J., Mascle, G., and Reibel, G., 1983: Geological studies in the Indus suture zone of Ladakh (Himalaya). *In*: Gupta, V. J. (ed), Stratigraphy and Structure of Kashmir and Ladakh, Himalaya. Hind. Publ. Co., Delhi, 96-124.

Bassoullet, J. P., Colchen, M., Marcoux, J., and Mascle, G., 1978: Une transversale de la zone de l'Indus de Khalsi à Phothaksar, Himalaya du Ladakh. Acad. Sci. Paris C. R., Sér. D, *286*, 563-566.

Bassoullet, J. P., Colchen, M., Marcoux, J., and Mascle, G., 1980: Les masses calcaires du flysch triasico-jurassique de Lamayuru (zone de la suture de l'Indus, Himalaya du Ladakh): Klippes sédimentaires et éléments de plate-forme remaniés. Riv. Ital. Paleont. Stratigr. *86*, 825-844.

Baud, A., 1983: The Zanskar side of the NW Himalaya (abstr.). Terra Cognita *3*, 263-264.

Baud, A., Arn, R., Bugnon, P., Crisinel, A., Dolivo, E., Escher, A., Hammerschlag, J. G., Marthaler, M., Masson, H., Steck, A., and Tièche, J. C., 1982: Le contact Gondwana—péri-Gondwana dans le Zanskar oriental (Ladakh, Himalaya). Soc. géol. France Bull., Sér. 7, *24*, 341-361.

Baud, A., Arn, R., Bugnon, P., Crisinel, A., Dolivo, E., Escher, A., Hammerschlag, J. G., Marthaler, M., Masson, H., Steck, A., and Tièche, J. C., 1983: Geological observations in the Eastern Zanskar Area, Ladakh (Himalaya). *In*: Gupta, V. J. (ed), Stratigraphy and Structure of Kashmir and Ladakh, Himalaya. Hind. Publ. Co., Delhi, 130-142.

Baud, A., Gaetani, M., Fois, E., Garzanti, D., Nicora, A., and Tintori, A., 1982: Les séries tibétaines de l'Himalaya sont-elles allochtones: Nouvelles observations dans le Zanskar oriental (Inde du N). Réunion ann. Sci. Terre, Paris, *9*, 33.

Baud, A., Gaetani, M., Garzanti, D., Fois, E., Nicora, A., and Tintori, A., 1984: Geological observations in southeastern Zanskar and adjacent Lahul area (northwestern Himalaya). Eclogae geol. Helvetiae *77*, 171-197.

Bhat, M. I., and Zainuddin, S. M., 1978: Geochemistry of the Panjal Traps of Mount Kayol. Lidderwat, Pahalgam, Kashmir. Geol. Soc. India Jour. *19*, 403-410.

Brookfield, M. E., and Andrews-Speed, C. P., 1984: Sedimentology, petrology and tectonic significance of the shelf, flysch and molasse clastic deposits across the Indus suture zone, Ladakh, NW India. Sedimentary Geology *40*, 249-286.

Brookfield, M. E., and Reynolds, P. H., 1981: Late Cretaceous emplacement of the Indus suture zone ophiolitic mélanges and an Eocene-Oligocene magmatic arc on the northern edge of the Indian plate. Earth Planet. Sci. Lett. *55*, 157-162.

Dietrich, V. J., Frank, W., and Honegger, K., 1983: A Jurassic-Cretaceous island arc in the Ladakh-Himalayas. Jour. Volcanology Geoth. Res. *18*, 405-433.

Frank, W., Gansser, A., and Trommsdorff, V., 1977: Geological observations in the Ladakh area (Himalayas). A preliminary report. Schweiz. Min. Pet. Mitt. *57*, 89-113.

Frank, W., Hoinkes, G., Miller, C., Purtscheller, F., Richter, W., and Thöni, M., 1973: Relations between metamorphism and orogeny in a typical section of the Indian Himalayas. Tschermaks Min. Pet. Mitt., Ser. 3, *20*, 303-332.

Frank, W., Thöni, M., and Purtscheller, F., 1977: Geology and petrography of Kulu–South Lahul area. C.N.R.S., Coll. Internat. *268*, 2, 147-172.

Fuchs, G., 1967: Zum Bau des Himalaya. Oester. Akad. Wiss., math.-naturw. Kl., Denkschr. *113*, 211 p.

Fuchs, G., 1975: Contributions to the geology of the north-western Himalayas. Geol. Bundesanstalt Wien, Abh. *32*, 59 p.

Fuchs, G., 1977: Traverse of Zanskar, from the Indus to the valley of Kashmir. A preliminary note. Geol. Bundesanstalt Wien, Jahrb. *120*, 219-229.

Fuchs, G., 1979: On the geology of western Ladakh. Geol. Bundesanstalt Wien, Jahrb. *122*, 513-540.

Fuchs, G., 1981: Outline of the geology of the Himalaya. Oesterr. Geol. Ges. Mitt. *74/75*, 101-127.

Fuchs, G., 1982a: The geology of the Pin valley in Spiti, H.P., India. Geol. Bundesanstalt Wien, Jahrb. *124*, 325-359.

Fuchs, G., 1982b: The geology of western Zanskar. Geol. Bundesanstalt Wien, Jahrb. *125*, 1-50.

Fuchs, G., and Frank, W., 1970: The geology of west Nepal between the rivers Kali Grandaki and Thulo Bheri. Geol. Bundesanstalt Wien, Jahrb., Sdbd. *18*, 103 p.

Gaetani, M., Nicora, A., and Premoli Silva, I., 1980: Uppermost Cretaceous and Paleocene in the Zanskar Range (Ladakh-Himalaya). Riv. Ital. Paleont. Stratigr. *86*, 127-166.

Gaetani, M., Nicora, A., Premoli Silva, I., Fois, E., Garzanti, E., and Tintori, A., 1983: Upper Cretaceous and Paleocene in Zanskar Range, NW-Himalaya. Riv. Ital. Paleont. Stratigr. *89*, 81-118.

Gansser, A., 1964: The geology of the Himalayas. Wiley-Interscience, London, 289 p.

Gansser, A., 1977: The great suture zone between Himalaya and Tibet. A preliminary account. C.N.R.S., Coll. Internat. *268*, 2, 181-191.

Gansser, A., 1980: The significance of the Himalayan suture zone. Tectonophysics *62*, 37-52.

Gupta, V. J., Gautam Mahajan, S., Kumar, D. K., Chadha, P. C., Bisaria, N. S., Virdi, N., Kichhar, N., and Kashyap, S., 1970: Stratigraphy along the Manali-Leh road. Panjab Univ., Chandigarh, Cent. Adv. Stud. Geol. Publ. *7*, 77-84.

Gupta, V. J., and Kumar, S., 1975: Geology of Ladakh, Lahaul and Spiti regions of Himalaya with special reference to the stratigraphic position of flysch deposits. Geol. Rundschau *64*, 540-563.

Heim, A., and Gansser, A., 1939: Central Himalaya: Geological observations of the Swiss expedition 1936. Soc. Helv. Sci. nat. Mém. *73*, no. 1, 245 p., Zürich.

Honegger, K., 1983: Strukturen und Metamorphose in Zanskar Kristallin (Ladakh-Kashmir, Indien). Dissertation ETH Zürich, Nr. 7456, 118 p.

Honegger, K., Dietrich, V., Frank, W., Gansser, A., Thöni, M., and Trommsdorff, V., 1982: Magmatism and metamorphism in the Ladakh-Himalayas (the Indus-Tsangpo suture zone). Earth Planet. Sci. Lett. *60*, 253-292.

Klootwijk, C. T., 1980: A summary of paleomagnetic data from extra-peninsular Indo-Pakistan and south central Asia: Implication for collision tectonics. *In*: Saklani, P. S. (ed), Structural Geology of the Himalaya, *2*. New Delhi, 307-360.

Klootwijk, C., Sharma, M. B., Gergan, J., Tirkey, B., Shah, S. K., and Agarval, V., 1979: The extent of greater India, II. Paleomagnetic data from the Ladakh intrusives at Kargil, northwestern Himalayas. Earth Planet. Sci. Lett. *44*, 47-64.

Le Fort, P., 1975: Himalayas: The collided range. Present knowledge of the continental arc. Am. Jour. Sci. *275A*, 1-44.

Molnar, P., Chen, W. P., Fitch, T. J., Tapponnier, P., Warsi, W.E.K., and Wu, F. T., 1977: Structure and tectonics of the Himalaya: A brief summary of relevant geophysical observations. C.N.R.S., Coll. Internat. *268*, 2, 269-294.

Mostler, H., 1980: Zur Mikrofauna des Unterkambriums in der Hazira Formation—Hazara, Pakistan. Naturh. Mus. Wien Ann. (Zapfe Festschrift) *83*, 245-257.

Pal, D., Srivastava, R.A.K., and Mathur, N. S., 1978: Tectonic framework of the miogeosynclinal sedimentation in the Ladakh Himalaya: A critical analysis. Himalayan Geology *8*, 500-523.

Pareek, H. S., 1976: On studies of the Agglomeratic Slate and Panjal Trap in the Jhelum, Liddar, and Sind valleys, Kashmir. India Geol. Survey Red. *107*, no. 2, 12-37.

Pecher, A., 1978: Déformations et métamorphisme associés à une zone de cisaillement. Exemple du grand chevauchement central himalayan (M.C.T.), transversale des Annapurnas et du Manaslu, Nepal. Thèse ès sciences, Univ. Grenoble, France, 354 p.

Sharma, K. K., and Kumar, S., 1978: Contribution to the geology of Ladakh, north western Himalaya. Himalayan Geology *8*, 252-287.

Srikantia, S. V., and Bhargava, O. N., 1978: The Indus tectonic belt of Ladakh Himalaya: Its geology, significance and evolution. *In*: Saklani, P. S. (ed), Tectonic Geology of the Himalaya. New Delhi, 43-62.

Srikantia, S. V., Ganesan, T. M., Rao, P. N., Sinha, P. K., and Tirkey, B., 1978: Geology of Zanskar area, Ladakh Himalaya. Himalayan Geology *8*, 1009-1033.

Srikantia, S. V., and Razdan, M. L., 1980: The ophiolite-sedimentary belt of the Indus tectonic zone of the Ladakh Himalaya: Its stratigraphic and tectonic significance. Internat. Ophiolite Symp., Cyprus 1979, Proc., 430-441.

Thöni, M., 1977: Geology, structural evolution and metamorphic zoning in the Kulu Valley (Himachal Himalayas, India) with special reference to the reversed metamorphism. Ges. Geol. Bergbaustud. Wien Mitt. *24*, 125-187.

Trommsdorff, V., Dietrich, V., and Honegger, K., 1983: The Indus suture zone: Paleotectonic and igneous evolution in the Ladakh-Himalayas. *In*: Hsü, K. J. (ed), Mountain Building Processes. Academic Press, London, 213-219.

Valdiya, K. S., 1970: Simla slates: The Precambrian flysch of the lesser Himalaya, its turbidites, sedimentary structures and paleocurrents. Geol. Soc. America Bull. *81*, 451-467.

Wadia, D. N., 1934: The Cambrian-Trias sequence of northwestern Kashmir (parts of Muzaffarabad and Baramula districts). India Geol. Survey Rec. *68*, 121-176.

Chapter 15

THE ACTIVE TAIWAN MOUNTAIN BELT

JOHN SUPPE

Princeton University

The rugged island of Taiwan (Figure 15-1) lies on the active boundary between the Philippine Sea plate and the Eurasian plate. This boundary extends eastward and northeastward along the Ryukyu Trench to southern Japan and southward along the Manila Trench to Luzon (Figure 15-2). The Philippine Sea plate is moving in a northwesterly direction relative to the Eurasian plate at about 70 km per million years, according to Seno (1977). Northeast of Taiwan, the Philippine Sea plate is subducting beneath the Ryukyu island arc of the Eurasian plate. The Okinawa Trough is an actively spreading back-arc basin (Letouzey and Kimura 1985). In contrast, south of Taiwan, the ocean floor of the South China Sea, which is part of the Eurasian plate, is subducting beneath the Luzon island arc along the Manila Trench. Thus Taiwan occupies the region in which the polarity of subduction changes.

The geology of Taiwan is in fact more closely related to the west-vergent Luzon arc than to the Ryukyu arc, a fact that is well displayed in the bathymetric continuity between Luzon and Taiwan (Figures 15-2 and 15-3). The Luzon volcanic arc, including the islands of Babuyan, Batan, Lanyu, and Lutao, continues northward into the low Coastal Range of eastern Taiwan where Neogene island-arc volcanic rocks are exposed (Ho 1975). The outer nonvolcanic arc or accretionary wedge, submerged to the south, expands continuously northward, rising above sea level at the southern tip of Taiwan to merge morphologically with the Central Mountains of Taiwan, which are constructed of west-vergent thrust sheets. The Luzon Trough, a fore-arc basin, lies between the Luzon volcanic arc to the east and the outer arc to the west; it narrows and shallows toward the north and projects directly into the southern Coastal Range and into the Longitudinal Valley, which lies between the Coastal Range and the Central Mountains (Figures 11-2 and 11-5). The Coastal Range contains uplifted and deformed sediments of the Luzon Trough, as well as of the Luzon volcanic arc (Chi et al. 1981, Page and Suppe 1981). The Manila Trench extends northward to become the foredeep of western Taiwan, which exists as the shallow bathymetric trough of the Taiwan Strait, less than 100 m deep, but more fundamentally it appears as the eastward-thickening wedge of Pleistocene orogenic sediments, reaching a thickness of over 4 km at the edge of the active fold-and-thrust belt. Thus Taiwan is part of the west-vergent Luzon arc system. The plate convergence changes polarity in northernmost Taiwan in a complex way that is beyond the scope of this article (see Suppe 1984). It is our purpose here to introduce the principal tectonic features of Taiwan and to point out the insights into orogenic processes that are revealed in this site of active arc-continent collision.

FIGURE 15-1. Cliffs of upper Miocene sandstones, Nanchuang Formation, in the Western Foothills of southern Taiwan near Alishan (2,500 m elevation), exemplifying the active mountain building in Taiwan. Cliffs are about 300 m high.

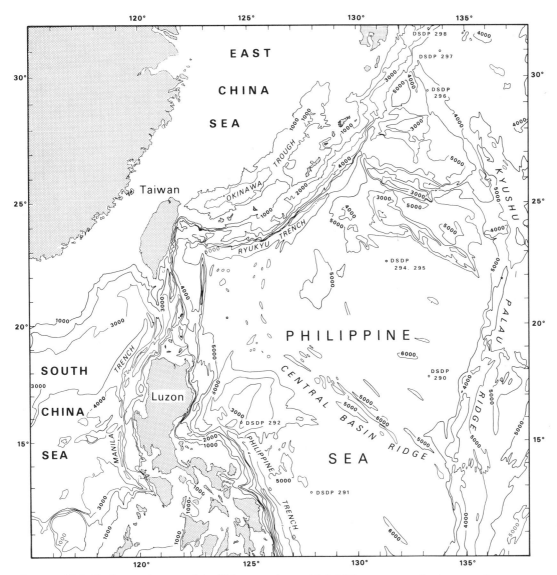

FIGURE 15-2. Bathymetric map of the Taiwan region, with 1,000 m contours (bathymetry from Mammerickx et al. 1976).

The ongoing arc-continent collision in Taiwan has been noted by many authors, for example, Biq (1972), Chai (1972), Jahn (1972), Yen (1973), Bowen et al. (1978), Tsai (1978), Wu (1978), Ho (1979), and Suppe (1981, 1984). The geologic literature on this region is extensive and includes the *Proceedings* and *Memoirs of the Geological Society of China, Petroleum Geology of Taiwan, Acta Geologica Taiwanica, Acta Oceanographica Taiwanica, Bulletin of the Institute of Earth Sciences Academia Sinica, Bulletin of the Geological Survey of Taiwan*, and *Bulletin of the Central Geological Survey*. Ho (1975, 1982) has written two book-length introductions to the geology and tectonics of Taiwan. Sun (1982) has outlined the offshore geology.

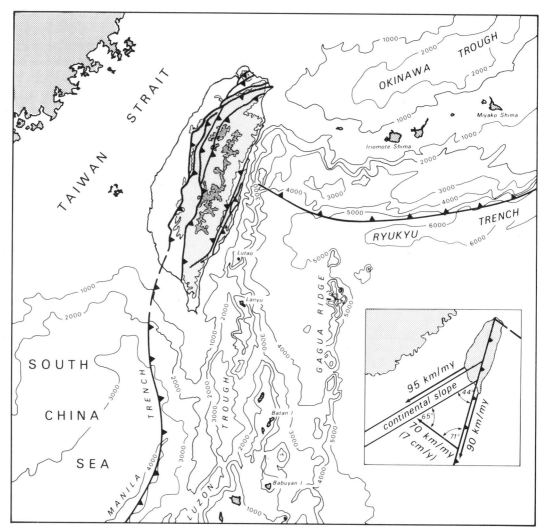

FIGURE 15-3. Tectonic and bathymetric setting of Taiwan, with 1,000 m contours (submarine and Taiwan only). Inset shows a velocity triangle for the propagation of the arc-continent collision, based on Seno's (1977) estimate of present-day plate motions (from Davis, Suppe, and Dahlen, Jour. Geophys. Research, v. 88, pp. 1153-1172, 1983, copyrighted by the American Geophysical Union).

KINEMATICS OF ARC-CONTINENT COLLISION

Present-day plate motions between the Philippine Sea and Eurasian plates are constrained largely by earthquake focal mechanisms and global plate motions. Seno (1977) estimates an instantaneous pole of rotation (1.2°/Ma, 45.5° N, 150.2° E) that predicts a rate of convergence near Taiwan of about 70 km/Ma in a west-northwesterly direction (305°). Minster and Jordan (1979) estimate a similar instantaneous pole of rotation (1.2°/Ma, 41.7° N, 162.5° E), as do Ranken et al. (1984). Theories of earlier motion of the Philippine Sea plate are given by Matsubara and Seno (1980) and Seno and Maruyama (1984).

    The arc-continent collision is oblique because the stable continental shelf of China is oriented northeast-southwest, whereas the Luzon island arc is oriented north-south. Northward from Luzon, the plate boundary impinges first on the abyssal plain of the South China Sea and then, in turn, on the Chinese continental rise, slope, and shelf (Figures 15-2 and 15-3). The arc-continent collision is beginning just now in southernmost Taiwan, whereas it began about 4 Ma ago in northern Taiwan, according to stratigraphic studies (Chi et al. 1981). Considering the oblique geometry, the site of collision should be propagating southward along the arc at about 90 km/Ma and southwestward along the continental margin at about 95 km/Ma (see inset, Figure 15-3).

Insofar as the island arc and continental margin have cylindrical symmetry, their collision should be similar at each point in time, shifted only in space. This time-space equivalence gives an important insight into the tectonics. To move 90 km south along the arc is equivalent to moving 1 million years back in time. Thus by moving northward along the Luzon arc we can observe the Taiwan mountain belt developing through time (Suppe 1981, Page and Suppe 1981).

The major bathymetric and topographic effect of arc-continent collision is the expansion of the accretionary wedge in both width and height as the plate boundary encounters and incorporates the thick sediments of the Chinese continental rise, slope, and shelf (Figure 15-3). More than 150 km south of Taiwan, the accretionary wedge is about 50 km wide with a crest 1,500 to 2,000 m below sea level. In the region 0 to 150 km south of Taiwan, the accretionary wedge expands to double this width as it encounters the Chinese continental rise and slope. With continued collision, the accretionary wedge rises above sea level to become the Central Mountains of Taiwan.

Using the time-space equivalence, we can observe the mountains growing with time as we move north from the southern tip of Taiwan. The mountains grow steadily in width and height for about 120 km (1.33 Ma), at which point they take on a constant width of about 90 km that persists over most of the island of Taiwan. Quantitative analysis indicates that this region of constant width is in a steady state topographically, balanced between growth by tectonic compression and erosion. This cross-sectional flux is about 500 km²/Ma (Suppe 1981).

The erosion and uplift rates in Taiwan are very high. Suspended and dissolved stream loads indicate a short-term erosion rate of 5 to 6 km/Ma (Li 1975). Fission-track dating of apatite, sphene, and zircon from the Central Mountains indicates a long-term (0.25 to 1.00 Ma) erosion rate of 5 to 9 km/Ma (Liu 1982). Dating of uplifted Holocene reefs and marine terraces indicates uplift rates of about 5 km/Ma (Peng et al. 1977).

STRATIGRAPHIC FRAMEWORK

The bulk of the subaerial mountain belt is the Central Mountains, Hsuehshan Range, and Western Foothills, 80 km wide in all, composed of parautochthonous Cenozoic clastic sediments and pre-Cenozoic basement rocks of the Chinese continental margin, together with an overlying Pliocene and Pleistocene sequence of orogenic sediments 4 km thick derived from the Taiwan mountain belt encroaching from the east (Figure 15-4). To the east is the ultra-allochthonous eastern Coastal Range, 10 km wide, composed of deformed rocks of the volcanic arc and fore-arc basin of the Luzon arc, together with an overlying Pliocene and

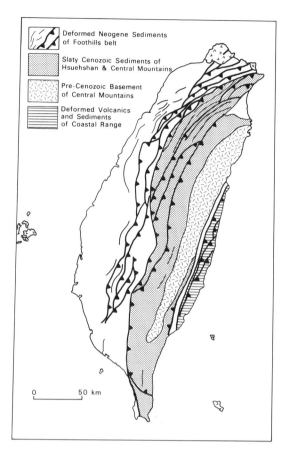

FIGURE 15-4. Simplified tectonic map of Taiwan showing the principal geologic provinces discussed in the text. Light, wavy lines show the trends of major folds; v's show Quaternary volcanic rocks.

Pleistocene sequence of orogenic sediments 4 to 5 km thick that records the rising Taiwan mountain belt to the west.

The stratigraphy of the stable Chinese continental margin is best known in the Taiwan region from petroleum and coal exploration in the Western Foothills, Coastal Plain, and Taiwan Strait. The stratigraphy is divided naturally into two sequences by a regional lower Oligocene unconformity (Huang 1982), which functions widely as the basal décollement in the Western Foothills fold-and-thrust belt. Consequently, pre-Oligocene stratigraphy is exposed in only one locality of the Foothills, although strata below the unconformity appear at the surface as quartzites and slates in more interior thrust sheets and folds of the Hsuehshan Range and Central Mountains. The Neogene sequence of the Foothills and Coastal Plain is in most areas thicker than 5 to 6 km; therefore the Paleogene is generally not reached in drilling for petroleum. Offshore to the west, however, farther from the deepest parts of the Pleistocene foredeep, the Neogene sequence is thinner and the Paleogene more accessible to drilling and seismic reflection profiling (Sun 1982). There, away from extensive Plio-Pleistocene deformation, the Paleogene sequence is seen to be deposited in a system of major normal-fault basins that strike roughly parallel to the trend of the continental margin of southeastern China and record the opening of the South China Sea. The normal-fault tectonics continued to a limited degree into the Miocene. Some of these faults have been reactivated in the Pleistocene compression in the Foothills as reverse faults or tear faults, depending on their orientation (Bonilla 1975; Suppe 1986).

The Mesozoic and Paleozoic basement of the continental margin is exposed in the coastal areas of the China mainland and has been encountered occasionally in drilling. This basement reappears as an allochthonous internal massif in the eastern Central Mountains, where it is overlain unconformably by Eocene through Miocene slates and phyllites at shallow greenschist grade (Suppe et al. 1976).

Immediately east of the exposures of basement rock in the Central Mountains is the low Coastal Range (Figure 15-5), 10 km wide and composed of ultra-allochthonous rocks of the Luzon volcanic arc and ophiolitic fragments, which are interpreted as pieces of the oceanic lithosphere of the South China Sea (Suppe et al. 1981). Thus the narrow Longitudinal Valley between the Central Mountains and the Coastal Range is commonly considered a suture zone between rocks of the Chinese continental margin and those of the Luzon volcanic arc and Philippine Sea. This suture zone marks the site of closure of the South China Sea.

The exposed stratigraphic sequence of the Coastal Range begins with extensive andesites and dacites of the Miocene Tuluanshan Formation (Figure 15-5) and associated volcanoclastic sediments and limestones. This arc volcanism and associated volcanoclastic sedimentation continues through the Pliocene and into the Pleistocene, but it is relatively less important because of the great influx of clastic sediment, whose petrography shows that it is derived from the rising Taiwan mountain belt (Teng 1979, Chi et al. 1981). The effect of arc-continent collision on the stratigraphy of the Coastal Range can be illustrated by plotting the stratigraphic sections of the Coastal Range, whose locations are shown in Figure 15-5, along their late Neogene tracks of plate motion relative to Eurasia (Figure 15-6). In this transformation we see that volcanic-arc sedimentation dominated the Coastal Range until about 4 Ma ago, when northern Taiwan began to impinge on the Eurasian continental margin, producing a clastic sediment source in the rising accretionary wedge to the west and north. We also observe that the arc-continent collision results in an enormous increase in sedimentation rate, from less than 500 m/Ma at the beginning of the Pliocene to 5 to 10 km/Ma in the early Pleistocene (Figure 15-7, Chi and Chang 1983, and Chi et al. 1981).

The stratigraphy of the Coastal Range records considerable paleogeographic complexity reflecting the bathymetric complexity of the volcanic arc and fore-arc basin. With the beginning of Taiwan-derived sedimentation, the Luzon Trough began to fill axially from the north and onlap up and over the volcanic arc; we see this same pattern of sedimentation continuing today at the northern end of the present Luzon Trough (Figure 15-3). Each stratigraphic section in the Coastal Range records a filling up of the basin of the Luzon Trough or the west flank of the Luzon volcanic arc.

One of the most interesting and well-known stratigraphic units of the Coastal Range is the Pliocene Lichi Mélange, which is a chaotic formation composed of blocks and slabs of deep-water sandstone, mudstone, and ophiolitic rocks ranging from millimeters to hundreds of meters in size, set in a scaly mudstone matrix. The Lichi Mélange has been shown to be composed of a sequence of submarine olistostromes and minor pebbly mudstones derived from the rising accretionary wedge to the west (Page and Suppe 1981). The Lichi olistostromes interfinger with bedded turbiditic sediments of the Takangkou Formation and repre-

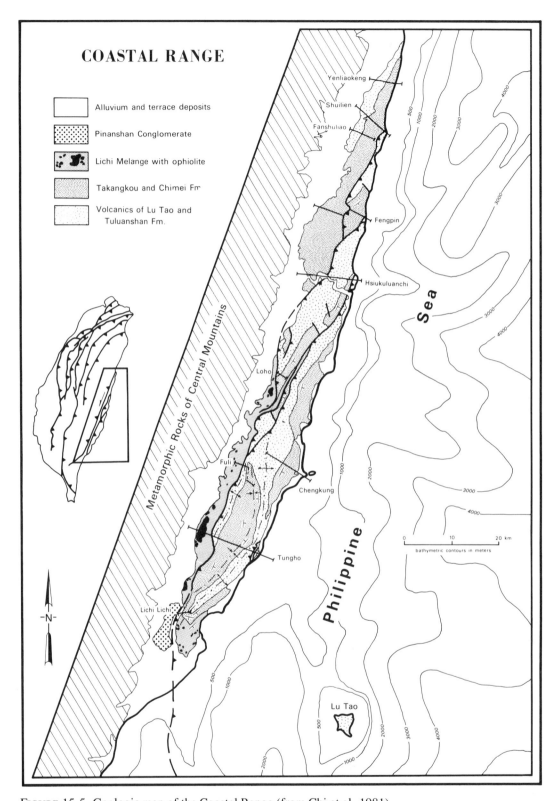

FIGURE 15-5. Geologic map of the Coastal Range (from Chi et al. 1981).

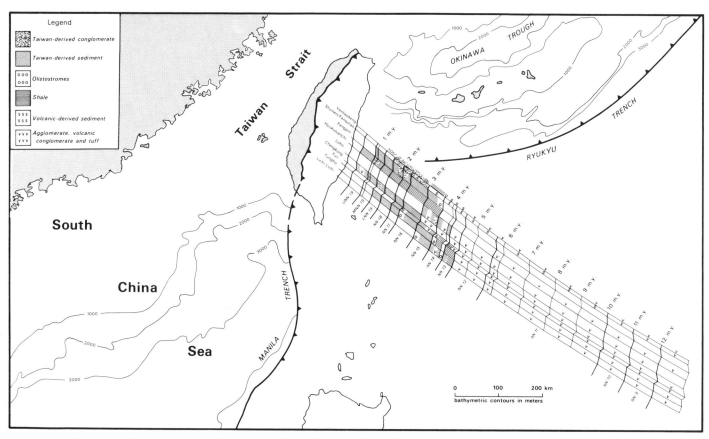

FIGURE 15-6. Paleogeographic tracks of the various stratigraphic sections of the Coastal Range shown in Figure 15-5, assuming Seno's (1977) instantaneous plate motions. Sediments derived from the rising island of Taiwan began to appear in the Coastal Range about 4 Ma ago, apparently recording the beginning of arc-continent collision (after Chi et al. 1981).

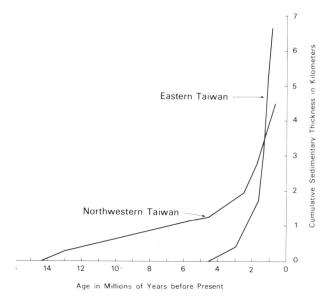

FIGURE 15-7. Typical increases in sedimentation rate in the last 4 Ma, reflecting the growth of the Taiwan mountain belt (data from Huang 1976, Chi and Huang 1981, Chi al. 1981).

sent a fault-scarp facies of the west flank of the Luzon Trough (Figure 15-8). Apparently the Lichi Mélange records east-vergent back thrusting of the accretionary wedge over the Luzon Trough fore-arc basin during arc-continent collision. This collapse of the fore-arc basin is apparently occurring today in the Luzon Trough about 200 km south of the Coastal Range (Page and Suppe 1981).

The ophiolitic fragments within the Lichi Mélange are called the East Taiwan Ophiolite, which exists as blocks and slabs up to 1 km across. The larger fragments contain an internal stratigraphy that indicates that the ophiolite probably formed along a major ridge-ridge transform fault about 15 Ma ago (Liou et al. 1977, Suppe et al. 1981). This transform fault was probably part of the China Basin Ridge system, a now inactive ridge system of the South China Sea (Taylor and Hays 1980).

WESTERN TAIWAN FOLD-AND-THRUST BELT

The western edge of the Taiwan mountain belt is marked by the western Taiwan foredeep, which displays flexure of the lithosphere beginning in the late Pliocene in response to the encroaching mountain load (Covey 1984). The Pleistocene orogenic sequence thickens eastward from less than 1 km in the Taiwan Strait to more than 4 km at the edge of the active thrusting (Figure 15-9). The rate of sedimentation increases from about 100 m/Ma under stable continental-margin conditions in the late Miocene, to 400 m/Ma in the Pliocene, to 1,500 to 2,000 m/Ma in the Pleistocene (Figure 15-7, Chi and Chang 1983, and Chi and Huang 1981). De-

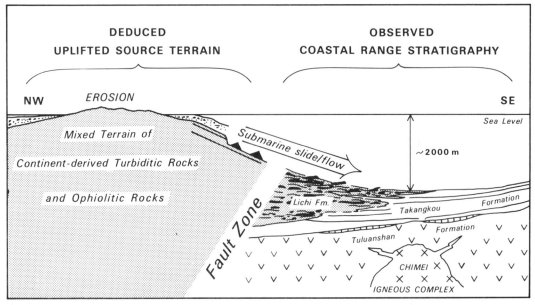

FIGURE 15-8. Schematic paleogeography during Pliocene deposition of the Lichi Mélange and Takangkou Formation. The basin of deposition is the Luzon Trough fore-arc basin, and the source terrain is the rising Taiwan accretionary wedge (from Page and Suppe 1981).

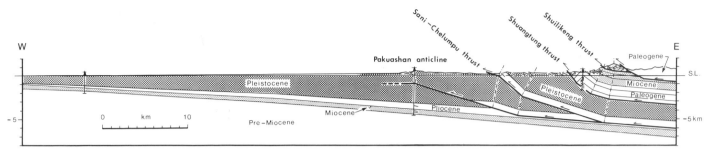

FIGURE 15-9. Cross section of the western Taiwan foredeep and the edge of the fold-and-thrust belt near Taichung in central Taiwan.

formation is actively encroaching on the foredeep; for example, the Pakuashan anticline in Figure 15-9 is growing up through the alluvium. Most of the frontal structures of the fold-and-thrust belt involve sediments dated at less than 0.9 to 0.4 Ma (Chi and Huang 1981), and some are known to be seismically active.

The structure of the western Taiwan fold-and-thrust belt is well known through extensive deep drilling, seismic reflection profiling, and detailed surface mapping. The structure involves displacement and folding above stratigraphically controlled décollement horizons within the Neogene section (Suppe and Namson 1979; Suppe 1980b; Namson 1981, 1983, 1984), with important imbricate slices (e.g., Figures 15-9 and 15-10). The overall shortening of the Chinese continental margin in northern Taiwan is not less than 160 to 200 km, based on retrodeformable cross sections of the observed structures of the fold-and-thrust belt (Suppe 1980a, 1986).

The presently active fold-and-thrust belt of western Taiwan offers an important opportunity to investigate the mechanical processes of fold-and-thrust belts. So far it has been

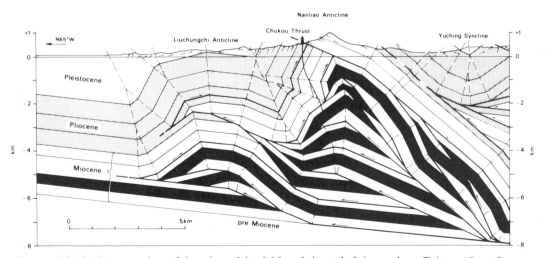

FIGURE 15-10. Cross section of the edge of the fold-and-thrust belt in southern Taiwan (from Suppe 1983).

learned that the important horizons of décollement show considerable stratigraphic variation over western Taiwan, reflecting important regional stratigraphic variation. The most through-going décollement lies at or near the base of the Neogene stratigraphic sequence. All the major horizons of décollement are observed to lie within stratigraphically controlled intervals of high pore-fluid pressure. In general the middle and lower Miocene is overpressured with a Hubbert and Rubey (1959) fluid-pressure coefficient of $\lambda = 0.7$ ($\lambda = P_f / \rho g z$), corresponding to a fluid-pressure gradient of about 17 kPa/m (Suppe and Wittke 1977).

The overall shape of the fold-and-thrust belt in cross section is a wedge-shaped compressive deformed zone analogous to the compressive wedge that develops in front of a bulldozer or snowplow (Figure 15-11). The dip of the basal décollement is about 6° to the east and surface slope is 2.9° to the west. This specific wedge shape is in agreement with that predicted theoretically for a wedge at brittle failure by horizontal compression throughout, for the observed fluid-pressure coefficient $\lambda = 0.7$ and laboratory frictional and fracture strengths (Davis et al. 1983, Dahlen et al. 1984). The best-fitting rock properties are a basal coefficient of friction of $\mu_b = 0.85$, corresponding to Byerlee's (1978) empirical ''law'' for most geo-

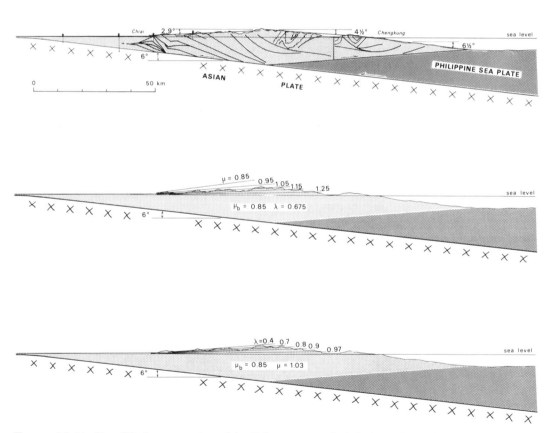

FIGURE 15-11. Simplified cross section of the Taiwan mountain belt through southern Taiwan showing the wedge-shaped deformed zone of western Taiwan. The lower two cross sections illustrate the sensitivity of the surface slope in Taiwan to coefficient of internal friction μ and fluid-pressure coefficient λ (from Davis, Suppe, and Dahlen, Jour. Geophys. Research, v. 88, pp. 1153-1172, 1983, copyrighted by the American Geophysical Union).

logical materials, and an internal strength of the wedge that is about 10% higher, corresponding to a cohesive strength of $S_o = 10$ to 15 MPa, in good agreement with rock-mechanics measurements for sedimentary rocks (e.g., Hoshino et al. 1972). Thus the western Taiwan mountain belt takes on a cross-sectional shape, through deformation, that is almost precisely what would be predicted for a compressive wedge in front of a bulldozer. The angle of initial step-up of thrust faults under these conditions is predicted to be 11.0 to 16.0°, which is in excellent agreement with the observed angle of 13.3° ± 2.4° in front of the fold-and-thrust belt (Dahlen et al. 1984).

METAMORPHISM

The Cenozoic rocks exposed in the Central Mountains and Hsuehshan Range of Taiwan display low-grade Plio-Pleistocene metamorphism (Chen et al. 1983). Much of the metamorphism occurred at relatively high CO_2 activity; consequently, the diagnostic calcium-aluminum silicates of low-grade regional metamorphism are not generally developed (Liou 1981). In the field, the rocks of the Foothills display moderate induration and low-rank bituminous coal (Ho 1959). In the Hsuehshan Range the rocks display local cleavage development and meta-anthracite coal. In the eastern Hsuehshan Range and Central Mountains the rocks display an increasingly strong development of slaty cleavage, locally developing phyllitic textures. Strong slaty cleavage is developed in rocks as young as Miocene. Shallow greenschist-facies mineral assemblages are displayed in rocks near the unconformity between basement and cover rocks in the eastern Central Mountains (Suppe et al. 1976, Liou 1981, Liou et al. 1981). Considering that arc-continent collision did not begin in northern Taiwan until early Pliocene (ca. 4 Ma) and that the rate of denudation is about 5 km/Ma, the metamorphism observed at the surface is apparently Pliocene.

Analogous metamorphism is presumably going on today at depth in the Taiwan mountain belt. Data from the western Taiwan fold-and-thrust belt indicate a geothermal gradient of 30 to 35°C/km to depths of 5 to 6 km. The bulldozer model of the Taiwan fold-and-thrust belt implies relatively large deviatoric stresses, such that the thermodynamic mean stress $\bar{\sigma} = (\sigma_1 + \sigma_2 + \sigma_3)/3$ is about twice the lithostatic overburden and the fluid-pressure/solid-pressure ratio $P_f/\bar{\sigma}$ is about 0.4 (Suppe 1981). Thus we expect that dehydration reactions are taking place at lower temperatures than would be expected from the common assumption of fluid pressure equals solid pressure ($P_f = \bar{\sigma}$ or $\lambda = 1$). The shape of the Taiwan mountain belt in cross section is inconsistent with fluid pressure equals solid pressure (see Figure 15-11 and Davis et al. 1983).

MAGMATISM

The Pliocene to Recent Taiwan orogeny has taken place with little magmatic activity. Minor volcanism is still occurring within the Luzon volcanic arc near Taiwan. Rare beds of tuff or volcanoclastic sandstone are observed in the Pliocene and Pleistocene stratigraphic sequences of the Coastal Range (Chi et al. 1981), but they are distinctly unimportant. Late Pleistocene andesitic volcanic and hypabyssal intrusive rocks exist at the northern tip of Taiwan and off-shore to the north and represent the southwestern end of the Ryukyu volcanic arc. Basalts of alkalic affinities are widely scattered through the Neogene and Paleogene stratigraphic section of western Taiwan and the Taiwan Strait and are possibly related to the normal-fault tectonics (Huang 1982).

DISCUSSION

Arc-continent collisions are widely inferred from the geological records of ancient mountain belts, for example, the Antler and Sonoma orogenies of the western North American Cordillera, the Taconic orogeny of the Appalachians, and the latest Cretaceous-Paleocene Caribbean orogeny of northern Venezuela. Arc-continent collisions are rare, however, among the active orogenic belts of the world; Taiwan and Timor are the principal examples. This rarity, relative to continent-continent orogenic belts and continental-margin arcs, is to be expected, because arc-continent collisions are short-lived; island arcs tend to move at 5 to 10 cm/year.

We observe these rapid orogenic processes in Taiwan. Several hundred kilometers of orogenic compression are accomplished in a few million years, topography reaches steady state in just over a million years, slates and phyllites form and reach the surface by erosion within 2 to 3 million years, and orogenic sediments are trapped adjacent to the mountain belt at rates of 1 to 10 km/Ma. According to Stanley et al. (1981), several generations of outcrop-scale metamorphic folds and associated cleavages have developed during the Plio-Pleistocene deformation. Furthermore, the orogenic activity associated with arc-continent collision may cease very soon after its onset. Arc-continent collision has already stopped in the northernmost quarter of Taiwan, because of southwestward propagation of the Ryukyu Trench subduction zone (Figure 15-3 and Suppe 1984). In this part of Taiwan the mountainous relief has decayed to a tenth of its initial value in less than a million years (Suppe 1981). Thus the time required for orogenic processes during arc-continent collision in Taiwan is short relative to the uncertainties in radiometric calibration of the pre-Cretaceous geological time scale. We might expect that ancient arc-continent collisions occurred with similar rapidity, even though it is more difficult to date them.

ACKNOWLEDGMENTS

I am grateful to many organizations for their contributions to my work in Taiwan, including the Chinese Petroleum Corporation, the Mining Research and Service Organization, and in particular at present, as I write this in Taipei, the Department of Geology of National Taiwan University and the National Science Council of the Republic of China. I also thank many individuals, including Stanley S. L. Chang, W. R. Chi, M. Covey, W. G. Ernst, C. S. Ho, J. G. Liou, C. Y. Meng, J. Namson, B. M. Page, R. S. Stanley, C. H. Tang, C. M. Wang-Lee, and Y. Wang. Support by the U.S. National Science Foundation includes grants EAR 80-18350 and 81-21196.

REFERENCES

Biq, C. C., 1972: Dual-trench structure in the Taiwan-Luzon region. Geol. Soc. China Proc. *15*, 65-75.

Bonilla, M. G., 1975: A review of recently active faults in Taiwan. U.S. Geol. Survey Open-File Report *75-41*, 72 p.

Bowen, C., Lu, R. S., Lee, C. S., and Schouten, H., 1978: Plate convergence and accretion in Taiwan-Luzon region. Am. Assoc. Petroleum Geologists Bull. *62*, 1645-1672.

Byerlee, J., 1978: Friction of rocks. Pure and Applied Geophys. *116*, 615-626.

Chai, B.H.T., 1972: Structure and tectonic evolution of Taiwan. Am. Jour. Sci. *272*, 389-422.

Chen, C. H., Chu, H. T., Liou, J. G., and Ernst, W. G., 1983: Explanatory notes for the metamorphic facies map of Taiwan. Central Geol. Survey (Taiwan) Special Publ. *2*, 32 p.

Chi, W. R., and Chang, S.S.L., 1983: Neogene nannoplankton biostratigraphy in Taiwan and the tectonic implications. Petroleum Geology Taiwan *19*, 93-147.

Chi, W. R., and Huang, H. M., 1981: Nannobiostratigraphy and paleo-environments of the late Neogene sediments and their tectonic implications in the Miaoli area, Taiwan. Petroleum Geology Taiwan *18*, 111-129.

Chi, W. R., Namson, J., and Suppe, J., 1981: Stratigraphic record of plate interactions in the Coastal Range of eastern Taiwan. Geol. Soc. China Mem. *4*, 155-194.

Covey, M., 1984: Lithofacies analysis and basin reconstruction, Plio-Pleistocene western Taiwan foredeep. Petroleum Geology Taiwan *20*, 53-83.

Dahlen, F. A., Suppe, J., and Davis, D., 1984: Mechanics of fold-and-thrust belts and accretionary wedges: Cohesive Coulomb theory. Jour. Geophys. Res. *89*, 10,087-10,101.

Davis, D., Suppe, J., and Dahlen, F. A., 1983: Mechanics of fold-and-thrust belts and accretionary wedges. Jour. Geophys. Res. *88*, 1153-1172.

Ho, C. S., 1959: Coal resources of Taiwan. Geol. Survey Taiwan Bull. *10*, 237 p.

Ho, C. S., 1975: An introduction to the geology of Taiwan. Explanatory text of the geologic map of Taiwan. Ministry of Economic Affairs, Republic of China, Taipei, 153 p.

Ho, C. S., 1979: Geologic and tectonic framework of Taiwan. Geol. Soc. China Mem. *3*, 57-72.

Ho, C. S., 1982: Tectonic evolution of Taiwan. Explanatory text of the tectonic map of Taiwan. Ministry of Economic Affairs, Republic of China, Taipei, 126 p.

Hoshino, K., Koide, H., Inami, K., Iwamura, S., and Mitsui, S., 1972: Mechanical properties of Japanese Tertiary sedimentary rocks under high confining pressures. Geol. Survey Japan Rept. *244*, 200 p.

Huang, T. C., 1976: Neogene calcareous nannoplankton biostratigraphy viewed from Chuhuangkeng section, northwestern Taiwan. Geol. Soc. China Proc. *19*, 7-24.

Huang, T. C., 1982: Tertiary calcareous nannofossil stratigraphy and sedimentation cycles in Taiwan. 2nd ASEAN Council on Petroleum (ASCOPE) Conference, Manila 1981, Proc., 873-886.

Hubbert, M. K., and Rubey, W. W., 1959: Role of fluid pressure in mechanics of overthrust faulting. I. Mechanics of fluid-filled porous solids and its application to overthrust faulting. Geol. Soc. America Bull. *70*, 115-166.

Jahn, B. M., 1972: Reinterpretation of geologic evolution of the Coastal Range, east Taiwan. Geol. Soc. America Bull. *83*, 241-247.

Letouzey, J., and Kimura, M., 1985: Okinawa Trough genesis: Structure and evolution of a backarc basin developed in a continent. Marine and Petroleum Geol. *2*, 111-130.

Li, Y. H., 1975: Denudation of Taiwan island since the Pliocene Epoch. Geology *4*, 105-107.

Liou, J. G., 1981: Recent high CO_2 activity and Cenozoic progressive metamorphism in Taiwan. Geol. Soc. China Mem. *4*, 551-581.

Liou, J. G., Ernst, W. G., and Moore, D. E., 1981: Geology and petrology of some polymetamorphosed amphibolites and associated rocks in northeastern Taiwan. Geol. Soc. America Bull. *93*, Part I, 219-224, Part II, 609-748.

Liou, J. G., Lan, C. Y., Suppe, J., and Ernst, W. G., 1977: The East Taiwan ophiolite. Taiwan, Mining Res. Serv. Org. (Taipei) Spec. Rept. *1*, 212 p.

Liu, Tsung-Kwei, 1982: Fission-track study of apatite, zircon, and sphene from Taiwan and its tectonic implication. Ph.D. thesis, National Taiwan University (in Chinese), 95 p.

Mammerickx, J., Fischer, R. L., Emmel, F. J., and Smith, S. M., 1976: Bathymetry of the east and southeast Asian seas. Geol. Soc. America, Map Chart Series, MC-17.

Matsubara, Y., and Seno, T., 1980: Paleogeographic reconstruction of the Philippine Sea at 5 m.y. B. P. Earth Planet. Sci. Lett. *51*, 406-414.

Minster, J. B., and Jordan, T. H., 1979: Rotation vectors for the Philippine and Rivera plates (abstr.). EOS (Am. Geophys. Union Trans.) *60*, 958.

Namson, J., 1981: Structure of the western foothills belt, Miaoli-Hsinchu area, Taiwan: I. Southern part. Petroleum Geology Taiwan *18*, 31-51.

Namson, J., 1983: Structure of the western foothills belt, Miaoli-Hsinchu area, Taiwan: II. Central part. Petroleum Geology Taiwan *19*, 51-76.

Namson, J., 1984: Structure of the western foothills belt, Miaoli-Hsinchu area, Taiwan: III. Northern part. Petroleum Geology Taiwan *20*, 35-52.

Page, B. M., and Suppe, J., 1981: The Pliocene Lichi mélange of Taiwan: Its plate-tectonic and olistostromal origin. Am. Jour. Sci. *281*, 193-227.

Peng, T. H., Li, Y. H., and Wu, F. T., 1977: Tectonic uplift rates of the Taiwan island since early Holocene. Geol. Soc. China Mem. *2*, 57-69.

Ranken, B., Cardwell, R. K., and Karig, D. E., 1984: Kinematics of the Philippine Sea plate. Tectonics *3*, 555-575.

Seno, T., 1977: The instantaneous rotation vector of the Philippine Sea plate relative to the Eurasian plate. Tectonophysics *42*, 209-226.

Seno, T., and Maruyama, S., 1984: Paleogeographic reconstruction and origin of the Philippine Sea. Tectonophysics *102*, 53-84.

Stanley, R. S., Hill, L. B., Chang, H. C., and Hu, H. N., 1981: A transect through the metamorphic core of the Central Mountains, southern Taiwan. Geol. Soc. China Mem. *4*, 443-473.

Sun, S. C., 1982: The Tertiary basins of offshore Taiwan. 2nd ASEAN Council on Petroleum (ASCOPE) Conference, Manila 1981, Proc., 125-135.

Suppe, J., 1980a: A retrodeformable cross section of northern Taiwan. Geol. Soc. China Proc. *23*, 46-55.

Suppe, J., 1980b: Imbricated structure of western foothills belt, south-central Taiwan. Petroleum Geology Taiwan *17*, 1-16.

Suppe, J., 1981: Mechanics of mountain building and metamorphism in Taiwan. Geol. Soc. China Mem. *4*, 67-89.

Suppe, J., 1983: Geometry and kinematics of fault-bend folding. Am. Jour. Sci. *283*, 684-721.

Suppe, J., 1984: Kinematics of arc-continent collision, flipping of subduction, and back-arc spreading near Taiwan. Geol. Soc. China Mem. *6*, 21-33.

Suppe, J., 1986: Reactivated normal faults in the western Taiwan fold-and-thrust belt. Geol. Soc. China Mem. *7*, 187-200.

Suppe, J., Liou, J. G., and Ernst, W. G., 1981: Paleogeographic origins of the Miocene East Taiwan ophiolite. Am. Jour. Sci. *281*, 228-246.

Suppe, J., and Namson, J., 1979: Fault-bend origin of frontal folds of the western Taiwan fold-and-thrust belt. Petroleum Geology Taiwan *16*, 1-18.

Suppe, J., Wang, Y., Liou, J. G., and Ernst, W. G., 1976: Observations of some contacts between basement and Cenozoic cover in the Central Mountains, Taiwan. Geol. Soc. China Proc. *19*, 59-70.

Suppe, J., and Wittke, J. H., 1977: Abnormal pore-fluid pressure in relation to stratigraphy and structure in the active fold-and-thrust belt of northwestern Taiwan. Petroleum Geology Taiwan *14*, 11-24.

Taylor, B., and Hays, D. E., 1980: The tectonic evolution of the South China Basin. Am. Geophys. Union Mono. *23*, 89-104.

Teng, L. S., 1979: Petrographical study of the Neogene sandstones of the Coastal Range, eastern Taiwan (I. Northern part). Acta Geol. Taiwanica *20*, 129-156.

Tsai, Y. B., 1978: Plate subduction and the Plio-Pleistocene orogeny in Taiwan. Petroleum Geology Taiwan *15*, 1-10.

Wu, F. T., 1978: Recent tectonics of Taiwan. Jour. Phys. Earth *26*, Suppl., S265-S299.

Yen, T. P., 1973: Plate tectonics in the Taiwan region. Geol. Soc. China Proc. *16*, 7-21.

INDEX

Library of Congress Cataloging-in-Publication Data

The Anatomy of mountain ranges.

   (Princeton series in geology and paleontology)
   Includes bibliographies and index.
   1. Orogeny. I. Schaer, Jean-Paul, 1928–
II. Rodgers, John, 1914–   . III. Series.
QE621.A53 1987   551.8   86–30686
ISBN 0–691–08452–1 (alk. paper)